T0235733

Lecture Notes in Computer Science 11972

More information about this series at http://www.springer.com/series/7407

Sergei Artemov · Anil Nerode (Eds.)

Logical Foundations of Computer Science

International Symposium, LFCS 2020
Deerfield Beach, FL, USA, January 4–7, 2020
Proceedings

 Springer

Editors
Sergei Artemov ⓘ
The Graduate Center, CUNY
New York, NY, USA

Anil Nerode
Cornell University
Ithaca, NY, USA

ISSN 0302-9743   ISSN 1611-3349 (electronic)
Lecture Notes in Computer Science
ISBN 978-3-030-36754-1   ISBN 978-3-030-36755-8 (eBook)
https://doi.org/10.1007/978-3-030-36755-8

LNCS Sublibrary: SL1 – Theoretical Computer Science and General Issues

This Springer imprint is published by the registered company Springer Nature Switzerland AG
The registered company address is: Gewerbestrasse 11, 6330 Cham, Switzerland

Preface

The Symposium on Logical Foundations of Computer Science (LFCS) series provides a forum for the fast-growing body of work in the logical foundations of computer science, e.g., those areas of fundamental theoretical logic related to computer science. The LFCS series began with "Logic at Botik," Pereslavl-Zalessky, 1989, which was co-organized by Albert R. Meyer (MIT) and Michael Taitslin (Tver). After that, organization passed to Anil Nerode.

Currently, LFCS is governed by a Steering Committee consisting of Anil Nerode (general chair), Samuel Buss, Stephen Cook, Dirk van Dalen, Yuri Matiyasevich, Gerald Sacks, Andre Scedrov, and Dana Scott.

LFCS 2020 took place at the Wyndham Deerfield Beach Resort, Deerfield Beach, Florida, USA, during January 4–7, 2020. This volume contains the extended abstracts of talks selected by the Program Committee for presentation at LFCS 2020.

The scope of the symposium is broad and includes constructive mathematics and type theory; homotopy type theory; logic, automata, and automatic structures; computability and randomness; logical foundations of programming; logical aspects of computational complexity; parameterized complexity; logic programming and constraints; automated deduction and interactive theorem proving; logical methods in protocol and program verification; logical methods in program specification and extraction; domain theory logics; logical foundations of database theory; equational logic and term rewriting; lambda and combinatory calculi; categorical logic and topological semantics; linear logic; epistemic and temporal logics; intelligent and multiple-agent system logics; logics of proof and justification; non-monotonic reasoning; logic in game theory and social software; logic of hybrid systems; distributed system logics; mathematical fuzzy logic; system design logics; and other logics in computer science.

We thank the authors and reviewers for their contributions. We acknowledge the support of the U.S. National Science Foundation, The Association for Symbolic Logic, Cornell University, the Graduate Center of the City University of New York, and Florida Atlantic University.

October 2019

Anil Nerode
Sergei Artemov

Organization

Steering Committee

Samuel Buss	University of California, San Diego, USA
Stephen Cook	University of Toronto, Canada
Yuri Matiyasevich	Steklov Mathematical Institute, St. Petersburg, Russia
Anil Nerode (General Chair)	Cornell University, USA
Gerald Sacks	Harvard University, USA
Andre Scedrov	University of Pennsylvania, USA
Dana Scott	Carnegie-Mellon University, USA
Dirk van Dalen	Utrecht University, The Netherlands

Program Committee

Sergei Artemov (Chair)	The City University of New York Graduate Center, USA
Eugene Asarin	Université Paris Diderot - Paris 7, France
Steve Awodey	Carnegie-Mellon University, USA
Lev Beklemishev	Steklov Mathematical Institute, Moscow, Russia
Andreas Blass	University of Michigan, USA
Samuel Buss	University of California, San Diego, USA
Robert Constable	Cornell University, USA
Thierry Coquand	University of Gothenburg, Sweden
Valeria de Paiva	Nuance Communications, Sunnyvale, USA
Nachum Dershowitz	Tel Aviv University, Israel
Melvin Fitting	The City University of New York, USA
Sergey Goncharov	Sobolev Institute of Mathematics, Novosibirsk, Russia
Denis Hirschfeldt	University of Chicago, USA
Rosalie Iemhoff	Utrecht University, The Netherlands
Hajime Ishihara	Japan Advanced Institute of Science and Technology, Japan
Bakhadyr Khoussainov	The University of Auckland, New Zealand
Roman Kuznets	Vienna University of Technology, Austria
Daniel Leivant	Indiana University Bloomington, USA
Robert Lubarsky	Florida Atlantic University, USA
Victor Marek	University of Kentucky, USA
Lawrence Moss	Indiana University Bloomington, USA
Anil Nerode	Cornell University, USA
Hiroakira Ono	Japan Advanced Institute of Science and Technology, Japan
Alessandra Palmigiano	Delft University of Technology, The Netherlands
Ruy de Queiroz	The Federal University of Pernambuco, Brazil

Ramaswamy Ramanujam	The Institute of Mathematical Sciences, Chennai, India
Michael Rathjen	University of Leeds, UK
Helmut Schwichtenberg	University of Munich, Germany
Sebastiaan Terwijn	Radboud University, The Netherlands

Additional Reviewers

Joachim Biskup
Hans van Ditmarsch
Peter Habermehl
Stepan Kuznetsov
Paul-André Melliès
Amaury Pouly
Katsuhiko Sano
S. P. Suresh
Yi Wang

Contents

Computability of Algebraic and Definable Closure

Nathanael Ackerman[1], Cameron Freer[2(✉)], and Rehana Patel[3]

[1] Harvard University, Cambridge, MA 02138, USA
`nate@aleph0.net`
[2] Massachusetts Institute of Technology, Cambridge, MA 02139, USA
`freer@mit.edu`
[3] African Institute for Mathematical Sciences, M'bour–Thiès, Senegal
`rpatel@aims-senegal.org`

Abstract. We consider computability-theoretic aspects of the algebraic and definable closure operations for formulas. We show that for φ a Boolean combination of Σ_n-formulas and in a given computable structure, the set of parameters for which the closure of φ is finite is Σ_{n+2}^0, and the set of parameters for which the closure is a singleton is Δ_{n+2}^0. In addition, we construct examples witnessing that these bounds are tight.

Keywords: Algebraic closure · Definable closure · Computable model theory

1 Introduction

An important step towards understanding the relationship between model theory and computability theory is to calibrate the effective content of concepts that are fundamental in classical model theory. There is a long history of efforts to understand this calibration within computable model theory; see, e.g., [4].

In this paper, we study the computability of two particular model-theoretic concepts, namely the related notions of *algebraic closure* and *definable closure*, which provide natural characterizations of a "neighborhood" of a set; for more details, see [5, §4.1]. In recent years, the property of a structure having *trivial* definable closure (i.e., the definable closure of every finite set is itself), or equivalently, trivial algebraic closure, has played an important role in combinatorial model theory and descriptive set theory; for some characterizations in terms of this property see, e.g., [1–3].

The standard notions of algebraic and definable closure can be refined by carrying out a formula-by-formula analysis. We consider the computational strength of the problem of identifying the algebraic or definable closure of a *formula* in a computable structure, and we give tight bounds on the complexity of both. Further, when the formula is quantifier-free, we achieve tightness of these bounds via structures that are model-theoretically "nice", namely, are $\aleph_0$-categorical or of finite Morley rank.

© Springer Nature Switzerland AG 2020
S. Artemov and A. Nerode (Eds.): LFCS 2020, LNCS 11972, pp. 1–11, 2020.
https://doi.org/10.1007/978-3-030-36755-8_1

1.1   Preliminaries

For standard notions from computability theory, see, e.g., [6]. We write $\{e\}(n)$ to represent the output of the eth Turing machine run on input n, if it converges, and in this case write $\{e\}(n)\downarrow$. Define $W_e := \{n \in \mathbb{N} : \{e\}(n)\downarrow\}$ and Fin $:= \{e \in \mathbb{N} : W_e \text{ is finite}\}$. Recall that Fin is Σ^0_2-complete [6, Theorem 4.3.2]).

In this paper we will focus on computable languages that are relational. Note that this leads to no loss of generality due to the standard fact that computable languages with function or constant symbols can be interpreted computably in relational languages where there is a relation for the graph of each function. For the definitions of languages, first-order formulas, and structures, see [5].

We will work with many-sorted languages and structures; for more details, see [7, §1.1]. Let $\mathcal{L}$ be a (many-sorted) language, let $\mathcal{A}$ be an $\mathcal{L}$-structure, and suppose that $\bar{a}$ is a tuple of elements of $\mathcal{A}$. We say that the **type** of $\bar{a}$ is $\prod_{i \leq n} X_i$ when $\bar{a} \in \prod_{i \leq n}(X_i)^{\mathcal{A}}$, where each of $X_0, \ldots, X_{n-1}$ is a sort of $\mathcal{L}$. The type of a tuple of variables is the product of the sorts of its constituent variables (in order). The type of a relation symbol is defined to be the type of the tuple of its free variables, and similarly for formulas. We write $(\forall \bar{x} : X)$ and $(\exists \bar{x} : X)$ to quantify over a tuple of variables $\bar{x}$ of type X (which includes the special case of a single variable of a given sort).

If we so desired, we could encode each sort using a unary relation symbol, and this would not affect most of our results. However, in Sect. 3 we are interested in how model-theoretically complicated the structures we build are, and if we do not allow sorts then the construction in Proposition 9 providing a lower bound on the complexity of algebraic closure will not yield an $\aleph_0$-categorical structure.

We now define computable languages and structures.

Definition 1. *Suppose* $\mathcal{L} = \big((X_j)_{j \in J}, (R_i)_{i \in I}\big)$ *is a language, where* $I, J \in \mathbb{N} \cup \{\mathbb{N}\}$ *and* $(X_j)_{j \in J}$ *and* $(R_i)_{i \in I}$ *are collections of sorts and relation symbols, respectively. Let* $\text{ty}_{\mathcal{L}}: I \to J^{<\omega}$ *be such that for all* $i \in I$*, we have* $\text{ty}_{\mathcal{L}}(i) = (j_0, \ldots, j_{n-1})$ *where the type of* R_i *is* $\prod_{k<n} X_{j_k}$*. We say that* $\mathcal{L}$ *is a* **computable language** *when* $\text{ty}_{\mathcal{L}}$ *is a computable function. For each computable language, we fix a computable encoding of all first-order formulas of the language.*

A **computable $\mathcal{L}$-structure** $\mathcal{A}$ *is an $\mathcal{L}$-structure with computable underlying set such that the sets* $\{(a, j) : a \in X_j^{\mathcal{A}}\}$ *and* $\{(\bar{b}, i) : \bar{b} \in R_i^{\mathcal{A}}\}$ *are computable subsets of the appropriate domains.*

We say that $c \in \mathbb{N}$ *is a* **code for a structure** *if* $\{c\}(0)$ *is a code for a computable language (via some fixed enumeration of functions of the form* $\text{ty}_{\mathcal{L}}$*) and* $\{c\}(1)$ *is a code for some computable structure in that language. In this case, we write* $\mathcal{L}_c$ *for the language that* $\{c\}(0)$ *codes,* $\mathcal{M}_c$ *for the structure that* $\{c\}(1)$ *codes, and* T_c *for the first-order theory of* $\mathcal{M}_c$*. Let* CompStr *be the collection of* $c \in \mathbb{N}$ *that are codes for structures.*

Note that these notions relativize in the obvious way. For more details on basic notions in computable model theory, see [4].

Towards defining algebraic closure and definable closure for formulas, we first describe when a formula is algebraic or definable at a given tuple.

Definition 2. *Let $\varphi(\overline{x}; \overline{y})$ be a first-order $\mathcal{L}$-formula, let $\mathcal{A}$ be an $\mathcal{L}$-structure, and suppose $\overline{a} \in \mathcal{A}$ has the same type as $\overline{x}$.*

- *The formula $\varphi(\overline{x}; \overline{y})$ is* **algebraic at** $\overline{a}$ *if*

$$\mathrm{cl}_{\varphi, \mathcal{A}}(\overline{a}) := \{\overline{b} \in \mathcal{A} : \mathcal{A} \models \varphi(\overline{a}; \overline{b})\}$$

  is finite (possibly empty).
- *The formula $\varphi(\overline{x}; \overline{y})$ is* **definable at** $\overline{a}$ *if* $\left| \mathrm{cl}_{\varphi, \mathcal{A}}(\overline{a}) \right| = 1$.

We now describe several sets that encode those formulas that are algebraic or definable at given tuples. These are our analogues, for individual formulas, of the standard notions of algebraic closure and definable closure. See [5, §4.1] for more details on these standard notions.

Definition 3.

- $\mathrm{CL} := \big\{ (c, \varphi(\overline{x}; \overline{y}), \overline{a}, k) \ : \ c \in \mathrm{CompStr}, \ \varphi(\overline{x}; \overline{y})$ a first-order $\mathcal{L}_c$-formula, $\overline{a} \in \mathcal{M}_c$ having the same type as $\overline{x}$, and $k \in \mathbb{N} \cup \{\infty\}$ with $|\mathrm{cl}_{\varphi, \mathcal{M}_c}(\overline{a})| = k \big\}$.
- $\mathrm{ACL} := \big\{ (c, \varphi(\overline{x}; \overline{y}), \overline{a}) \ : \$ there exists $k \in \mathbb{N}$ with $(c, \varphi(\overline{x}; \overline{y}), \overline{a}, k) \in \mathrm{CL} \big\}$.
- $\mathrm{DCL} := \big\{ (c, \varphi(\overline{x}; \overline{y}), \overline{a}) \ : \ (c, \varphi(\overline{x}; \overline{y}), \overline{a}, 1) \in \mathrm{CL} \big\}$.
- *For $Y \in \{\mathrm{CL}, \mathrm{ACL}, \mathrm{DCL}\}$ and $n \in \mathbb{N}$ let*

$$Y_n := \{t \in Y \ : \ \text{the second coordinate of } t \text{ is a Boolean combination of}$$
$$\Sigma_n\text{-formulas}\}.$$

- *For $Y \in \{\mathrm{CL}, \mathrm{ACL}, \mathrm{DCL}\} \cup \{\mathrm{CL}_n, \mathrm{ACL}_n, \mathrm{DCL}_n\}_{n \in \mathbb{N}}$ and $c \in \mathrm{CompStr}$, let $Y^c := \{u \ : \ (c)^\wedge u \in Y\}$, i.e., select those elements of Y whose first coordinate is c, and then remove this first coordinate.*

Note that CompStr is a Π_2^0 class. Hence even before we consider the complexity of whether formulas are algebraic or definable at various tuples, the sets $\mathrm{CL}, \mathrm{ACL}, \mathrm{DCL}$ are already complicated computability-theoretically. As such, we will mainly be interested in the question of how complex $\mathrm{CL}^c, \mathrm{ACL}^c, \mathrm{DCL}^c$ can be, when c is a code for a structure. The next three lemmas connect these sets.

Lemma 4. *Uniformly in the parameters $c \in \mathrm{CompStr}$ and $n \in \mathbb{N}$, the set*

$$\{(\varphi(\overline{x}; \overline{y}), \overline{a}, k) \in \mathrm{CL}_n^c \ : \ k \in \mathbb{N}, \ k \geq 1\}$$

is computably enumerable from DCL_n^c.

Proof. Suppose $\varphi(\overline{x}; \overline{y})$ is a Boolean combination of Σ_n-formulas, and let $k \geq 1$. For each $j < k$, choose a tuple of new variables $\overline{z}^j$ of the same type as $\overline{y}$. Define the formula

$$\Phi_{\varphi(\overline{x}; \overline{y}), k} := \bigwedge_{k_0 < k_1 < k} (\overline{z}^{k_0} \neq \overline{z}^{k_1}) \wedge \bigwedge_{k_1 < k} \varphi(\overline{x}; \overline{z}^{k_1})$$

which specifies k-many distinct realizations of the tuple $\overline{y}$ in $\varphi(\overline{x}; \overline{y})$, given an instantiation of $\overline{x}$. Note that $\Phi_{\varphi(\overline{x};\overline{y}),k}$ is also a Boolean combination of Σ_n-formulas.

For $j < k$, let $\tau_j := \overline{x}\,\overline{z}^0 \cdots \overline{z}^{j-1}\overline{z}^{j+1} \cdots \overline{z}^{k-1}$, and write $\Phi_{\varphi(\overline{x};\overline{y}),k}(\tau_j; \overline{z}^j)$ to mean $\Phi_{\varphi(\overline{x};\overline{y}),k}$ considered as a formula whose free variables are partitioned as $(\tau_j, \overline{z}^j)$. Note that $(\varphi(\overline{x}; \overline{y}), \overline{a}, k) \in \mathrm{CL}_n^c$ if and only if

$$\left(\Phi_{\varphi(\overline{x};\overline{y}),k}(\tau_j; \overline{z}^j), \overline{a}\,\overline{b}^0 \cdots \overline{b}^{j-1}\overline{b}^{j+1} \cdots \overline{b}^{k-1} \right) \in \mathrm{DCL}_n^c$$

for some $j < k$ and $\overline{b}^0, \ldots, \overline{b}^{j-1}, \overline{b}^{j+1}, \ldots, \overline{b}^{k-1} \in \mathcal{M}_c$. By enumerating over all such parameters, and enumerating over all choices of φ and k, we see that the desired set is c.e. from DCL_n^c. □

Lemma 5. *Uniformly in the parameters $c \in \mathrm{CompStr}$ and $n \in \mathbb{N}$, the set*

$$\left\{ (\varphi(\overline{x}; \overline{y}), \overline{a}, k) \in \mathrm{CL}_n^c \ : \ k = 0 \right\}$$

is computably enumerable from DCL_n^c.

Proof. Suppose $\varphi(\overline{x}; \overline{y})$ is a Boolean combination of Σ_n-formulas. Let $\overline{z}$ be a tuple of variables having the same type as $\overline{y}$ and disjoint from $\overline{x}\,\overline{y}$. Let $\Psi_{\varphi(\overline{x};\overline{y})}(\overline{x}\,\overline{z}; \overline{y}) := \varphi(\overline{x}; \overline{y}) \vee (\overline{y} = \overline{z})$. Note that $\Psi_{\varphi(\overline{x};\overline{y})}(\overline{x}\,\overline{z}; \overline{y})$ is also a Boolean combination of Σ_n-formulas.

Now suppose $\overline{b}_0$ and $\overline{b}_1$ are distinct tuples of elements of $\mathcal{M}_c$ having the same type as $\overline{z}$. Then the following are equivalent:

- $\left(\Psi_{\varphi(\overline{x};\overline{y})}(\overline{x}\,\overline{z}; \overline{y}), \overline{a}\,\overline{b}_0 \right) \in \mathrm{DCL}_n^c$ and $\left(\Psi_{\varphi(\overline{x};\overline{y})}(\overline{x}\,\overline{z}; \overline{y}), \overline{a}\,\overline{b}_1 \right) \in \mathrm{DCL}_n^c$;
- $\left\{ \overline{b} : \mathcal{M}_c \models \varphi(\overline{a}; \overline{b}) \right\} = \emptyset$, i.e., $(\varphi(\overline{x}; \overline{y}), \overline{a}, 0) \in \mathrm{CL}_n^c$.

The result is then immediate. □

Lemma 6. *Uniformly in the parameters $c \in \mathrm{CompStr}$ and $n \in \mathbb{N}$, there are computable reductions in both directions between $\mathrm{ACL}_n^c \coprod \mathrm{DCL}_n^c$ and CL_n^c.*

Proof. It is immediate from the definitions that DCL_n^c is computable from CL_n^c. Further, ACL_n^c is computable from CL_n^c as

$$\mathrm{ACL}_n^c = \left\{ (\varphi(\overline{x}; \overline{y}), \overline{a}) \ : \ (\exists k) \ (\varphi(\overline{x}; \overline{y}), \overline{a}, k) \in \mathrm{CL}_n^c \text{ and } k \neq \infty \right\}$$

and $(\varphi(\overline{x}; \overline{y}), \overline{a}, k) \in \mathrm{CL}_n^c$ holds for a unique $k \in \mathbb{N} \cup \{\infty\}$.

Lemmas 4 and 5 together tell us that $\{(\varphi(\overline{x}; \overline{y}), \overline{a}, k) \in \mathrm{CL}_n^c \ : \ k \in \mathbb{N}\}$ is computably enumerable from DCL_n^c. But $(\varphi(\overline{x}; \overline{y}), \overline{a}, \infty) \in \mathrm{CL}_n^c$ if and only if $(\varphi(\overline{x}; \overline{y}), \overline{a}) \notin \mathrm{ACL}_n^c$. Therefore when $\varphi(\overline{x}; \overline{y})$ is a Boolean combination of Σ_n-formulas, and given $\overline{a} \in \mathcal{M}_c$, we can compute from ACL_n^c whether or not $(\varphi(\overline{x}; \overline{y}), \overline{a}, \infty) \in \mathrm{CL}_n^c$. Further, if $(\varphi(\overline{x}; \overline{y}), \overline{a}, \infty) \notin \mathrm{CL}_n^c$, then we can compute from DCL_n^c the (unique) value of k such that $(\varphi(\overline{x}; \overline{y}), \overline{a}, k) \in \mathrm{CL}_n^c$. Hence CL_n^c is computable from $\mathrm{ACL}_n^c \coprod \mathrm{DCL}_n^c$. □

Note that by Lemma 6 we are justified, from a computability-theoretic perspective, in restricting our attention to ACL and DCL (and their variants), as opposed to CL.

2   Upper Bounds for Quantifier-Free Formulas

We now provide straightforward upper bounds on the complexity of ACL_0^c and DCL_0^c for $c \in \mathrm{CompStr}$.

Proposition 7. *Uniformly in the parameter* $c \in \mathrm{CompStr}$, *the set* ACL_0^c *is a* Σ_2^0 *class.*

Proof. Uniformly in $c \in \mathrm{CompStr}$, a quantifier-free $\mathcal{L}_c$-formula $\varphi(\overline{x}; \overline{y})$, and tuple $\overline{a} \in \mathcal{M}_c$ of the same type as $\overline{x}$, we can computably find an $e \in \mathbb{N}$ such that W_e equals $\mathrm{cl}_{\varphi, \mathcal{M}_c}(\overline{a})$ (where the elements of $\mathrm{cl}_{\varphi, \mathcal{M}_c}(\overline{a})$ are encoded in $\mathbb{N}$ in a standard way).

Further, $(\varphi(\overline{x}; \overline{y}), \overline{a}) \in \mathrm{ACL}_0^c$ if and only if $\mathrm{cl}_{\varphi, \mathcal{M}_c}(\overline{a})$ is finite. Therefore ACL_0^c is Σ_2^0 as Fin is Σ_2^0. □

Proposition 8. *Uniformly in the parameter* $c \in \mathrm{CompStr}$, *the set* DCL_0^c *is the intersection of a* Π_1^0 *and a* Σ_1^0 *class (in particular, it is a* Δ_2^0 *class).*

Proof. Uniformly in $c \in \mathrm{CompStr}$, the set of all tuples $(\varphi(\overline{x}; \overline{y}), \overline{a})$ such that

$$\mathcal{M}_c \models (\forall \overline{y_0}, \overline{y_1}) \left(((\varphi(\overline{a}; \overline{y_0}) \wedge \varphi(\overline{a}; \overline{y_1})) \rightarrow (\overline{y_0} = \overline{y_1}) \right)$$

holds is a Π_1^0 class. Likewise, uniformly in $c \in \mathrm{CompStr}$, the set of all tuples $(\varphi(\overline{x}; \overline{y}), \overline{a})$ such that there exists $\overline{b}$ with $\mathcal{M}_c \models \varphi(\overline{a}; \overline{b})$ is a Σ_1^0 class. □

As a consequence, DCL_0^c is computable from $\mathbf{0}'$.

3   Lower Bounds for Quantifier-Free Formulas

We now show that the upper bounds in Sect. 2 are tight. Further, we do so using structures that have nice model-theoretic properties.

We first show that the upper bound in Proposition 7 is tight.

Proposition 9. *There is a parameter* $c \in \mathrm{CompStr}$ *such that the following hold.*

(a) $\mathcal{L}_c$ *has no relation symbols, i.e.,* $\mathcal{L}_c$ *consists only of sorts.*
(b) *For each ordinal* α, *the theory* T_c *has* $(|\alpha + 1|^\omega)$-*many models of size* $\aleph_\alpha$. *In particular,* T_c *is* $\aleph_0$-*categorical.*
(c) $\mathrm{ACL}_0^c \equiv_1 \mathrm{Fin}$. *In particular,* ACL_0^c *is a* Σ_2^0-*complete set.*

Proof. Let $\big((e_i, n_i)\big)_{i \in \mathbb{N}}$ be a computable enumeration without repetition of

$$\{(e, n) : e, n \in \mathbb{N} \text{ and } \{e\}(n)\downarrow\}.$$

Note that for each $\ell \in \mathbb{N} \cup \{\infty\}$, there are infinitely many programs that halt on exactly ℓ-many inputs, and so there are infinitely many $e \in \mathbb{N}$ that are equal to e_i for exactly ℓ-many i.

Let $c \in \mathrm{CompStr}$ be such that

- $\mathcal{L}_c$ consists of infinitely many sorts $(X_i)_{i \in \mathbb{N}}$ and no relation symbols,
- the underlying set of $\mathcal{M}_c$ is $\mathbb{N}$, and
- for each $i \in \mathbb{N}$, the element i is of sort X_{e_i} in $\mathcal{M}_c$.

A model of T_c is determined up to isomorphism by the number of elements in the instantiation of each sort. Hence there are $\aleph_0$-many sorts of each finite size and $\aleph_0$-many that are infinite (each of which may have size $\aleph_\beta$ for arbitrary $\beta \le \alpha$, in a model of size $\aleph_\alpha$), and so (b) holds.

Now $|W_e| = |(X_e)^{\mathcal{M}_c}|$ and so Fin is 1-equivalent to $\{e : (X_e)^{\mathcal{M}_c}$ is finite$\}$. Recall that each variable in a many-sorted language is assigned a single sort, and so no non-trivial Boolean combination of instantiations of sorts is definable. Since there are no relation symbols in $\mathcal{L}_c$, every quantifier-free definable set is contained in some product of instantiations of sorts, and is itself the product of finite or cofinite subsets of instantiations of sorts. Therefore ACL_0^c is 1-equivalent to $\{e : (X_e)^{\mathcal{M}_c}$ is finite$\}$ as well, establishing (c).   □

We now show that the upper bound in Proposition 8 is tight.

Proposition 10. *There is a parameter $c \in$ CompStr such that the following hold.*

(a) The language $\mathcal{L}_c$ has one sort and a single binary relation symbol E.

(b) The structure $\mathcal{M}_c$ is a countable saturated model of T_c with underlying set $\mathbb{N}$.

(c) For each ordinal α, the theory T_c has $(|\alpha + \omega|)$-many models of size $\aleph_\alpha$, and has finite Morley rank.

(d) There is a computable array $(U_{k,\ell})_{k,\ell \in \mathbb{N}}$ of subsets of $\mathbb{N}$ such that every countable model of T_c is isomorphic to the restriction of $\mathcal{M}_c$ to underlying set $U_{k,\ell}$ for exactly one pair (k, ℓ).

(e) If $\mathcal{N} \cong \mathcal{M}_c$ then uniformly in $\mathcal{N}$ we can compute $\mathbf{0}'$ from

$$\{a : |\{b : \mathcal{N} \models E(a; b)\}| = 1\}.$$

(f) The set

$$\{a : (E(x; y), a) \in \mathrm{DCL}_0^c\}$$

has Turing degree $\mathbf{0}'$.

Proof. Let $g : \mathbb{N} \to \{0, 1\}$ be the characteristic function of $\mathbf{0}'$, i.e., such that $g(n) = 1$ if and only if $n \in \mathbf{0}'$. As $\mathbf{0}'$ is a Δ_2^0 set, there is some computable function $f : \mathbb{N} \times \mathbb{N} \to \{0, 1\}$ such that $\lim_{s \to \infty} f(n, s) = g(n)$ for all $n \in \mathbb{N}$.

We will construct $\mathcal{M}_c$ in the language specified in (a) so as to satisfy the following axioms.

- $(\forall x)\, \neg E(x, x)$
- $(\forall x, y)\, (E(x, y) \to E(y, x))$
- $(\forall x)(\exists y)\, E(x, y)$
- $(\forall x)(\exists^{\le 2} y)\, E(x, y)$

By a *graph* we mean a structure with a single undirected irreflexive binary relation. A *chain* in a graph is a connected component of the graph each of whose vertices has degree 1 or 2; hence a chain either is finite with at least two vertices, or is infinite on one side (an N-chain), or is infinite on both sides (a Z-chain). By the *order* of a chain we mean its number of vertices.

The above axioms specify that $\mathcal{M}_c$ will be a graph (with edge relation E) that is the union of chains. In fact, we will construct $\mathcal{M}_c$ so as to have infinitely many chains of certain finite orders, infinitely many N-chains, and infinitely many Z-chains.

For $n \in \mathbb{N}$, let p_n denote the nth prime number. We now construct $\mathcal{M}_c$ with underlying set $\mathbb{N}$, in stages.

Stage 0:
Let $\{N_i\}_{i \in \mathbb{N}} \cup \{Z_i\}_{i \in \mathbb{N}} \cup \{F\}$ be a uniformly computable partition of $\mathbb{N}$ into infinite sets.

For each $i \in \mathbb{N}$, let the induced subgraph on N_i be an N-chain, and let the induced subgraph on Z_i be a Z-chain. The only other edges will be between elements of F (to be determined in later stages).

Stage $2s + 1$:
Let a_s be the least element of F that is not yet part of an edge. Create a finite chain of order $(p_s)^{2+f(s,s)}$ consisting of a_s and other elements of F not yet in any edge.

Stage $2s + 2$:
For each $n \leq s$, we have two cases, based on the values of f. If $f(n, s) = f(n, s + 1)$, do nothing.

Otherwise, if $f(n, s) \neq f(n, s + 1)$, consider the (unique) chain whose order so far is $(p_n)^k$ for some positive k. Extend this chain by $((p_n)^{k+1} - (p_n)^k)$-many elements of F which are not yet in any edge, so that the resulting chain has order $(p_n)^{2\ell + f(n,s+1)}$ for some $\ell \in \mathbb{N}$.

The resulting graph is computable, as every vertex participates in at least one edge, and whether or not there is an edge between a given pair of vertices is determined by the first stage at which each vertex of the pair becomes part of some edge.

Observe that every element of F is part of a chain of elements of F whose order is some positive power of a prime, which moreover is the only chain in $\mathcal{M}_c$ whose order is a power of that prime.

Now, every model of T_c is determined by the number of N-chains and the number of Z-chains in it. In a model of size $\aleph_\alpha$, there must be either $\aleph_\alpha$-many N-chains and 0-, 1-, …, $\aleph_0$-, …, or $\aleph_\alpha$-many Z-chains, or vice-versa. Condition (b) holds because the countable saturated models of T_c have $\aleph_0$-many N-chains and $\aleph_0$-many Z-chains, as does $\mathcal{M}_c$. Condition (c) holds because none of these N-chains or Z-chains are first-order definable.

For condition (d), let $U_{k,\ell} := \bigcup_{i<k} N_i \cup \bigcup_{i<\ell} Z_i \cup F$.

Towards condition (e), note that for each $n \in \mathbb{N}$, there is a unique chain of order a power of p_n. Writing $(p_n)^{j_n}$ for this order, we have $j_n \equiv g(n) \pmod 2$. An element $a \in \mathcal{N}$ is one of the two ends of a finite chain or the beginning of an $\mathbb{N}$-chain if and only if $|\{b : \mathcal{N} \models E(a;b)\}| = 1$. So, from the set $\{a : |\{b : \mathcal{N} \models E(a;b)\}| = 1\}$ we can enumerate the orders of all finite chains, and hence can compute $g(n)$ for all n.

Finally, recall that DCL_0^c is computable from $\mathbf{0}'$ and so the set $\{a : (E(x;y),a) \in \mathrm{DCL}_0^c\}$ is also computable from $\mathbf{0}'$. Hence (f) follows from (e). □

4   Boolean Combinations of Σ_n-Formulas

We now study the complexity of ACL^c and DCL^c with respect to Boolean combinations of Σ_n-formulas.

The following lemma captures a computable version of the standard process known as *Morleyization*. The proof is straightforward.

Lemma 11. *Let $\mathcal{L}$ be a computable language and $\mathcal{A}$ a computable $\mathcal{L}$-structure. For each $n \in \mathbb{N}$ there is a computable language $\mathcal{L}_n$ and a $\mathbf{0}^{(n)}$-computable $\mathcal{L}_n$-structure $\mathcal{A}_n$ such that*

- $\mathcal{L} \subseteq \mathcal{L}_n \subseteq \mathcal{L}_{n+1}$,
- $\mathcal{A}$ *is the reduct of $\mathcal{A}_n$ to the language $\mathcal{L}$,*
- *for each first-order $\mathcal{L}_n$-formula φ there is a first-order $\mathcal{L}$-formula ψ_φ (of the same type as φ) such that*

$$\mathcal{A}_n \models (\forall x_0, \ldots, x_{k-1})\ \varphi(x_0, \ldots, x_{k-1}) \leftrightarrow \psi_\varphi(x_0, \ldots, x_{k-1}),$$

  where k is the number of free variables of φ, and
- *for each first-order $\mathcal{L}$-formula ψ, if ψ is a Boolean combination of Σ_n-formulas then there is a first-order quantifier-free $\mathcal{L}_n$-formula φ_ψ (of the same type as ψ) such that*

$$\mathcal{A}_n \models (\forall x_0, \ldots, x_{k-1})\ \psi(x_0, \ldots, x_{k-1}) \leftrightarrow \varphi_\psi(x_0, \ldots, x_{k-1}),$$

  where k is the number of free variables of ψ.

Lemma 11 tells us that the methods used earlier in this paper to study quantifier-free algebraic and definable closures can be applied to more complicated formulas, provided that we allow the structures that we build to have greater complexity, as we now illustrate.

Corollary 12. *For every $n \in \mathbb{N}$ and $c \in \mathrm{CompStr}$,*

- ACL_n^c *is a Σ_{n+2}^0 class, and*
- DCL_n^c *is a Δ_{n+2}^0 class.*

Proof. By Lemma 11, we know that ACL_n is equivalent to the relativization of ACL_0 to the class of structures computable in $\mathbf{0}^{(n)}$, and that DCL_n is equivalent to the relativization of DCL_0 to the class of structures computable in $\mathbf{0}^{(n)}$.

Therefore by Propositions 7 and 8, ACL_n^c is a $\Sigma_2^0(\mathbf{0}^{(n)})$ class and DCL_n^c is a $\Delta_2^0(\mathbf{0}^{(n)})$ class. □

In Theorem 15 we will show that these bounds are tight. Towards this, we will need the next two results.

Suppose that $\mathcal{L}$ is a language containing a sort N and a relation symbol S of type $N \times N$. Let $\mathcal{A}$ be an $\mathcal{L}$-structure. We call $(N^{\mathcal{A}}, S^{\mathcal{A}})$ a **directed N-chain** when it is isomorphic to a single-sorted structure with underlying set $\mathbb{N}$ in a language consisting of the binary relation symbol S, in which $S(k, \ell)$ holds precisely when $\ell = k + 1$. In other words, $(N^{\mathcal{A}}, S^{\mathcal{A}})$ is a directed $\mathbb{N}$-chain if there is an isomorphism between it and $\mathbb{N}$ with its successor function viewed as a directed graph. Note that this isomorphism is necessarily unique. Given $\ell \in \mathbb{N}$, we write $\widehat{\ell}$ to denote the corresponding element of $N^{\mathcal{A}}$ according to this isomorphism.

Lemma 13. *Let $\mathcal{L}$ be a language containing a sort N and a relation symbol S of type $N \times N$ (and possibly other sorts and relation symbols). Let $\mathcal{A}$ be an $\mathcal{L}$-structure such that $(N^{\mathcal{A}}, S^{\mathcal{A}})$ is a directed $\mathbb{N}$-chain. Let $k \in \mathbb{N}$ and let $h(\overline{x}, m)$ be an $\mathcal{L}$-formula that is a Boolean combination of Σ_k-formulas, where $\overline{x}$ is of some type X, and m has sort N.*

Suppose that

$$\mathcal{A} \models (\forall \overline{x} : X)(\exists^{\leq 1} m : N)(\exists p : N) \; S(m, p) \wedge \big(h(\overline{x}, m) \leftrightarrow \neg h(\overline{x}, p) \big).$$

Let $H : X^{\mathcal{A}} \times \mathbb{N} \to \{\mathrm{True}, \mathrm{False}\}$ be the function where $H(\overline{a}, \ell) = \mathrm{True}$ if and only if $\mathcal{A} \models h(\overline{a}, \widehat{\ell})$. Note that $\lim_{\ell \to \infty} H(\overline{a}, \ell)$ exists for all $\overline{a} \in X^{\mathcal{A}}$.

There is an $\mathcal{L}$-formula $h'(\overline{x})$, where $\overline{x}$ is of type X, such that h' is a Boolean combination of Σ_{k+1}-formulas and for all $\overline{a} \in X^{\mathcal{A}}$,

$$\mathcal{A} \models h'(\overline{a}) \quad \text{if and only if} \quad \lim_{m \to \infty} H(\overline{a}, m) = \mathrm{True}.$$

Proof. Define the formula h' by

$$h'(\overline{x}) := \big[(\forall m : N) \, h(\overline{x}, m) \big] \quad \vee \quad \big[(\exists m, p : N) \, \big(\neg h(\overline{x}, m) \wedge h(\overline{x}, p) \wedge S(m, p) \big) \big].$$

Clearly h' is a Boolean combination of Σ_{k+1}-formulas and has the desired property. □

Proposition 14. *Let $n \in \mathbb{N}$ and let $\mathcal{L}$ be a language containing a sort N and a relation symbol S of type $N \times N$ (and possibly other sorts and relation symbols). Suppose $\mathcal{A}$ is an $\mathcal{L}$-structure that is computable in $\mathbf{0}^{(n)}$ and such that $(N^{\mathcal{A}}, S^{\mathcal{A}})$ is a computable directed $\mathbb{N}$-chain. Then there is a computable language $\mathcal{L}^+$ and a computable $\mathcal{L}^+$-structure $\mathcal{A}^+$ such that for every relation symbol $R \in \mathcal{L}$ other than S, there is an $\mathcal{L}^+$-formula φ_R that is a Boolean combination of Σ_n-formulas for which $R^{\mathcal{A}} = (\varphi_R)^{\mathcal{A}^+}$.*

Proof. We begin by defining, for relation symbols in $\mathcal{L}$ other than S, certain auxiliary functions. Let R be a relation symbol in $\mathcal{L}$ that is not S, and let X be its type. For every $k \in \mathbb{N}$ such that $0 \leq k \leq n$, there is some $\mathbf{0}^{(n-k)}$-computable function $F_{R,k} \colon X^{\mathcal{A}} \times \mathbb{N}^k \to \{\text{True}, \text{False}\}$ such that for all $\bar{a} \in X^{\mathcal{A}}$, the following hold:

- $F_{R,0}(\bar{a}) = 1$ if and only if $\mathcal{A} \models R(\bar{a})$.
- Suppose $k \geq 1$ and let $\ell_0, \ldots, \ell_{k-2} \in \mathbb{N}$. There is at most one $s \in \mathbb{N}$ for which

$$F_{R,k}(\bar{a}, \ell_0, \ldots, \ell_{k-2}, s) \neq F_{R,k}(\bar{a}, \ell_0, \ldots, \ell_{k-2}, s+1).$$

Further,

$$F_{R,k-1}(\bar{a}, \ell_0, \ldots, \ell_{k-2}) = \lim_{\ell_{k-1} \to \infty} F_{R,k}(\bar{a}, \ell_0, \ldots, \ell_{k-2}, \ell_{k-1}).$$

Next we define the computable language $\mathcal{L}^+$ as follows:

- $\mathcal{L}^+$ has the same sorts as $\mathcal{L}$.
- For each relation symbol $R \in \mathcal{L}$ other than S, there is a relation symbol $R^+ \in \mathcal{L}^+$ of type $X \times N^n$, where X is the type of R.

Now define the computable $\mathcal{L}^+$-structure $\mathcal{A}^+$ as follows:

- $\mathcal{A}^+$ has the same underlying set as $\mathcal{A}$, and sorts are instantiated on the same sets in $\mathcal{A}^+$ as in $\mathcal{A}$.
- $S^{\mathcal{A}^+}$ is the same relation as $S^{\mathcal{A}}$.
- For each $R \in \mathcal{L}$ other than S, each tuple $\bar{a} \in X^{\mathcal{A}^+}$ where X is the type of R, and any $\ell_0, \ldots, \ell_{n-1} \in \mathbb{N}$, we have

$$\mathcal{A}^+ \models R^+(\bar{a}, \widehat{\ell_0}, \ldots, \widehat{\ell_{n-1}}) \text{ if and only if } F_{R,n}(\bar{a}, \ell_0, \ldots, \ell_{n-1}) = \text{True}.$$

(Recall that for $\ell \in \mathbb{N}$, we have defined $\widehat{\ell} \in N^{\mathcal{A}^+}$ to be the ℓth element of the directed $\mathbb{N}$-chain.)

Finally, we build, for each relation symbol $R \in \mathcal{L}$ other than S, an $\mathcal{L}^+$-formula φ_R. First apply Lemma 13 (with $k = 0$) to $\mathcal{A}^+$ and the $\mathcal{L}^+$-formula

$$h_0(\bar{x}y_0 \cdots y_{n-2}, y_{n-1}) := R^+(\bar{x}, y_0, \ldots, y_{n-1})$$

(where $\bar{x}$ has type X and each y_i has type N) to obtain an $\mathcal{L}^+$-formula $h_0'(\bar{x}y_0 \cdots y_{n-2})$ that is a Boolean combination of Σ_1-formulas. Next apply Lemma 13 again (with $k = 1$) to $\mathcal{A}^+$ and the $\mathcal{L}^+$-formula

$$h_1(\bar{x}y_0 \cdots y_{n-3}, y_{n-2}) := h_0'(\bar{x}y_0 \cdots y_{n-2})$$

to obtain an $\mathcal{L}^+$-formula $h_1'(\bar{x}y_0 \cdots y_{n-3})$ that is a Boolean combination of Σ_2-formulas. Proceed in this way for $k = 2, \ldots, n-1$, to obtain an $\mathcal{L}^+$-formula $\varphi_R(\bar{x}) := h_{n-1}'(\bar{x})$ that is a Boolean combination of Σ_n-formulas for which $R^{\mathcal{A}} = (\varphi_R)^{\mathcal{A}^+}$. □

Combining this with results from Sect. 3, we obtain the following.

Theorem 15. *For each $n \in \mathbb{N}$,*

(a) *there is an element $a \in$ CompStr such that ACL_n^a is a $\Sigma_2^0(\mathbf{0}^{(n)})$-complete set, and*

(b) *there is an element $b \in$ CompStr such that $\mathrm{DCL}_n^b \equiv_{\mathrm{T}} \mathbf{0}^{(n+1)}$.*

Proof. Let $\mathcal{P}$ be the structure constructed in the proof of Proposition 9, relativized to the oracle $\mathbf{0}^{(n)}$, i.e., so that $\mathcal{P}$ is computable from $\mathbf{0}^{(n)}$. Let the structure $\mathcal{P}^*$ be $\mathcal{P}$ augmented with a sort N (instantiated on a new set of elements) along with a relation symbol S of type $N \times N$, such that $(N^{\mathcal{P}^*}, S^{\mathcal{P}^*})$ is a computable directed $\mathbb{N}$-chain. Part (a) then follows by applying Proposition 14 to $\mathcal{P}^*$ to obtain some computable structure, namely $\mathcal{M}_a$ for some $a \in$ CompStr. Then ACL_n^a is a $\Sigma_2^0(\mathbf{0}^{(n)})$-complete set.

Let $\mathcal{Q}$ be the structure constructed in the proof of Proposition 10 relativized to the oracle $\mathbf{0}^{(n)}$, i.e., so that $\mathcal{Q}$ is computable from $\mathbf{0}^{(n)}$. Let the structure $\mathcal{Q}^*$ be obtained from $\mathcal{Q}$ by similarly augmenting it by N and S, so that $(N^{\mathcal{Q}^*}, S^{\mathcal{Q}^*})$ is a new computable directed $\mathbb{N}$-chain. Part (b) then follows by applying Proposition 14 to $\mathcal{Q}^*$ to obtain a computable structure $\mathcal{M}_b$ for some $b \in$ CompStr. Then $\mathrm{DCL}_n^b \equiv_{\mathrm{T}} \mathbf{0}^{(n+1)}$. □

Note that the structures constructed in Theorem 15 do not obviously have the nice model-theoretic properties ($\aleph_0$-categoricity or finite Morley rank) that those constructed in Propositions 9 and 10 do, because the application of Proposition 14 makes their theories more elaborate.

Question 16. Is there some $c \in$ CompStr such that ACL_n^c is a $\Sigma_2^0(\mathbf{0}^{(n)})$-complete set or $\mathrm{DCL}_n^c \equiv_{\mathrm{T}} \mathbf{0}^{(n+1)}$ and $\mathcal{M}_c$ is nice model-theoretically (e.g., $\aleph_0$-categorical, strongly minimal, stable, etc.)?

References

1. Ackerman, N., Freer, C., Patel, R.: Invariant measures concentrated on countable structures. Forum Math. Sigma **4**(e17), 59 (2016)
2. Chen, R., Kechris, A.S.: Structurable equivalence relations. Fund. Math. **242**(2), 109–185 (2018)
3. Cherlin, G., Shelah, S., Shi, N.: Universal graphs with forbidden subgraphs and algebraic closure. Adv. Appl. Math. **22**(4), 454–491 (1999)
4. Harizanov, V.S.: Pure computable model theory. In: Handbook of recursive mathematics, Logic and the Foundations of Mathematics, North-Holland, vol. 138, pp. 3–114 (1998)
5. Hodges, W.: Model theory, Encyclopedia of Mathematics and its Applications, vol. 42. Cambridge University Press, Cambridge (1993)
6. Soare, R.I.: Turing Computability. Theory and Applications of Computability. Springer, Berlin (2016)
7. Tent, K., Ziegler, M.: A Course in Model Theory. Lecture Notes in Logic, vol. 40. Cambridge University Press (2012)

Observable Models

Sergei Artemov$^{(\boxtimes)}$ (ID)

The City University of New York, The Graduate Center, 365 Fifth Avenue,
New York City, NY 10016, USA
sartemov@gc.cuny.edu

Abstract. Epistemic reading of Kripke models relies on a hidden
assumption of *common knowledge of the model* which is too restrictive in
epistemic contexts since agents may have different views of the situation.
We explore possible worlds models in their full generality without common knowledge assumptions. Our starting point is a collection of possible
worlds with accessibility relations *"whatever is known in u is true in v."*
We call such a structure an *observable model* since, contrary to the popular belief, it is not generally a Kripke model but rather an "observable
section" of some Kripke model. We sketch a theory of observable models
and argue that they bring a new conceptual clarity to epistemic modeling. In practical terms, observable models are as manageable as Kripke
models and have advantages over the latter in representing (un)awareness
and ignorance. Similar analysis applies to intuitionistic models.

Keywords: Modal logic · Epistemic logic · Intuitionistic logic · Kripke
models

1 Preliminaries

In this note we will try to present things at both levels, conceptual and technical.
On the formal side, we will focus on the propositional n-agent epistemic logic
$\mathsf{S5}^n$, cf. [4] though all the major findings and suggestions apply to other modal
logics as well. Similar considerations apply to other classes of observable models,
e.g., Aumann structures [1].

Informally, by a global state[1] of a multi-agent system we understand a complete description of epistemic states of agents along with the state of nature,
represented as a set of propositions in an appropriate epistemic language.

A *global state* is a maximal consistent (over a given logic base, among which
$\mathsf{S5}^n$ is the default) set of formulas. A set W of global states and truth assignment
to formulas at each world

$$w \models F \quad iff \quad F \in w$$

determines relations of epistemic accessibility uR_iv:

whatever agent i knows in u is true in v.

---

[1] In this text we will also be using terms *state* or *world* for global states, when convenient.

S. Artemov and A. Nerode (Eds.): LFCS 2020, LNCS 11972, pp. 12–26, 2020.
https://doi.org/10.1007/978-3-030-36755-8_2

Each such structure $(W, \models)$ has an *induced Kripke model* $(W, R_1, \ldots, R_n, \Vdash)$ with the same atomic evaluation $\Vdash$ as in $(W, \models)$

$$u \Vdash p \qquad \textit{iff} \qquad p \in u.$$

Truth assignments $\models$ and in $\Vdash$ coincide for the atomic propositions, but can differ for compound formulas, cf. Example 1.

For a preliminary version of this paper, cf. [2].

1.1 Motivations

We quote [6] for the standard approach to motivate Kripke models in epistemology[2]:

> *Informally, we interpret W as a set of mutually exclusive, jointly exhaustive worlds or states, ... R is a relation of epistemic accessibility: a world w has R to a world x if and only if ... whatever the agent knows in w is true in x. We define a function $\mathbf{K}$ from propositions to propositions by the following equation for all propositions p:*

$$\mathbf{K}p = \{w \in W : \forall x \in W, wRx \Rightarrow x \in p\}. \tag{1}$$

> *In other words, $\mathbf{K}p$ is true at a world if and only if p is true at every world epistemically accessible from that one. Informally, $\mathbf{K}p$ is interpreted as the proposition that the agent knows p.*

Formally, the characterization of R via knowledge at the states in W is

$$R(w) = \{x \in W : \forall p, \ \mathbf{K}p \in w \Rightarrow p \in x\}. \tag{2}$$

This yields

$$\mathbf{K}F \in w \quad \Rightarrow \quad \textit{for all } x \in R(w), \ F \in x. \tag{3}$$

However, this does not guarantee the converse:

$$(\textit{for all } x \in R(w), \ F \in x) \quad \Rightarrow \quad \mathbf{K}F \in w, \tag{4}$$

which is built into Definition (1).

Conceptually, the fact that F holds at some designated set of states should not automatically yield knowledge of F at a given state.

Technically, Eqs. (1) and (2) do not match. Given knowledge assertions at states of the model, we indeed can find accessibility relation R by (2), and then determine the knowledge modality $\mathbf{K}$ by (1). The problem is that this $\mathbf{K}p$ is different from the original knowledge assertion "p is known."

Example 1 (technical). Consider S5 with a single propositional letter p. Consider also a structure $\mathcal{M}_1$ consisting of one state w generated by $\Gamma = \{p, \neg \mathbf{K}p\}$.[3]

---

[2] Analyzing the role of knowing the model, normally assumed and not acknowledged in formal epistemology, has been long overdue. The paper that prompted completing this study was [6].

[3] State w is constructive: one can check that Γ is a complete set of formulas, i.e., for each F, either Γ proves F or Γ proves $\neg F$, and w is the set of formulas derivable from Γ, cf. also Sect. 7.

Fig. 1. Model $\mathcal{M}_1$.

The induced accessibility relation R, (2), is wRw, the truth in the model is membership in w: F holds iff $F \in w$. In particular, $\mathbf{K}p$ is false at w. On the other hand, by (1), $\mathbf{K}p$ ought to be true at w. So Definitions (2) and (1) do not match in $\mathcal{M}_1$.

Example 2 (conceptual). Kripke models are a convenient vehicle for specifying worlds: each node in a model yields a specific maximal consistent set of formulas. The downside of Kripke specification is that in order to model ignorance of a fact F, one has to commit to a hypothetic world at which F is false. Such a world may not exist.

> *An agent knows the axioms of Peano Arithmetic* PA, *but does not know a theorem F. In a Kripke model, we have to have a world v deemed possible by the agent at which $\neg F$ holds. However, there cannot be such a world v because all axioms of* PA *should be true at v and* PA $\cup \{\neg F\}$ *is inconsistent.*

A possible way out of this predicament is by epistemic models which naturally allow F to be true at each possible world but yet remain unknown.

2   Observable Models

Definition 1. *An* observable model, OM, *over an n-agent logic with knowledge modalities* $\mathbf{K}_1, \ldots, \mathbf{K}_n$ *is a tuple* $(W, R_1, \ldots, R_n, \models)$ *in which*

- W *is a nonempty set elements of which are called states (possible worlds);*
- $\models$ *is a complete truth relation at each world respecting the base logic: for each $u \in W$ the set of formulas true at u, $\{F \mid u \models F\}$, is a maximal consistent set over the base logic;*
- $uR_i v$ *yields for all F ($u \models \mathbf{K}_i F \Rightarrow v \models F$).*

By $R_i(w)$ we understand the set $\{x \in W \mid wR_i x\}$. Obviously,

$$w \models \mathbf{K}_i F \quad \Rightarrow \quad R_i(w) \models F.$$

A Kripke model $(W, R_1, \ldots, R_n, \Vdash)$ associated with $(W, R_1, \ldots, R_n, \models)$ is a Kripke model with the frame $(W, R_1, \ldots, R_n)$ and atomic forcing relation "$\Vdash$":

$$u \Vdash p \quad iff \quad u \models p.$$

Definition 2. *A model $(W, R_1, \ldots, R_n, \models)$ is* fully observable *if for each i, w, F,*

$$R_i(w) \models F \quad \Rightarrow \quad w \models \mathbf{K}_i F.$$

Example 3. Model $\mathcal{M}_1$ from Example 1 is not fully observable.

Proposition 1. *An observable model $\mathcal{M}$ coincides with its associated Kripke model iff $\mathcal{M}$ is fully observable*

Proof. Any Kripke model is a fully observable model over the same frame. Let an observable model $(W, R_1, \ldots, R_n, \models)$ be fully observable and $(W, R_1, \ldots, R_n, \Vdash)$ be the associated Kripke model. By induction on formula F we check that for each $u \in W$,

$$u \models F \quad \Leftrightarrow \quad u \Vdash F.$$

The claim is secured by definitions for atomic F's and obvious for the Boolean steps. Let F be $\mathbf{K}_i X$. If $u \models \mathbf{K}_i X$ then $R_i(w) \models X$. By IH, $R_i(w) \Vdash X$, hence $u \Vdash \mathbf{K}_i X$. If $u \Vdash \mathbf{K}_i X$ then $R_i(w) \Vdash X$. By IH, $R_i(w) \models X$, hence, by full observability, $u \models \mathbf{K}_i X$.

So, Kripke models are exactly **fully observable models**. Here is an informal[4] sufficient condition under which an observable model is a Kripke model (a fully observable model):

Kripke models are observable models **commonly known to all agents**.

Once F holds everywhere in $R(u)$, the agent knows this and, knowing R, can conclude that F holds at all states epistemically possible in u, thus coming to justified (by virtue of this argument) knowledge of F. So, in Kripke models, knowledge of F at u, given that F holds in $R(u)$, does not appear from nowhere. A justification for such a knowledge

$$u \Vdash \mathbf{K}F$$

is merely assumed knowledge of the model itself relativized to a specific state u.

Definition 3. *An observable model $(W, R_1, \ldots, R_n, \models)$ is induced if*

$$uR_i v \quad \Leftrightarrow \quad for\ all\ F\ (u \models \mathbf{K}_i F \Rightarrow v \models F). \tag{5}$$

Informally, an induced *OM* $(W, R_1, \ldots, R_n, \models)$ has all possible accessibility relations given $(W, \models)$. Since, in canonical models accessibility relations R_i's satisfy (5), all canonical models are induced.

We will further discuss the ontological status of observable models vs. Kripke models in Sect. 6.

3   On the Structure of Induced Observable Models

Let $(W, R_1, \ldots, R_n, \models)$ be an induced observable model over $\mathsf{S5}^n$.

---

[4] It appears that a natural formalization of this condition leads us beyond the current level of propositional modal logic.

Proposition 2. *Each R_i is an equivalence relation on W.*

Proof. Let $\mathbf{K}$ and R denote $\mathbf{K}_i$ and R for any i. Reflexivity and transitivity are immediate. Let us check symmetry. Let wRx, and suppose $x \models \mathbf{K}F$. We have to prove that $w \models F$. Suppose $w \not\models F$, then, by reflexivity in $\mathsf{S5}^n$, $w \not\models \mathbf{K}F$, hence $w \models \neg\mathbf{K}F$. By negative introspection, $w \models \mathbf{K}\neg\mathbf{K}F$. By definition of R, $x \models \neg\mathbf{K}F$, which is impossible since x is consistent and $x \models \mathbf{K}F$.

The intuition of indistinguishability for states from $R(w)$ in observable models is similar to Kripke models: we can interpret $R(w)$ as some set of states indistinguishable from w by facts known to the agent. Apparently, $w \Vdash \mathbf{K}F$ yields $R(w) \Vdash F$ and all facts known at w are true everywhere in $R(w)$. So, a state $x \in R(w)$ cannot be distinguished from w by any fact known to the agent.

The principal difference between observable models and Kripke models is that in the former, a validity of F in $R(w)$ does not yield knowledge of F: there is room for ignorance of agents about valid facts[5]. In particular, it is possible to have F throughout $R(w)$, but $\neg\mathbf{K}F$ at each state in $R(w)$.

The following proposition shows that knowledge assertions respect indistinguishability: either $\mathbf{K}F$ holds everywhere in $R(w)$, or $\neg\mathbf{K}F$ holds everywhere in $R(w)$.

Corollary 1. $R(w) \models \mathbf{K}F$ *or* $R(w) \models \neg\mathbf{K}F$.

Indeed, suppose $w \models \mathbf{K}F$, but for some $x \in R(w)$, $x \models \neg\mathbf{K}F$. By negative introspection, $x \models \mathbf{K}\neg\mathbf{K}F$, hence $R(x) \models \neg\mathbf{K}F$. Since, by Proposition 2, $w \in R(x)$, $w \models \neg\mathbf{K}F$. A contradiction.

4   Derivations from Hypotheses in Modal Logic

The standard formulation of modal logics postulates the Necessitation rule:

$$\vdash F \;\Rightarrow\; \vdash \mathbf{K}_i F.$$

However, this rule is not valid in a general setting for derivations from assumptions: for some Γ, $\Gamma \vdash F$ does not yield $\Gamma \vdash \mathbf{K}_i F$. Therefore, when speaking about derivations from hypotheses in $\mathsf{S5}^n$, we do not postulate Necessitation.

Definition 4. *For a given set of formulas Γ (here called "hypotheses" or "assumptions") we consider derivations from Γ: assume all $\mathsf{S5}^n$-theorems together with Γ and use classical propositional reasoning (rule Modus Ponens). The notation*

$$\Gamma \vdash A$$

represents 'A is derivable from Γ.'

For some "good" Γ's, Necessitation is a valid rule.

---

[5] We regard this as a feature that makes observable models more flexible and realistic.

5   Canonical Models for S5 with a single letter

In this section we will offer a useful elaborate example of canonical model constructions associated to S5 with a single propositional letter p, S5(p). Such models are defined by their possible worlds W and truth relations $\models$ since the accessibility relations are induced and can be recovered from $(W, \models)$.

We first note that the modal-free fragment generated by $\{p\}$, i.e., the usual classical propositional logic with a single propositional letter p, admits two possible worlds: one generated by $\{p\}$ and the other generated by $\{\neg p\}$.

We claim that S5(p) admits exactly four possible worlds (maximal consistent sets):

- A, generated by $\{\mathbf{K}p\}(= \{p, \mathbf{K}p\})$;
- B, generated by $\{p, \neg \mathbf{K}p\}$;
- C, generated by $\{\neg p, \neg \mathbf{K}\neg p\}$;
- D, generated by $\{\mathbf{K}\neg p\}(= \{\neg p, \mathbf{K}\neg p\})$.

Consistency of each of A–D is straightforward since each has an easy Kripke model.

Now we check that each of A–D is complete, i.e., that each proves F or $\neg F$ for any formula F in the language of S5(p).

Completeness of A. First we note that A is closed under Necessitation: $A \vdash F$ yields $A \vdash \mathbf{K}F$. Standard induction on derivations of F. The key point here is that $A \vdash \mathbf{K}A$. Once we establish Necessitation in A, we proceed to proving that for each F, $A \vdash F$ or $A \vdash \neg F$. Induction on F. Obvious for atomic formulas and Boolean connectives. Let $F = \mathbf{K}X$. If $A \vdash X$, then, by Necessitation, $A \vdash \mathbf{K}X$. If $A \vdash \neg X$, then, by reflexivity, $A \vdash \neg \mathbf{K}X$.

Completeness of B. Here Necessitation is not admissible since $B \vdash p$, but $B \not\vdash \mathbf{K}p$. We will use the S5-normal forms, cf. [5].

Lemma 1 (S5 normal forms). *In* S5, *every formula is provably equivalent to a formula in normal form which is a disjunction of conjunctions of type*

$$\alpha \wedge \mathbf{K}\beta \wedge \neg \mathbf{K}\gamma_1 \wedge \ldots \wedge \neg \mathbf{K}\gamma_m \tag{6}$$

where $\alpha, \beta, \gamma_1, \ldots, \gamma_m$ *are all purely propositional formulas. For* S5(p) *we may assume that each of them is one of* $\top, \bot, p, \neg p$.

It now suffices to check that for each formula F of type (6), $B \vdash F$ or $B \vdash \neg F$.

If $\alpha = \bot, \neg p$, then $B \vdash \neg F$. If $\alpha = \top, p$, then $B \vdash \alpha$ and we proceed to β.

If $\beta = \bot, \neg p$, then, by reflexivity, $B \vdash \neg F$. If $\beta = p$, then again, $B \vdash \neg F$. If $\beta = \top$, then $B \vdash \mathbf{K}\beta$ and we proceed to γ_i.

If at least one of γ_i is $\top$, then $B \vdash \neg F$. Otherwise, all conjuncts in F are provable in B. Indeed, for $\gamma_i = \bot$, use $B \vdash \neg \mathbf{K}\bot$. For $\gamma_i = p$ use the fact that $\neg \mathbf{K}p \in B$. For $\gamma_i = \neg p$, use reflexivity $p \rightarrow \neg \mathbf{K}\neg p$. In either case, $B \vdash \neg \mathbf{K}\gamma_i$.[6]

---

[6] A similar normal form-based proof of completeness can be given for each of A–D, but we have opted for Necessitation-based proof for A and D to underline the fact that both A and D enjoy Necessitation.

Completeness of C. Similar to B

Completeness of D. Similar to A, since D also enjoys Necessitation.

The collection of A–D exhausts all logical possibilities for states over $\mathsf{S5}(p)$. Indeed, for the remaining four logical possibilities for p and knowledge assertion about p, $\{p, \mathbf{K}\neg p\}$ and $\{\neg p, \mathbf{K}p\}$ are inconsistent and so are any of its extensions. The last two options: $\{p, \neg \mathbf{K}\neg p\} \subset A$ and $\{\neg p, \neg \mathbf{K}p\} \subset C$, hence they generate no new states.

Finally, we describe the induced accessibility relation R on $W = \{A, B, C, D\}$. By Proposition 2, R is an equivalence relation on W. Consider all six possible pairs of different states in W and rule out the ones that are not accessible from each other.

By definition of R, for ARX, p should be in X, which rules out ARC, ARD. Likewise, B and D are not related by R.

$\{A, B\}$ and $\{C, D\}$ are not connected due to positive introspection, e.g., since $\mathbf{K}p \in A$, $\mathbf{KK}p \in A$ too, hence ARX yields $\mathbf{K}p \in X$; this rules out ARB.

The only remaining possibility for R-connection is pair $\{B, C\}$, and they are related! From the Kripke canonical model perspective, there should be a state in W accessible from B in which $\neg p$ holds, and C is the only remaining possibility, hence BRC. The resulting picture of the canonical model for $\mathsf{S5}(p)$ is

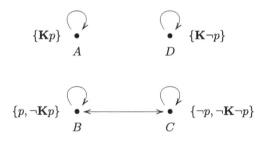

Fig. 2. Canonical model for $\mathsf{S5}(p)$.

Proposition 3. *Each of fifteen non-empty subsets of* $\{A, B, C, D\}$ *is an observable model. Seven of them are fully observable and hence Kripke models:*

$$\{A\}, \ \{D\}, \ \{A, D\}, \ \{B, C\}, \ \{A, B, C\}, \ \{B, C, D\}, \ \{A, B, C, D\}.$$

The remaining eight are observable models which are non-fully observable and thus not Kripke models.

Among OM's here is $\{A, B\}$ – all states at which p holds; this may be regarded as the canonical observable model of $\Gamma = \{p\}$ which is therefore not a Kripke model.

For a general theory of canonical models for sets of assumptions Γ cf. Sect. 7.

6 From Observable Models to Kripke Models

Each Kripke model is an observable model, hence observable models constitute a well-principled generalization of Kripke models covering more epistemic situations.

How to build observable models in general? To define an observable, we have to specify complete truth relations at each world, and this is the task which Kripke models do naturally. Can we combine the generality of observable models and the convenience of Kripke models?

In this section we show that each observable model is a fragment of an appropriate Kripke model. This suggests a general method, "scaffolding," of defining observable models: build an umbrella Kripke model, and carve out of it the desired observable fragment.

Theorem 1. *For any observable model* $(W, R_1, \ldots, R_n, \models)$ *there is a Kripke model* $(\widetilde{W}, \widetilde{R}_1, \ldots, \widetilde{R}_n, \Vdash)$ *such that*

(a) $W \subseteq \widetilde{W}$;
(b) $R_i \subseteq \widetilde{R}_i$;
(c) for each $u \in W$ *and each* F, $u \models F$ *iff* $u \Vdash F$.

Proof. Take

$$(\widetilde{W}, \widetilde{R}_1, \ldots, \widetilde{R}_n, \Vdash)$$

to be the canonical Kripke model $CM(\mathsf{S5}^n)$: each $u \in W$ is also a world in $CM(\mathsf{S5}^n)$ and in both models, $(W, R_1, \ldots, R_n, \models)$ and $CM(\mathsf{S5}^n)$, F holds at u iff $F \in u$. For (b) it suffices to note that uR_iv yields $u\widetilde{R}_iv$.

Example 4. Here is the finite example of such an embedding. The aforementioned observable model $\mathcal{M}_1$ from Example 1 is naturally embedded into Kripke model $\mathcal{M}_2$. A singleton model $\mathcal{M}_1$, as a world, coincides with world w in $\mathcal{M}_2$.

Fig. 3. Model $\mathcal{M}_2$.

We see here that the Kripke model requires two worlds to emulate a singleton observable model. It is easy to build a singleton observable model for which the corresponding Kripke model is necessarily infinite.

Definition 5. *Let* Γ *be a set of formulas. By* $\Gamma \models F$ *we understand the situation when for each observable model* $\mathcal{M}$ *and its state* u,

$$\mathcal{M}, u \models \Gamma \quad \Rightarrow \quad \mathcal{M}, u \models F.$$

Proposition 4. *Soundness and completeness of* $S5^n$ *w.r.t. observable models:*

$$\Gamma \vdash F \quad iff \quad \Gamma \models F.$$

Proof. Soundness. If $\Gamma \vdash F$, then F belongs to any maximal consistent extension of Γ and hence is true at each state of each observable model of Γ.

Completeness follows from the saturation lemma: if $\Gamma \nvdash F$, then $\Gamma \cup \{\neg F\}$ is consistent and its maximal consistent extension Δ is a desired singleton observable model: $\Delta \nvDash F$.

7   Canonical Observable Models in a General Setting

Kripke models constitute a comprehensive semantical tool in modal logic: every consistent configuration is realized in an appropriate node of the canonical model. However, in modal logic we don't normally care how "possible" the states in a Kripke model are, their formal consistency is sufficient. Epistemic scenarios are different. We can imagine a situation in which some propositions Γ (e.g., reflecting the state of nature) should hold at all possible states. Furthermore, if this Γ is not common knowledge the corresponding set of states is, generally speaking, not a Kripke model, and requires a broader approach.

Definition 6. *A canonical observable model of a set of formulas Γ, $CM(\Gamma)$ is, by definition, the collection of all worlds containing Γ with induced accessibility relations.*

We show that for many (intuitively, most) Γ's, the corresponding canonical model $CM(\Gamma)$ is not fully observable, hence not a Kripke model. We give a criterion of when the canonical model of Γ is a Kripke model:

Γ *is closed under Necessitation* iff Γ *proves its own common knowledge.*

Example 5. Canonical model $CM(p)$ for $\Gamma = \{p\}$ in S5 has been described in Sect. 5. There are two possible worlds in $CM(p)$, generated by $\{\mathbf{K}p\}$ (world A) and $\{p, \neg\mathbf{K}p\}$ (world B). Worlds A and B are not connected by the undistinguishability relation R: p holds at both worlds, but is not known in B. The canonical model $CM(p)$ is not a Kripke model, since p holds in $CM(p)$, but $\mathbf{K}p$ does not.

7.1   Common Knowledge and Necessitation

In this section we consider a representative case of two agents. We will use abbreviations: for "everybody's knowledge"

$$\mathbf{E}X = \mathbf{K}_1 X \wedge \mathbf{K}_2 X,$$

and "common knowledge"

$$\mathbf{C}X = \{X, \ \mathbf{E}X, \ \mathbf{E}^2 X, \ \mathbf{E}^3 X, \ \ldots\}.$$

As one can see, $\mathbf{C}X$ is an infinite (though easy to describe and decidable) set of formulas. Since modalities $\mathbf{K}_i$ commute with the conjunction $\wedge$, $\mathbf{C}X$ is logically equivalent to the set of all formulas which are X prefixed by iterated knowledge modalities:

$$\{P_1 P_2 \ldots P_k X \mid k = 0, 1, 2, \ldots, \quad P_i \in \{\mathbf{K}_1, \mathbf{K}_2\}\}$$

and we regard this as an alternative definition of $\mathbf{C}X$. Naturally,

$$\mathbf{C}\Gamma = \bigcup \{\mathbf{C}F \mid F \in \Gamma\}$$

represents "Γ is common knowledge." The following proposition states that the rule of Necessitation corresponds to derivable common knowledge of assumptions.

Proposition 5. *A set of formulas Γ is closed under Necessitation if and only if $\Gamma \vdash \mathbf{C}\Gamma$, i.e., Γ proves its own common knowledge.*

Proof. Direction 'if.' Assume $\Gamma \vdash \mathbf{C}\Gamma$ and prove by induction on derivations that $\Gamma \vdash X$ yields $\Gamma \vdash \mathbf{K}_i X$. For X being from $\mathsf{S5}_n$, this follows from the rule of Necessitation in $\mathsf{S5}_n$. For $X \in \Gamma$, it follows from the assumption that $\Gamma \vdash \mathbf{C}X$, hence $\Gamma \vdash \mathbf{K}_i X$. If X is obtained from *Modus Ponens*, $\Gamma \vdash Y \to X$ and $\Gamma \vdash Y$. By IH, $\Gamma \vdash \mathbf{K}_i(Y \to X)$ and $\Gamma \vdash \mathbf{K}_i Y$. By the distributivity principle of $\mathsf{S5}_n$, $\Gamma \vdash \mathbf{K}_i X$.

For 'only if,' suppose that Γ is closed under Necessitation and $F \in \Gamma$, hence $\Gamma \vdash F$. Using appropriate instances of the Necessitation rule in Γ we can derive $P_1 P_2 P_3, \ldots, P_k F$ for each prefix $P_1 P_2 P_3, \ldots, P_k$ with P_i is one of $\mathbf{K}_1, \mathbf{K}_2$. Therefore, $\Gamma \vdash \mathbf{C}F$ and $\Gamma \vdash \mathbf{C}\Gamma$.

7.2 Canonical Models of Sets of Assumptions

We answer the question of when the canonical model of Γ is a Kripke model.

Theorem 2. *The following are equivalent:*

(a) $CM(\Gamma)$ is fully observable (i.e., a Kripke model);
(b) Γ admits Necessitation;
(c) Γ proves its own common knowledge.

Proof. The fact that (b) is equivalent to (c) has already been established in Proposition 5. We now check that (a) and (b) are equivalent.

If Γ does not admit Necessitation, there is a formula F such that $\Gamma \vdash F$, but $\Gamma \nvdash \mathbf{K}F$[7]. Hence F holds everywhere in $CM(\Gamma)$ and $\neg\mathbf{K}F$ is consistent with Γ. Consider a maximal consistent extension u of $\Gamma \cup \{\neg\mathbf{K}F\}$. Obviously, $\neg\mathbf{K}F$ holds in u and F holds in $R(u)$ which makes $CM(\Gamma)$ not fully observable and hence not a Kripke model.

---

[7] Here again, we ignore indices i in $\mathbf{K}_i$ and R_i.

Suppose Γ admits Necessitation. We have to prove that observable model $CM(\Gamma)$ with the canonical evaluation

$$u \models F \quad \Leftrightarrow \quad F \in u$$

and the induced R as in (2) is fully observable, i.e.

$$R(u) \models F \quad \Rightarrow \quad u \models \mathbf{K}F.$$

Put

$$u^{\mathbf{K}} = \{F \mid \mathbf{K}F \in u\}.$$

Note that $\Gamma \subseteq u^{\mathbf{K}}$: $X \in \Gamma$ yields, by Necessitation in Γ, that $\mathbf{K}X \in \Gamma$ hence $\mathbf{K}X \in u$ and $X \in u^{\mathbf{K}}$. Suppose $u \not\models \mathbf{K}F$. Then $u^{\mathbf{K}} \cup \{\neg F\}$ is consistent. Indeed, otherwise

$$\Gamma \vdash X_1 \wedge \ldots \wedge X_n \to F$$

for some $X_i \in u^{\mathbf{K}}$. Since Γ is closed under Necessitation,

$$\Gamma \vdash \mathbf{K}(X_1 \wedge \ldots \wedge X_n \to F).$$

By standard S5-reasoning,

$$\Gamma \vdash \mathbf{K}X_1 \wedge \ldots \wedge \mathbf{K}X_n \to \mathbf{K}F.$$

Since $\mathbf{K}X_i \in u$, $\mathbf{K}F$ should be in u as well – a contradiction.

Now consider a maximal consistent set v extending $u^{\mathbf{K}} \cup \{\neg F\}$. Since $\Gamma \subseteq u^{\mathbf{K}}$, $v \in CM(\Gamma)$. Since $u^{\mathbf{K}} \subseteq v$, uRv. Since $\neg F \in v$, $F \notin v$ hence $v \not\models F$, which yields $u \not\models \mathbf{K}F$.

Example 6. Consider worlds A, B, C, D from the canonical model for S5(p), Sect. 5, Fig. 2. The canonical model of $\Gamma = \{p\}$, $CM(p)$[8], is the set of worlds at which p holds, i.e., $W = \{A, B\}$, Fig. 4. Model $CM(p)$ is not fully observable, not a Kripke model, and is an illustration of Theorem 2, since $\{p\}$ is not closed under Necessitation.

The canonical model of $\Gamma = \{\mathbf{K}p\}$ is the set of worlds at which $\mathbf{K}p$ holds, i.e., $W = \{A\}$. This Γ enjoys Necessitation, its canonical observable model is fully observable, i.e., is a Kripke model.

The canonical model of $\Gamma = \{\neg\mathbf{K}p, \neg\mathbf{K}\neg p\}$. By negative and positive introspection, Γ is closed under Necessitation, hence $CM(\Gamma)$ should also be fully observable. Worlds A and D are not compatible with Γ and hence are not in $CM(\Gamma)$. Since none of B and C is fully observable, $CM(\Gamma) = \{B, C\}$. It is an easy exercise to derive $\neg\mathbf{K}p$ in C and $\neg\mathbf{K}\neg p$ in B. This is a Kripkean situation, i.e., $CM(\{\neg\mathbf{K}p, \neg\mathbf{K}\neg p\})$ is fully observable.

---

[8] We drop brackets in $CM(\{p\})$ and similar cases for better readability.

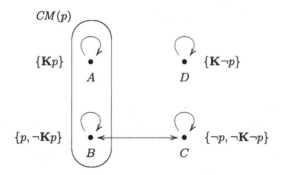

Fig. 4. Canonical model $CM(p)$ (in the oval).

8   *OM* for epistemic modeling

Kripke models represent ignorance by building special "possible worlds": these worlds, however, can be epistemically impossible for the agent, cf. Examples 2 and 4. In this respect, *OM*'s offer a more refined epistemic picture than Kripke models alone.

Another area in which *OM*'s can have advantage is avoiding unnecessary *pro forma* worlds.

8.1   Ignorant Agent Example

Here is a natural example, *Ignorant Agent*, of a two state *OM* for which scaffolding Kripke models are infinite and nonsensical. Consider S4 with a single propositional letter p and a situation in which the agent has no specific knowledge about the state of nature, i.e., the agent knows only logical truths derived in S4. Consider two sets of formulas:

$$\Gamma^+ = \{p\} \cup \{\neg \mathbf{K} F \mid \text{S4} \not\vdash F\}, \quad \Gamma^- = \{\neg p\} \cup \{\neg \mathbf{K} F \mid \text{S4} \not\vdash F\}$$

Proposition 6. Γ^+ *and* Γ^- *are consistent, complete, and decidable.*

Proof. Consider Γ^+, the case of Γ^- is similar.

Consistency. Suppose otherwise: $\Gamma^+ \vdash \bot$. Then $p, \neg \mathbf{K} F_1, \ldots, \neg \mathbf{K} F_n \vdash \bot$ for some F_i's not provable in S4. By S4-reasoning, $\text{S4} \vdash p \rightarrow (\mathbf{K} F_1 \vee \ldots \vee \mathbf{K} F_n)$. Since each of F_i's has a countermodel, by a standard S4-model construction, we build an S4 Kripke model, at the root node r of which $r \Vdash p$ and $r \not\Vdash \mathbf{K} F_1 \vee \ldots \vee \mathbf{K} F_n$, a contradiction.

Completeness: For each F, $\Gamma^+ \vdash F$ or $\Gamma^+ \vdash \neg F$. Induction on F. Atomic and Boolean cases are trivial. Let F be $\mathbf{K} X$ for some X. Case 1. $\text{S4} \vdash X$. Then $\text{S4} \vdash \mathbf{K} X$, $\Gamma^+ \vdash \mathbf{K} X$. Case 2. $\text{S4} \not\vdash X$. Then $\Gamma^+ \vdash \neg \mathbf{K} X$.

Decidability of Γ^+ is immediate due to its completeness and decidability of S4.

Fig. 5. Model $\mathcal{M}_3$.

A natural *OM* over S4 for Ignorant Agent is $\mathcal{M}_3$:

It suffices to show that worlds Γ^+ and Γ^- are indeed indistinguishable, i.e., $\Gamma^+ \vdash \mathbf{K}F$ yields $\Gamma^- \vdash F$ (and symmetrically for Γ^- and Γ^+). Suppose $\Gamma^+ \vdash \mathbf{K}F$. Then S4 $\vdash F$, since otherwise S4 $\nvdash F$ and $\neg\mathbf{K}F \in \Gamma^+$ which would make Γ^+ inconsistent. Then $\Gamma^- \vdash F$.

Any scaffolding Kripke model for $\mathcal{M}_3$ is infinite and meaningless. It must explicitly contain counter-worlds for all F's s.t. S4 $\nvdash F$ which does not make much sense since S4 is decidable and there is no need to carry all these additional worlds to define the situation epistemically.

9   Intuitionistic Observable Models

Consider the case of one agent. Nodes in an intuitionistic Kripke model $(W, \preceq, \Vdash)$ are information states in a discovery process and $u \preceq v$ means that v has more truths than u. In order to use $(W, \preceq, \Vdash)$ as a specification device, **we have to assume the knowledge of the model**. Without this assumption, the fact that F does not hold at all states accessible from w can be unknown and cannot constitute a sufficient ground for claiming $\neg F$ at w. Although

$$w \Vdash \neg F \quad \Rightarrow \quad v \nVdash F \;\; for\; all\; v \; s.t.\; w \preceq v,$$

the converse "$\Leftarrow$" is not well-principled: the fact that $v \nVdash F \; for\; all\; v \; s.t.\; w \preceq v$ alone, without the knowledge of the model assumption, does not produce enough evidence at w to conclude $\neg F$.

We propose an appropriate adjustment: an **observable fragment** of an intuitionistic Kripke model $\mathcal{K} = (W, \preceq, \Vdash)$ is $\mathcal{O} = (W_O, \preceq_O, \models)$ with

- $W_O \subseteq W$, states from W_O are called "observable states";
- "observable accessibility" $\preceq_O$ is a subset of $\preceq$ restricted to W_O: $\preceq_O \subseteq \preceq\!\restriction_{W_O}$;
- for $v \in W_O$, $v \models F$ is defined as $v \Vdash F$
  (the whole model $\mathcal{K}$ is needed to define "$\models$" in $\mathcal{O}$).

An intuitionistic observable model is, by definition, an observable fragment $\mathcal{O}$ of some intuitionistic Kripke model $\mathcal{K}$.

Whereas the whole scenario can be represented traditionally, by a Kripke model $\mathcal{K}$, only a part $\mathcal{O}$ of $\mathcal{K}$ is observable by the agent. Within the traditional Kripkean approach, we assume that a model $\mathcal{K}$ is known and recover the truths at each world. Within the observable intuitionistic model $\mathcal{O}$, we are given whole truths at each world consistent with the intuitionistic semantics. A scaffolding Kripke model $\mathcal{K}$ is one of possible methods of defining $\mathcal{O}$.

$2 \bullet p$

$1 \bullet$

Example 7. Intuitionistic Kripke model $\mathcal{K}$: $W = \{1, 2\}$, $1 \preceq 2$, $1 \nVdash p$, $2 \Vdash p$. *Intuitionistic observable model* $\mathcal{O}$ (circled): $W_O = \{1\}$, $1 \nvDash p$. In $\mathcal{O}$, p does not hold, but yet $1 \nVdash \neg p$. $\mathcal{O}$ itself is not a Kripke model!

Observable intuitionistic models can help to ease the aforementioned tension between Kripkean intuitionistic semantics and the intended Brouwer-Heyting-Kolmogorov semantics of proofs for intuitionistic logic, cf. [3]. The standard Kripkean condition

$$w \Vdash \neg F \quad \Leftrightarrow \quad v \nVdash F \ for \ all \ v \ s.t. \ w \preceq v$$

appears to require complete knowledge of the model, which is neither well-principled, nor realistic. Observable models provide more conceptual clarity and flexibility.

10   Findings and Suggestions

The principal contributions of this paper are conceptual: we suggest streamlining the foundations of observable modeling.

1. When epistemic logic starts from a given set of possible worlds and defines an accessibility relation in the standard way, the result might not be a Kripke model: only those structures that are fully observable are Kripke models. The full observability property is a propositional equivalent to common knowledge of the model, which is quite restrictive but has been hiddenly assumed in epistemic modeling. This assumption should be acknowledged, made explicit, and properly studied.
2. We have sketched a basic theory of observable models in a general setting without the common knowledge of the model constraint. This is just a step towards a general theory of observable modeling with possible worlds; more expressive tools are needed to capture partial and asymmetric knowledge of the model for multiple agents.
3. A similar suggestion is made for intuitionistic models. A well-known discon-nection between intuitionistic Kripke models and the intended constructive semantics of intuitionistic logic can be mitigated by viewing intuitionistic possible worlds models as intuitionistic observable models.

References

1. Aumann, R.: Backward induction and common knowledge of rationality. Games Econ. Behav. **8**(1), 6–19 (1995)
2. Artemov, S.: Knowing the Model. arXiv:1610.04955 [math.LO] (2016)
3. Artemov, S., Fitting, M.: Justification Logic: Reasoning with Reasons. Cambridge University Press, Cambridge (2019)
4. Fagin, R., Halpern, J., Moses, Y., Vardi, M.: Reasoning About Knowledge. MIT Press, Cambridge (1995)
5. Meyer, J.-J.C., van der Hoek, W.: Epistemic Logic for AI and Computer Science. CUP, Cambridge (1995)
6. Williamson, T.: A note on Gettier cases in epistemic logic. Philos. Stud. **172**(1), 129–140 (2015)

Countermodel Construction via Optimal Hypersequent Calculi for Non-normal Modal Logics

Tiziano Dalmonte[1]([✉])⬤, Björn Lellmann[2]⬤, Nicola Olivetti[1]⬤,
and Elaine Pimentel[3]⬤

[1] Aix Marseille Univ, Université de Toulon, CNRS, LIS, Marseille, France
tiziano.dalmonte@etu.univ-amu.fr, nicola.olivetti@lis-lab.fr
[2] Technische Universität Wien, Vienna, Austria
lellmann@logic.at
[3] Universidade Federal do Rio Grande do Norte, Natal, Brazil
elaine.pimentel@gmail.com

Abstract. We develop semantically-oriented calculi for the cube of non-normal modal logics and some deontic extensions. The calculi manipulate hypersequents and have a simple semantic interpretation. Their main feature is that they allow for direct countermodel extraction. Moreover they provide an optimal decision procedure for the respective logics. They also enjoy standard proof-theoretical properties, such as a syntactical proof of cut-admissibility.

Keywords: Hypersequents · Non-normal modal logics · Countermodel construction · Cut elimination

1 Introduction

Non-normal modal logics–NNMLs for short–have a long history, going back to the seminal works by Kripke, Montague, Segeberg, Scott, and Chellas (see [3] for an introduction). They are "non-normal" as they do not contain all axioms of minimal *normal* modal logic **K**. NNMLs find an interest in several areas: in epistemic reasoning they offer a simple (although partial) solution to the problem of logical omniscience (see [19]); in deontic logic, they allow avoiding well-known paradoxes (such as Ross's Paradox) and to represent conflicting obligations (see [8]); NNMLs are needed also when $\Box A$ is interpreted as "A is true in most of the cases" [1]; finally NNMLs naturally arise in game-theoretical interpretation of $\Box A$ as "the agent has a winning strategy to bring about A" (indeed, non-normal monotonic logic **M** can be seen as a 2-agent case of coalition logic with determinacy [18]). In this work, we consider the *classical cube* on

Supported by WWTF project MA16-28 and by Project TICAMORE ANR-16-CE91-0002-01.

S. Artemov and A. Nerode (Eds.): LFCS 2020, LNCS 11972, pp. 27–46, 2020.
https://doi.org/10.1007/978-3-030-36755-8_3

NMMLs, given by the extensions of the minimal modal logic **E**, containing only the *congruence rule*, with axioms C, M and N.

NNMLs have a well-understood semantics defined in terms of neighbourhood models [16]: in these models each world w has an associated set of neighbourhoods $\mathcal{N}(w)$, each one of them being a set of worlds/states. If we accept the traditional interpretation of a "proposition" as a set of worlds (= its truth set), we can think of each neighbourhood in $\mathcal{N}(w)$ as a proposition: a formula $\Box A$ is true in a world w if "the proposition" A, *i.e.* the truth-set of A, belongs to $\mathcal{N}(w)$. The classical cube can be modelled by imposing additional closure properties of the set of neighbourhoods. In this work we adopt a variant of neighbourhood semantics defined in terms of bi-neighbourhood models [4]: in these structures each world has associated a set of *pairs* of neighbourhoods. The intuition is that the two components of a pair provide positive and negative support for a modal formula, being more natural for "non-monotonic" logics (*i.e.* not containing axiom M). The reason is that, instead of specifying exactly the truth sets in $\mathcal{N}(w)$, the pairs of neighbourhoods specify just lower and upper bounds of truth sets, so that the same pair may be a "witness" for several propositions. This makes the generation of *countermodels* easier, as shown in this paper.

It is curious to note that, although some proof-systems for NNMLs have been proposed in the past, countermodel extraction has been rarely addressed and complexity is seldom analysed. Indeed, the works [4,7,11,14] propose countermodel extraction, but all of them require either a complicated procedure or an extended language with labels. [12] presents a nested sequent calculus for a logic combining normal and monotone non-normal modal logic that supports countermodel extraction, but the nested sequent structure is not suitable for logics lacking monotonicity. On the other hand, cut-free sequent/linear nested calculi for the classical cube and its extensions with standard axioms of normal modal logics (the non-normal counterpart of logics from **K** to **S5**) are studied in [9,10,13,15]; however neither semantic completeness and countermodel extraction, nor complexity are studied.

In this work, we intend to fill this gap by proposing *modular* calculi for the classical cube (and also some deontic extensions) that provide *direct countermodel extraction* and are of *optimal complexity*. Our calculi are semantically based on the bi-neighbourhood models, and have two syntactic features: they manipulate hypersequents and sequents may contain blocks of $\Box$-ed formulas in the antecedent. A hypersequent [2] is just a multiset of sequents and can be understood as a (meta-logical) disjunction of sequents. Sequents within hypersequents can be read as formulas of the logic. Blocks of formulas are interpreted as conjunction of positive $\Box$-ed formulas. Intuitively each block represents a neighbourhood satisfying one or more $\Box$-ed formulas, and this allows for the formulation of modular calculi for the whole cube. The advantage of using hypersequents is that all rules become invertible, thus there is no need for backtracking in proof search. For the same reason, the hypersequent calculi provide directly countermodel extraction: from *one* failed proof, it can be extracted *directly* a countermodel in the bi-neighbourhood semantics of the sequent/formula at the root of the derivation. For logics without the C axiom, our calculi also provide an *optimal* NP/CoNP decision procedure. For logics including C, we can still obtain an optimal PSPACE decision procedure but, as usual in modal logic, at the cost of sacrificing the invertibility of some rules. Finally, the calculi have also good proof-theoretical properties, as they support a syntactic proof of cut admissibility.

It turns out that our calculi can be modularly extended by adding modal axioms. We illustrate this by extending them with axioms T, P, D, the last two are of interest in deontic logic.

All in all, we believe that the structure of our calculi, namely hypersequents with blocks, is adequate for NNMLs from a semantical, computational and a proof-theoretical point of view since it: (i) has a semantic interpretation; (ii) allows direct countermodel generation; (iii) has optimal complexity; and (iv) has good proof-theoretical properties.

2   Non-normal Modal Logics

In this section, we present the classical cube of NNMLs, both axiomatically and semantically in terms of neighbourhood models. We also present bi-neighbourhood models, a variant of the neighbourhood semantics introduced in [4]. The propositional language $\mathcal{L}$ contains *formulas* given by the following grammar: $A:: = p \mid \top \mid \bot \mid A \to A \mid \Box A$, where $p \in$ Atm, the set of propositional variable symbols. Other propositional connectives are defined by the standard equivalences. The minimal logic $\mathbf{E}$ in the language $\mathcal{L}$ is given Hilbert-style by extending classical propositional logic with only the *congruence rule*

$$\text{RE} \ \frac{A \to B \quad B \to A}{\Box A \to \Box B}$$

The *classical cube* (below on the right) is formed by extending $\mathbf{E}$ with any combination of axioms M, C, and N (below on the left).

M	$\Box(A \land B) \to \Box A$
C	$\Box A \land \Box B \to \Box(A \land B)$
N	$\Box\top$

As usual for logics containing M we omit E, *e.g.*, we write $\mathbf{MC}$ for $\mathbf{EMC}$. In the following, for any system $\mathbf{L}$ of the cube we denote with $\mathbf{L}^*$ any extension of $\mathbf{L}$ obtained by adding one or more of these axioms. We recall that axioms M and N are, respectively, equivalent to the rules RM ($A \to B$ / $\Box A \to \Box B$) and RN (A / $\Box A$), and that axiom K ($\Box(A \to B) \to \Box A \to \Box B$) is derivable from M and C. As a consequence, the top system $\mathbf{MCN}$ is equivalent to $\mathbf{K}$, the weakest normal modal logic.

The standard semantics for NNMLs is defined in terms of so-called neighbourhood (or minimal) models [3].

Definition 1. *A* standard neighbourhood model *is a tuple* $\mathcal{M} = \langle W, \mathcal{N}_s, \mathcal{V} \rangle$, *where* W *is a non-empty set,* $\mathcal{V}$ *is a valuation function, and* $\mathcal{N}_s$ *is a neighbourhood function* $W \longrightarrow \mathcal{P}(\mathcal{P}(W))$. *A model* $\mathcal{M}$ *is* supplemented *if* $\alpha \in \mathcal{N}_s(w)$ *and* $\alpha \subseteq \beta \subseteq W$ *implies* $\beta \in \mathcal{N}_s(w)$, *it* contains the unit *if* $W \in \mathcal{N}_s(w)$ *for all* $w \in W$, *and it is* closed under intersection *if* $\alpha, \beta \in \mathcal{N}_s(w)$ *implies* $\alpha \cap \beta \in \mathcal{N}_s(w)$. *The* standard forcing relation *for boxed formulas is:* $\mathcal{M}, w \Vdash_s \Box A$ *iff* $[A]_{\mathcal{M}} \in \mathcal{N}_s(w)$, *where* $[A]_{\mathcal{M}}$ *denotes the set* $\{v \in W \mid \mathcal{M}, v \Vdash_s A\}$ *of the worlds* v *that force* A, *also called the* truth set *of* A.

Neighbourhood models characterise modularly the classical cube of NNMLs [3] in the sense that a formula A is a theorem of **E** if and only if it is valid in all neighbourhood models. Furthermore, A is a theorem of **E**+(M/C/N) iff it is valid, respectively, in all models that are supplemented (M), closed under intersection (C), and contain the unit (N) (and any combination of these previous axioms/conditions).

Here we consider bi-neighbourhood semantics [4], a variant of neighbourhood semantics. In this semantics, neighbourhoods come in pairs, the two components provide, so to say, "positive" and "negative" support for a modal formula.

Definition 2. *A* bi-neighbourhood model *is a tuple* $\mathcal{M} = \langle W, \mathcal{N}_b, \mathcal{V} \rangle$, *where* W *is a non-empty set,* $\mathcal{V}$ *is a valuation function, and* $\mathcal{N}_b$ *is a bi-neighbourhood function* $W \longrightarrow \mathcal{P}(\mathcal{P}(W) \times \mathcal{P}(W))$. *We say that* $\mathcal{M}$ *is a M-model if* $(\alpha, \beta) \in \mathcal{N}_b(w)$ *implies* $\beta = \emptyset$, *it is a N-model if for all* $w \in W$ *there is* $\alpha \subseteq W$ *such that* $(\alpha, \emptyset) \in \mathcal{N}_b(w)$,[1] *and it is a C-model if* $(\alpha_1, \beta_1), (\alpha_2, \beta_2) \in \mathcal{N}_b(w)$ *implies* $(\alpha_1 \cap \alpha_2, \beta_1 \cup \beta_2) \in \mathcal{N}_b(w)$. *The forcing relation for boxed formulas is as follows:*

$$\mathcal{M}, w \Vdash_b \Box A \quad \textit{iff} \quad \textit{there is } (\alpha, \beta) \in \mathcal{N}_b(w) \textit{ s.t. } \alpha \subseteq [A]_{\mathcal{M}} \textit{ and } \beta \subseteq [\neg A]_{\mathcal{M}}.$$

Every standard model gives rise to a bi-neighbourhood model, by taking for each neighbourhood $\alpha \in \mathcal{N}_s(w)$, the pair $(\alpha, W \setminus \alpha)$. Conversely, every bi-neighbourhood model can be transformed into a standard model by assigning to $\mathcal{N}_s(w)$, for each pair $(\alpha, \beta) \in \mathcal{N}_b(w)$, the subsets γ such that $\alpha \subseteq \gamma \subseteq W \setminus \beta$. In this sense α and β give upper and lower bounds for neighbourhoods in standard models. For the non-monotonic case there is also a finer transformation which depends on a set $\mathcal{S}$ of formulas [4].

Proposition 1. *Let* $\mathcal{M} = \langle W, \mathcal{N}_b, \mathcal{V} \rangle$ *be a bi-neighbourhood model and* $\mathcal{S}$ *be a set of formulas closed under subformulas. We define the standard neighbourhood model* $\mathcal{M}' = \langle W', \mathcal{N}_s, \mathcal{V}' \rangle$ *by taking* $W' = W$, $\mathcal{V}' = \mathcal{V}$ *and for all* $w \in W$, $\mathcal{N}_s(w) = \{[A]_{\mathcal{M}} \mid w \Vdash_b \Box A \textit{ and } \Box A \in \mathcal{S}\}$. *Then for all* $w \in W$, $A \in \mathcal{S}$,

$$\mathcal{M}', w \Vdash_s A \quad \textit{iff} \quad \mathcal{M}, w \Vdash_b A.$$

Further, if $\mathcal{M}$ *is a N-model and* $\Box\top \in \mathcal{S}$, *then* $\mathcal{M}'$ *contains the unit, and if* $\mathcal{M}$ *is a C-model and* $\Box A, \Box B \in \mathcal{S}$ *implies* $\Box(A \wedge B) \in \mathcal{S}$, *then* $\mathcal{M}'$ *is closed under intersection.*

---

[1] The N-model condition in [4] was slightly different. However, it is easy to verify that the two conditions are equivalent with respect to the validity of formulas.

The transformation in Proposition 1 produces models with a smaller neighbourhood function. Whereas in models produced by the first transformation the size of $\mathcal{N}_s(w)$ can be exponential with respect to $\mathcal{N}_b(w)$, here the size of $\mathcal{N}_s(w)$ is linearly bounded by the number of boxed formulas in $\mathcal{S}$. As a paradigmatic case, suppose there is $(\emptyset, \emptyset) \in \mathcal{N}_b(w)$. By the rough transformation $\mathcal{N}_s$ would contain all subsets of $\mathcal{W}$, whereas by the finer one it would contain only the truth sets of some boxed formulas.

While the two semantics characterise equally well the classical cube, we shall see in Sect. 5 that bi-neighbourhood semantics is more suited for direct countermodels extraction from failed proofs.

3   Hypersequent Calculi

We now move our attention to proof theory. We will construct our calculi in the *hypersequent* framework (see, *e.g.*, [2]). This choice is motivated mainly by the possibility of getting direct countermodel extraction, as detailed in the following. Moreover, our calculi will contain additional structures, called *blocks*, which are used to collect boxed formulas.

Our calculi are built as follows: A *block* is a structure $\langle \Sigma \rangle$, where Σ is a multiset of formulas of $\mathcal{L}$. A *sequent* is a pair $\Gamma \Rightarrow \Delta$, where Γ is a multiset of formulas and blocks, and Δ is a multiset of formulas. We sometimes consider $\mathrm{set}(\Gamma)$: the *support* of a multiset Γ, *i.e.* the set of its elements disregarding multiplicities. A *hypersequent* is a multiset $S_1 \mid ... \mid S_n$, where $S_1, ..., S_n$ are sequents. $S_1, ..., S_n$ are called *components* of the hypersequent. Single sequents can be interpreted as formulas of the logic in the following manner:

$$i(A_1, ..., A_n, \langle \Sigma_1 \rangle, ..., \langle \Sigma_m \rangle \Rightarrow B_1, ..., B_k) = \bigwedge_{i \leq n} A_i \wedge \bigwedge_{j \leq m} \Box \bigwedge \Sigma_j \to \bigvee_{\ell \leq k} B_\ell .$$

We say that a sequent S is *valid* in a (bi-)neighbourhood model $\mathcal{M}$ ($\mathcal{M} \models S$) if for all $w \in \mathcal{M}$, $\mathcal{M}, w \Vdash i(S)$; and that a hypersequent H is *valid* in $\mathcal{M}$ ($\mathcal{M} \models H$) if $\mathcal{M} \models S$ for some $S \in H$. Finally, we say that H is valid in (M/C/N-)models if it is valid in all models of that kind.

The hypersequent calculi $\mathcal{H}_{\mathsf{E}*}$ are defined by the rules in Fig. 1, in particular: $\mathcal{H}_{\mathsf{E}} :=$ propositional rules $+ \Box_{\mathsf{L}} + \Box_{\mathsf{R}}$; $\mathcal{H}_{\mathsf{EN}} := \mathcal{H}_{\mathsf{E}} + \mathsf{N}$; $\mathcal{H}_{\mathsf{EC}} := \mathcal{H}_{\mathsf{E}} + \mathsf{C}$; $\mathcal{H}_{\mathsf{ECN}} := \mathcal{H}_{\mathsf{E}} + \mathsf{C} + \mathsf{N}$; $\mathcal{H}_{\mathsf{M}} :=$ propositional rules $+ \Box_{\mathsf{L}} + {}_{\mathsf{M}}\Box_{\mathsf{R}}$; $\mathcal{H}_{\mathsf{MN}} := \mathcal{H}_{\mathsf{M}} + \mathsf{N}$; $\mathcal{H}_{\mathsf{MC}} := \mathcal{H}_{\mathsf{M}} + \mathsf{C}$; and $\mathcal{H}_{\mathsf{MCN}} := \mathcal{H}_{\mathsf{M}} + \mathsf{C} + \mathsf{N}$.

Rules are given in their *kleene'd* versions, *i.e.*, where the principal formula (or structure) is copied into every premiss. As usual, initial sequents init are restricted to propositional variables, but it is easy to see that $G \mid A, \Gamma \Rightarrow \Delta, A$ is derivable for any A. Note that the only rule which expands blocks is C, then in absence of this rule the blocks occurring in a proof for a single formula contain only one formula. Examples of derivations are the following.

$$\text{init} \; \frac{}{G \mid p, \Gamma \Rightarrow \Delta, p} \qquad\qquad \bot_\mathsf{L} \; \frac{}{G \mid \bot, \Gamma \Rightarrow \Delta} \qquad\qquad \top_\mathsf{R} \; \frac{}{G \mid \Gamma \Rightarrow \Delta, \top}$$

$$\rightarrow_\mathsf{L} \; \frac{G \mid A \rightarrow B, \Gamma \Rightarrow \Delta, A \qquad G \mid B, A \rightarrow B, \Gamma \Rightarrow \Delta}{G \mid A \rightarrow B, \Gamma \Rightarrow \Delta} \qquad \rightarrow_\mathsf{R} \; \frac{G \mid A, \Gamma \Rightarrow \Delta, A \rightarrow B, B}{G \mid \Gamma \Rightarrow \Delta, A \rightarrow B}$$

$$\wedge_\mathsf{L} \; \frac{G \mid A, B, A \wedge B, \Gamma \Rightarrow \Delta}{G \mid A \wedge B, \Gamma \Rightarrow \Delta} \qquad \wedge_\mathsf{R} \; \frac{G \mid \Gamma \Rightarrow \Delta, A \wedge B, A \qquad G \mid \Gamma \Rightarrow \Delta, A \wedge B, B}{G \mid \Gamma \Rightarrow \Delta, A \wedge B}$$

$$\Box_\mathsf{L} \; \frac{G \mid \langle A \rangle, \Box A, \Gamma \Rightarrow \Delta}{G \mid \Box A, \Gamma \Rightarrow \Delta} \qquad {}_\mathsf{M}\Box_\mathsf{R} \; \frac{G \mid \langle \Sigma \rangle, \Gamma \Rightarrow \Delta, \Box B \mid \Sigma \Rightarrow B}{G \mid \langle \Sigma \rangle, \Gamma \Rightarrow \Delta, \Box B}$$

$$\Box_\mathsf{R} \; \frac{G \mid \langle \Sigma \rangle, \Gamma \Rightarrow \Delta, \Box B \mid \Sigma \Rightarrow B \qquad \{G \mid \langle \Sigma \rangle, \Gamma \Rightarrow \Delta, \Box B \mid B \Rightarrow A\}_{A \in \Sigma}}{G \mid \langle \Sigma \rangle, \Gamma \Rightarrow \Delta, \Box B}$$

$$\mathsf{N} \; \frac{G \mid \langle \top \rangle, \Gamma \Rightarrow \Delta}{G \mid \Gamma \Rightarrow \Delta} \qquad\qquad \mathsf{C} \; \frac{G \mid \langle \Sigma, \Pi \rangle, \langle \Sigma \rangle, \langle \Pi \rangle, \Gamma \Rightarrow \Delta}{G \mid \langle \Sigma \rangle, \langle \Pi \rangle, \Gamma \Rightarrow \Delta}$$

Fig. 1. Rules of $\mathcal{H}_{\mathsf{E}^\star}$.

Example 1. Axioms M, N, C are derivable in $\mathcal{H}_{\mathsf{M}^\star}$, $\mathcal{H}_{\mathsf{EN}^\star}$, and $\mathcal{H}_{\mathsf{EC}^\star}$, respectively

$$\cfrac{\cfrac{\langle A \wedge B \rangle, \Box(A \wedge B) \Rightarrow \Box A \mid A \wedge B \Rightarrow A}{\langle A \wedge B \rangle, \Box(A \wedge B) \Rightarrow \Box A} \; {}_\mathsf{M}\Box_\mathsf{R}}{\Box(A \wedge B) \Rightarrow \Box A} \; \Box_\mathsf{L}$$

$$\cfrac{\cfrac{\langle \top \rangle \Rightarrow \Box \top \mid \top \Rightarrow \top \qquad \dots \mid \top \Rightarrow \top}{\langle \top \rangle \Rightarrow \Box \top} \; \Box_\mathsf{R}}{\Rightarrow \Box \top} \; \mathsf{N}$$

$$\cfrac{\cfrac{\cfrac{\cfrac{\langle A, B \rangle, \dots \mid A, B \Rightarrow A \wedge B \qquad \langle A, B \rangle, \dots \mid A \wedge B \Rightarrow A \qquad \langle A, B \rangle, \dots \mid A \wedge B \Rightarrow B}{\langle A, B \rangle, \langle A \rangle, \langle B \rangle, \Box A, \Box B, \Box A \wedge \Box B \Rightarrow \Box(A \wedge B)} \; \Box_\mathsf{R}}{\langle A \rangle, \langle B \rangle, \Box A, \Box B, \Box A \wedge \Box B \Rightarrow \Box(A \wedge B)} \; \mathsf{C}}{\Box A, \Box B, \Box A \wedge \Box B \Rightarrow \Box(A \wedge B)} \; \Box_\mathsf{L} \times 2}{\Box A \wedge \Box B \Rightarrow \Box(A \wedge B)} \; \wedge_\mathsf{L}$$

We then have the following soundness theorem.

Theorem 1. *If H is derivable in $\mathcal{H}_{\mathsf{E(M/C/N)}}$, then it is valid in (M/C/N-) models.*

Proof. We show that whenever the premises of a rule are valid, so is the conclusion. For propositional rules the proof is standard.

$\Box_\mathsf{L}$) Assume $G \mid \langle A \rangle, \Gamma \Rightarrow \Delta$ valid. Then for all $\mathcal{M}$, $\mathcal{M} \models S$ for some $S \in G$, or $\mathcal{M} \models \langle A \rangle, \Box A, \Gamma \Rightarrow \Delta$. In the second case, for all $w \in \mathcal{M}$, $w \Vdash i(\langle A \rangle, \Box A, \Gamma \Rightarrow \Delta) = i(\Box A, \Box A, \Gamma \Rightarrow \Delta)$, which is equivalent to $i(\Box A, \Gamma \Rightarrow \Delta)$. Then $G \mid \Box A, \Gamma \Rightarrow \Delta$ is valid.

$\Box_\mathsf{R}$) Let $\Sigma = C_1, ..., C_n$ and assume $G \mid \langle \Sigma \rangle, \Gamma \Rightarrow \Delta, \Box B \mid \Sigma \Rightarrow B$ and $G \mid \langle \Sigma \rangle, \Gamma \Rightarrow \Delta, \Box B \mid B \Rightarrow C_i$ valid for all $1 \leq i \leq n$. Then for all $\mathcal{M}$, $\mathcal{M} \models S$ for some $S \in G$, or $\mathcal{M} \models \langle \Sigma \rangle, \Gamma \Rightarrow \Delta, \Box B$. Otherwise $\Sigma \Rightarrow B, B \Rightarrow C_1, ..., B \Rightarrow C_n$

are valid in $\mathcal{M}$. In the last case, for all $w \in \mathcal{M}$, $w \Vdash (\bigwedge \Sigma \rightarrow B) \wedge (B \rightarrow C_1) \wedge \dots \wedge (B \rightarrow C_n)$, that is $\mathcal{M} \models \bigwedge \Sigma \leftrightarrow B$. Then $\mathcal{M} \models \Box \bigwedge \Sigma \leftrightarrow \Box B$, so $\mathcal{M} \models \langle \Sigma \rangle \Rightarrow \Box B$. Therefore $G \mid \langle \Sigma \rangle, \Gamma \Rightarrow \Delta, \Box B$ is valid.

$_{\text{M}}\Box_{\text{R}}$) Analogous to $\Box_{\text{R}}$, by considering that in M-models $\mathcal{M} \models \bigwedge \Sigma \rightarrow B$ implies $\mathcal{M} \models \Box \bigwedge \Sigma \rightarrow \Box B$.

N) Assume $G \mid \langle \top \rangle, \Gamma \Rightarrow \Delta$ valid. Then for all $\mathcal{M}$, $\mathcal{M} \models S$ for some $S \in G$, or $\mathcal{M} \models \langle \top \rangle, \Gamma \Rightarrow \Delta$. In the second case, for all $w \in \mathcal{M}$, $w \Vdash i(\langle \top \rangle, \Gamma \Rightarrow \Delta)$, which is equivalent to $\Box \top \rightarrow i(\Gamma \Rightarrow \Delta)$. Since $\Box \top$ is valid in N-models, $\mathcal{M} \models \Gamma \Rightarrow \Delta$.

C) Assume $G \mid \langle \Sigma \rangle, \langle \Pi \rangle, \langle \Sigma, \Pi \rangle, \Gamma \Rightarrow \Delta$ valid. Then for all $\mathcal{M}$, $\mathcal{M} \models S$ for some $S \in G$, or $\mathcal{M} \models \langle \Sigma \rangle, \langle \Pi \rangle, \langle \Sigma, \Pi \rangle, \Gamma \Rightarrow \Delta$. In the second case, for all $w \in \mathcal{M}$, $w \Vdash (\Box \bigwedge \Sigma \wedge \Box \bigwedge \Pi \wedge \Box (\bigwedge \Sigma \wedge \bigwedge \Pi)) \rightarrow i(\Gamma \Rightarrow \Delta)$. Since $\Box \bigwedge \Sigma \wedge \Box \bigwedge \Pi \rightarrow \Box (\bigwedge \Sigma \wedge \bigwedge \Pi)$ is valid in C-models, $w \Vdash (\Box \bigwedge \Sigma \wedge \Box \bigwedge \Pi) \rightarrow i(\Gamma \Rightarrow \Delta)$. Then $\mathcal{M} \models \langle \Sigma \rangle, \langle \Pi \rangle, \Gamma \Rightarrow \Delta$. $\square$

We make clear that to the purpose of having sound and complete calculi for NNMLs the hypersequent framework is not necessary, as for instance the sequent calculi in [11] show. Moreover, for the calculi $\mathcal{H}_{\text{E}*}$ whenever a hypersequent is derivable there is a component which is derivable. But as we shall see, the hypersequent framework is very adequate to extract countermodels from a single failed proof, ensuring at the same time good computational and structural properties. As a matter of fact, even in the bi-neighbourhood semantics, non-normal modal logics without monotonicity ultimately need to consider truth sets of formulas. Hence, to make our calculi suitable for a reasonably straightforward counter-model construction, we need to be able to represent essentially all worlds of a possible model in the data structure used by the calculus. While this could also be accomplished by, *e.g.* nested sequents, for obtaining small countermodels in non-monotonic logics it is crucial that every world (represented by a component of the hypersequent) has access to *all* other worlds which have been constructed so far. This very strongly suggests a *flat* structure, as given by hypersequents.

Structural Properties and Syntactic Completeness. We now investigate the structural properties of our calculi. We first show that weakening and contraction are height-preserving (*hp* for short) admissible, both in their internal and in their external variants. Then, we prove that the cut rule is admissible, which allows us to directly prove the completeness of the calculi with respect to the corresponding axiomatisations. In the proofs we use the following definition of weight of formulas and blocks.

Definition 3. *The weight* w *of a formula is recursively defined as* $\mathsf{w}(\bot) = \mathsf{w}(\top) = \mathsf{w}(p) = 0$, $\mathsf{w}(A \rightarrow B) = \mathsf{w}(A) + \mathsf{w}(B) + 1$, $\mathsf{w}(\langle A_1, \dots, A_n \rangle) = max_i \{\mathsf{w}(A_i)\} + 1$, $\mathsf{w}(\Box A) = \mathsf{w}(A) + 2$.

Proposition 2. *The following structural rules are hp-admissible in* $\mathcal{H}_{\text{E}*}$:

$$\text{wk}_\text{L} \frac{G \mid \Gamma \Rightarrow \Delta}{G \mid A, \Gamma \Rightarrow \Delta} \qquad \text{wk}_\text{R} \frac{G \mid \Gamma \Rightarrow \Delta}{G \mid \Gamma \Rightarrow \Delta, A} \qquad \text{wk}_{\langle\rangle} \frac{G \mid \Gamma \Rightarrow \Delta}{G \mid \langle \Sigma \rangle, \Gamma \Rightarrow \Delta} \qquad \text{ctr}_\text{L} \frac{G \mid A, A, \Gamma \Rightarrow \Delta}{G \mid A, \Gamma \Rightarrow \Delta}$$

$$\text{ctr}_\text{R} \frac{G \mid \Gamma \Rightarrow \Delta, A, A}{G \mid \Gamma \Rightarrow \Delta, A} \qquad \text{ctr}_{\langle\rangle} \frac{G \mid \langle \Sigma \rangle, \langle \Sigma \rangle, \Gamma \Rightarrow \Delta}{G \mid \langle \Sigma \rangle, \Gamma \Rightarrow \Delta} \qquad \text{ew} \frac{G}{G \mid \Gamma \Rightarrow \Delta} \qquad \text{ec} \frac{G \mid \Gamma \Rightarrow \Delta \mid \Gamma \Rightarrow \Delta}{G \mid \Gamma \Rightarrow \Delta}$$

Note that, since weakening is admissible, invertibility of all rules is immediate.

We now move our attention to the admissibility of the following cut rule

$$\mathsf{cut} \; \frac{G \mid \Gamma \Rightarrow \Delta, A \qquad G \mid A, \Gamma \Rightarrow \Delta}{G \mid \Gamma \Rightarrow \Delta}$$

In order do to this, we prove simultaneously the admissibility of cut and of the following rule sub, which states that a formula A inside one or more blocks can be replaced by any equivalent set of formulas Σ:

$$\mathsf{sub} \; \frac{G \mid \Sigma \Rightarrow A \qquad \{G \mid A \Rightarrow B\}_{B \in \Sigma} \qquad G \mid \langle A^{n_1}, \Pi_1 \rangle, ..., \langle A^{n_k}, \Pi_k \rangle, \Gamma \Rightarrow \Delta}{G \mid \langle \Sigma^{n_1}, \Pi_1 \rangle, ..., \langle \Sigma^{n_k}, \Pi_k \rangle, \Gamma \Rightarrow \Delta}$$

where A^{n_i} (resp. Σ^{n_i}) is a compact way to denote n_i occurrences of A (resp. Σ). In the monotonic case we need to consider, instead of sub, the rule

$$\mathsf{sub_M} \; \frac{G \mid \Sigma \Rightarrow A \qquad G \mid \langle A^{n_1}, \Pi_1 \rangle, ..., \langle A^{n_k}, \Pi_k \rangle, \Gamma \Rightarrow \Delta}{G \mid \langle \Sigma^{n_1}, \Pi_1 \rangle, ..., \langle \Sigma^{n_k}, \Pi_k \rangle, \Gamma \Rightarrow \Delta}$$

Theorem 2. *If $\mathcal{H}_{\mathsf{E}^\star}$ contains $\square_\mathsf{R}$, then the rules cut and sub are admissible in $\mathcal{H}_{\mathsf{E}^\star}$, otherwise cut and sub$_\mathsf{M}$ are admissible in $\mathcal{H}_{\mathsf{E}^\star}$.*

Sketch of Proof. Let $Cut(c, h)$ mean that all applications of cut of height h on a cut formula of weight c are admissible, and $Sub(c)$ mean that all applications of sub where A has weight c are admissible. Then the theorem is a consequence of the following claims (for all $\Sigma, \Pi_1, ..., \Pi_k$): **(A)** $\forall c. Cut(c, 0)$; **(B)** $\forall h. Cut(0, h)$; **(C)** $\forall c. (\forall h. Cut(c, h) \rightarrow Sub(c))$; **(D)** $\forall c. \forall h. ((\forall c' < c. (Sub(c') \wedge \forall h'. Cut(c', h')) \wedge \forall h'' < h. Cut(c, h'')) \rightarrow Cut(c, h))$. Proof in the Appendix. $\square$

Theorem 3 (Syntactic completeness). *If $i(\Gamma \Rightarrow \Delta)$ is derivable in $\mathbf{E}^\star$, then $\Gamma \Rightarrow \Delta$ is derivable in $\mathcal{H}_{\mathsf{E}^\star}$.*

Proof. By deriving the axioms, simulating MP using cut, and checking that RE is derivable using ew, $\square_\mathsf{R}$ and $\square_\mathsf{L}$. $\square$

4 Complexity of Proof Search

We would like to use the calculus to obtain an optimal decision procedure for the considered logics. As established in [20], logics without axiom C are coNP-complete, whereas logics with C are in PSPACE (although we are not aware of a proof of the lower bound). We accordingly distinguish cases whether axiom C is present or not.

Extensions without C. The decision procedures for the logics **E**, **M**, **EN** and **MN** implement backwards proof search on a polynomially bounded nondeterministic Turing machine with universal choices to handle the branching caused by rules with several premises, as shown in Algorithm 1. To prevent loops, we employ a *local loop checking* strategy, stating that a rule is not applied (bottom-up) to a hypersequent G, if for at least one of its premises H we have that

---

Algorithm 1. Decision procedure for the derivability problem in $\mathcal{H}_{E^*}$

Input: a hypersequent G and the code of a logic $\mathbf{L} \in \{\mathbf{E}, \mathbf{M}, \mathbf{EN}, \mathbf{MN}\}$
Output: Is G derivable in $\mathcal{H}_L$?

1 **if** *there is a component* $\Gamma \Rightarrow \Delta$ *in* G *with* $\bot \in \Gamma$, *or* $\top \in \Delta$, *or* $\Gamma \cap \Delta \neq \emptyset$ **then**
2 $\quad|\quad$ halt and accept;
3 pick the next applicable rule R from $\mathcal{H}_L$, a matching component $\Gamma \Rightarrow \Delta$ and principal formula (and block) from $\Gamma \Rightarrow \Delta$;
4 universally choose a premiss H of this rule application ;
5 check recursively whether H is derivable, output the answer and halt;
6 halt and reject;

---

for every component $\Gamma \Rightarrow \Delta$ in H there is a component $\Sigma \Rightarrow \Pi$ in G with set$(\Gamma) \subseteq$ set(Σ) and set$(\Delta) \subseteq$ set(Π). The order of applications of the rules is arbitrary but fixed, and once a rule has been applied the algorithm universally chooses one of its premisses and then recursively checks that this premiss is derivable.

It is easy to see that Algorithm 1 is correct and complete. In particular, completeness follows from admissibility of the structural rules of Proposition 2, and the fact that in view of this it suffices to search for *minimal derivations*, i.e., derivations where none of the branches can be shortened.

Theorem 4. *Algorithm 1 runs in* coNP, *whence for the logics without* C *the calculi provide a complexity-optimal decision procedure.*

Proof. Since the procedure is in the form of a non-deterministic Turing machine with universal choices, it suffices to show that every computation of this machine has polynomial length. Every application of a rule adds either a subformula of its conclusion or a new block to one of the components, or adds a new component. Due to local loop checking it never adds a formula, block or component which is already in the premiss, so it suffices to calculate the maximal size of a hypersequent occurring in proof search for G. Suppose that the size of G is n. Then both the number of components and the number of subformulas of G are bounded by n. Since the local loop check prevents the duplication of formulas, each component contains at most n formulas in the antecedent and n formulas in the succedent. Moreover, since we only consider logics without the axiom C, every newly created block contains exactly one formula. Again, due to the local loop checking condition no block is duplicated, so every component contains at most n blocks. Thus every component has size at most $3n$. The procedure creates new components from a block and a formula of an already existing component, hence there are only n^2 many different components which can be created without violating the local loop checking condition. Thus every hypersequent occurring in the proof contains at most $n + n^2$ many components, each of size at most $3n$, giving a total size and thus running time of $\mathcal{O}(n^3)$. $\qquad\square$

Extensions with C. For the logics with axiom C we would like to use our calculi to obtain PSPACE decision procedures. It can be easily shown that Algorithm 1 works properly also for these logics, ensuring in particular termination. However, because of C-rule, hypersequents can be exponentially large, and therefore PSPACE upper bound *cannot* be achieved. In order to obtain PSPACE decision procedures we adopt a different strategy: Instead of the rules in Fig. 1, we consider their *unkleene'd* version, *i.e.*, the ones with all principal formulas and structures deleted from the premises. For instance $_\mathsf{M}\Box_\mathsf{R}$, $\Box_\mathsf{R}$ and C are replaced respectively with

$$\frac{G|\Gamma \Rightarrow \Delta|\Sigma \Rightarrow B}{G|\langle\Sigma\rangle,\Gamma \Rightarrow \Delta,\Box B} \qquad \frac{G|\Gamma \Rightarrow \Delta|\Sigma \Rightarrow B \quad \{G|\Gamma \Rightarrow \Delta|B \Rightarrow A\}_{A\in\Sigma}}{G|\langle\Sigma\rangle,\Gamma \Rightarrow \Delta,\Box B} \qquad \frac{G|\langle\Sigma,\Pi\rangle,\Gamma \Rightarrow \Delta}{G|\langle\Sigma\rangle,\langle\Pi\rangle,\Gamma \Rightarrow \Delta}$$

Call the resulting calculus $\mathcal{H}_{\mathsf{E}^\ast}{}^-$. Backwards proof search is then implemented on an alternating Turing machine by existentially guessing the last applied rule except for N, and universally checking that all of its premises are derivable. To ensure that N is applied if it is present in the system, we stipulate that it is applied once to every component of the input, and that if the existentially guessed rule is one of $\Box_\mathsf{R}$ or $_\mathsf{M}\Box_\mathsf{R}$, the rule N is applied immediately afterwards to each of its premises. Since no rule application keeps the principal formulas in the premises, and since the rule N if present is applied exactly once to every component, there is no need for any loop checking condition.

The calculi $\mathcal{H}_{\mathsf{E}^\ast}{}^-$ are sound and complete. Soundness is obvious, since we can add the missing formulas and structures and recover derivations in $\mathcal{H}_{\mathsf{E}^\ast}$. Completeness can be proved syntactically by a cut elimination argument similar to the one in the previous section, or alternatively by simulating the calculi in [11]. While it is easy to see that the calculi in [11] give a PSPACE upper bound, this is less obvious for $\mathcal{H}_{\mathsf{E}^\ast}{}^-$ because of the hypersequent structure. Nonetheless we obtain the following result.

Theorem 5. *Backwards proof search in* $\mathcal{H}_{\mathsf{E}^\ast}{}^-$ *is in* PSPACE.

Proof. We need to show that every run of the procedure terminates in polynomial time. Assume that the size of the input is n. Let the *weight* of a component in a hypersequent be the sum of the weights of the formulas and blocks occurring in it according to Definition 3, and suppose that the maximal weight of components in the input is w. Then every rule apart from N decreases the weight of the component active in its conclusion. Moreover, a new component is only introduced in place of a subformula of the input, hence any hypersequent occurring in the proof search has at most $n + n$ components. The weight of each of these components is at most the maximal weight of a component of the input (plus one in the cases with N). Since the rule N is applied at most once to each component, it is thus applied at most n times in the total proof search. Thus the runtime in total is $\mathcal{O}(n^2 \cdot w)$, hence polynomial in the size of the input. Thus the procedure runs in alternating polynomial time, and thus in PSPACE.   □

5   Countermodel Extraction

We now prove semantic completeness of the calculus, *i.e.*: every valid hypersequent is derivable in the calculus. This amounts to show that a non-provable hypersequent has a countermodel. Countermodels are found in the bi-neighbourhood semantics, as it is more suited for direct countermodels extraction from failed proofs than the standard semantics. The reason is that in order to define a neighbourhood model we need to determine *exactly* the truth sets of formulas: If we want a world w to force $\Box A$, then we have to make sure that $[A]$ belongs to $\mathcal{N}(w)$, thus $[A]$ must be computed. On the contrary, in bi-neighbourhood models it suffices to find a suited pair (α, β) such that $\alpha \subseteq [A]$ and $\beta \subseteq [\neg A]$. As we shall see, such a pair can be extracted direclty from the failed proof even without knowing exactly the extension of $[A]$.

In order to prove semantic completeness we make use of the backwards proof search strategy based on local loop checking already considered in Sect. 4 (Algorithm 1). This strategy amounts to consider the following notion of saturation, stating that a bottom-up application of a rule R is not allowed to a hypersequent G if G already fulfills the corresponding saturation condition (R).

Definition 4 (Saturated hypersequent). *Let $H = \Gamma_1 \Rightarrow \Delta_1 \mid ... \mid \Gamma_n \Rightarrow \Delta_n$ be a hypersequent occurring in a proof for H'. The saturation conditions associated to each application of a rule of $\mathcal{H}_{\mathsf{E}^*}$ are as follows:* (init) $\Gamma_i \cap \Delta_i = \emptyset$; $(\bot_{\mathsf{L}})$ $\bot \notin \Gamma_i$; $(\top_{\mathsf{R}})$ $\top \notin \Delta_i$; $(\rightarrow_{\mathsf{L}})$ *If $A \rightarrow B \in \Gamma_i$, then $A \in \Delta_i$ or $B \in \Gamma_i$;* $(\rightarrow_{\mathsf{R}})$ *If $A \rightarrow B \in \Delta_i$, then $A \in \Gamma_i$ and $B \in \Delta_i$;* $(\wedge_{\mathsf{L}})$ *If $A \wedge B \in \Gamma_i$, then $A \in \Gamma_i$ and $B \in \Gamma_i$;* $(\wedge_{\mathsf{R}})$ *If $A \wedge B \in \Delta_i$, then $A \in \Delta_i$ or $B \in \Delta_i$;* $(\Box_{\mathsf{L}})$ *If $\Box A \in \Gamma_i$, then $\langle A \rangle \in \Gamma_i$;* (N) $\langle \top \rangle \in \Gamma_i$; (C) *If $\langle \Sigma \rangle, \langle \Pi \rangle \in \Gamma_i$, then there is $\langle \Omega \rangle \in \Gamma_i$ such that $\mathsf{set}(\Sigma, \Pi) = \mathsf{set}(\Omega)$.* $(\mathsf{M}\Box_{\mathsf{R}})$ *If $\langle \Sigma \rangle, \Gamma \Rightarrow \Delta, \Box B$ is in H, then there is $\Sigma, \Gamma' \Rightarrow \Delta', B$ in H.* $(\Box_{\mathsf{R}})$ *If $\langle \Sigma \rangle, \Gamma \Rightarrow \Delta, \Box B$ is in H, then there is $\Sigma, \Gamma' \Rightarrow \Delta', B$ in H or there is $B, \Gamma' \Rightarrow \Delta', A$ in H for some $A \in \Sigma$.*

We say that H is saturated with respect to an application of a rule R if it satisfies the corresponding saturation condition (R) for that particular rule application, and that it is saturated with respect to $\mathcal{H}_{\mathsf{E}^*}$ if it is saturated with respect to all possible applications of any rule of $\mathcal{H}_{\mathsf{E}^*}$.

Given a saturated hypersequent H we can construct directly a countermodel for H in the bi-neighbourhood semantics in the following way.

Definition 5 (Countermodel construction). *Let H be a saturated hypersequent occurring in a proof for H'. Moreover, let $e : \mathbb{N} \longrightarrow H$ be an enumeration of the components of H. Given e, we can write H as $\Gamma_1 \Rightarrow \Delta_1 \mid ... \mid \Gamma_k \Rightarrow \Delta_k$. Model $\mathcal{M} = \langle \mathcal{W}, \mathcal{N}, \mathcal{V} \rangle$ is defined as follows:*

- $\mathcal{W} = \{n \mid \Gamma_n \Rightarrow \Delta_n \in H\}$.
- $\mathcal{V}(p) = \{n \mid p \in \Gamma_n\}$.
- *For all blocks $\langle \Sigma \rangle$ appearing in a component $\Gamma_m \Rightarrow \Delta_m$ of H,*
  $\Sigma^+ = \{n \mid \mathsf{set}(\Sigma) \subseteq \Gamma_n\}$ *and* $\Sigma^- = \{n \mid \Sigma \cap \Delta_n \neq \emptyset\}$.
  - *Non-monotonic case: $\mathcal{N}(n) = \{(\Sigma^+, \Sigma^-) \mid \langle \Sigma \rangle \in \Gamma_n\}$.*
  - *Monotonic case: $\mathcal{N}(n) = \{(\Sigma^+, \emptyset) \mid \langle \Sigma \rangle \in \Gamma_n\}$.*

Lemma 1. *Let $\mathcal{M}$ be defined as in Definition 5. Then for every A, $\langle \Sigma \rangle$ and every $n \in \mathcal{W}$, we have: If $A \in \Gamma_n$, then $n \Vdash A$; if $\langle \Sigma \rangle \in \Gamma_n$, then $n \Vdash \Box \bigwedge \Sigma$;*

and if $A \in \Delta_n$, then $n \nVdash A$. Moreover, (a) $\mathcal{M}$ is a M-model if $\mathcal{H}_{E^*}$ contains rule $_M\square_R$; (b) $\mathcal{M}$ is a N-model if $\mathcal{H}_{E^*}$ contains rule N; and (c) $\mathcal{M}$ is a C-model if $\mathcal{H}_{E^*}$ contains rule C.

Proof. The first claim is proved by mutual induction on A and $\langle \Sigma \rangle$. We only consider the cases of modal formulas, the other are similar and simpler. ($\langle \Sigma \rangle \in \Gamma_n$) In the non-monotonic case, by def $(\Sigma^+, \Sigma^-) \in \mathcal{N}(n)$. We show that $\Sigma^+ \subseteq [\bigwedge \Sigma]$ and $\Sigma^- \subseteq [\neg \bigwedge \Sigma]$, which implies $n \Vdash \square \bigwedge \Sigma$. If $m \in \Sigma^+$, then $\mathsf{set}(\Sigma) \subseteq \Gamma_m$. By i.h. $m \Vdash A$ for all $A \in \Sigma$, then $m \Vdash \bigwedge \Sigma$. If $m \in \Sigma^-$, then there is $B \in \Sigma \cap \Delta_m$. By i.h. $m \nVdash B$, then $m \nVdash \bigwedge \Sigma$. In the monotonic case the proof is analogous. ($\square B \in \Gamma_n$) By saturation of $\square_L$, $\langle B \rangle \in \Gamma_n$. Then by i.h. $n \Vdash \square B$. ($\square B \in \Delta_n$) In the non-monotonic case, assume $(\Sigma^+, \Sigma^-) \in \mathcal{N}(n)$. Then there is $\langle \Pi \rangle \in \Gamma_n$ s.t. $\Pi^+ = \Sigma^+$ and $\Pi^- = \Sigma^-$. By saturation of $\square_R$, there is $m \in \mathcal{W}$ s.t. $\Pi \subseteq \Gamma_m$ and $B \in \Delta_m$, or there is $m \in \mathcal{W}$ s.t. $B \in \Gamma_m$ and $\Pi \cap \Delta_m \neq \emptyset$. In the first case, $m \in \Pi^+$ and by i.h. $m \nVdash B$. In the second case, $m \in \Pi^-$ and by i.h. $m \Vdash B$. That is $\Sigma^+ \nsubseteq [B]$ or $\Sigma^- \nsubseteq [\neg B]$. Then $n \nVdash \square B$. The monotonic case is analogous.

The model conditions are proved as follows: (a) By definition of $\mathcal{N}$ in monotonic case. (b) By saturation $\langle \top \rangle \in \Gamma_n$ for all $n \in \mathcal{W}$. Then $(\top^+, \top^-) \in \mathcal{N}(n)$, where by saturation of $\top_R$, $\top^- = \emptyset$. (c) Assume $(\Sigma^+, \Sigma^-), (\Pi^+, \Pi^-) \in \mathcal{N}(n)$. Then there are $\langle \Lambda \rangle, \langle \Theta \rangle \in \Gamma_n$ s.t. $\Sigma^+ = \Lambda^+, \Sigma^- = \Lambda^-, \Pi^+ = \Theta^+$ and $\Pi^- = \Theta^-$. By saturation, there is $\langle \Omega \rangle \in \Gamma_n$ s.t. $\mathsf{set}(\Omega) = \mathsf{set}(\Lambda, \Theta)$, thus $(\Omega^+, \Omega^-) \in \mathcal{N}(n)$. We show that (i) $\Omega^+ = \Sigma^+ \cap \Pi^+$ and (ii) $\Omega^- = \Sigma^- \cap \Pi^-$. (i) If $m \in \Omega^+$, then $\mathsf{set}(\Omega) \subseteq \Gamma_m$, then $\mathsf{set}(\Lambda, \Theta) \subseteq \Gamma_m$, then $\mathsf{set}(\Lambda) \subseteq \Gamma_m$ and $\mathsf{set}(\Theta) \subseteq \Gamma_m$, then $m \in \Lambda^+$ and $m \in \Theta^+$. If $m \in \Lambda^+ \cap \Theta^+$, then $m \in \Lambda^+$ and $m \in \Theta^+$, then $\mathsf{set}(\Lambda) \subseteq \Gamma_m$ and $\mathsf{set}(\Theta) \subseteq \Gamma_m$, then $\mathsf{set}(\Lambda, \Theta) \subseteq \Gamma_m$, then $\mathsf{set}(\Omega) \subseteq \Gamma_m$, then $m \in \Omega^+$. (ii) If $m \in \Omega^-$, then $\Omega \cap \Delta_m \neq \emptyset$, then $\Lambda, \Theta \cap \Delta_m \neq \emptyset$, then $\Lambda \cap \Delta_m \neq \emptyset$ or $\Theta \cap \Delta_m \neq \emptyset$, then $m \in \Lambda^-$ or $m \in \Theta^-$. If $m \in \Lambda^- \cup \Theta^-$, then $m \in \Lambda^-$ or $m \in \Theta^-$, then $\Lambda \cap \Delta_m \neq \emptyset$ or $\Theta \cap \Delta_m \neq \emptyset$, then $\Lambda, \Theta \cap \Delta_m \neq \emptyset$, then $\Omega \cap \Delta_m \neq \emptyset$, then $m \in \Omega^-$. $\square$

Observe that, since all rules are cumulative, $\mathcal{M}$ is a countermodel for the root hypersequent H'. Moreover, since every proof built in accordance to the strategy either provides a derivation of the root hypersequent or contains a saturated hypersequent, this allows us to prove the following theorem.

Theorem 6. *If H is valid in (M/C/N-)models, then it is derivable in $\mathcal{H}_{E(M/C/N)}$.*

As the above construction shows, we can directly extract a bi-neighbourhood countermodel from any failed proof. If we want to obtain a countermodel in the stardard semantics we then need to apply the transformations presented in Sect. 2. In principle, the rough transformation can be embedded into the countermodel construction in order to get immediately a neighbourhood model. However, as illustrated in Sect. 2 we might obtain a much larger model than needed. On the other hand, there is no obvious way to integrate the finer transformation of Proposition 1 since it rests on the evaluation in an already existing model.

An alternative way to obtain countermodels in the neighbourhood semantics is proposed in [11]: It basically consists in forcing the proof search procedure to determine exactly the truth set of each formula. To this aim, whenever a sequent representing a new world is created, the sequent is saturated with respect to all disjunctions $A \lor \neg A$ such that A is a subformula of the root sequent. This solution is equivalent to use *analytic cut* and makes the proof search procedure significantly more complex than the one presented here (moreover it makes use of a more complex data structure than hypersequents).

Below we show some examples of countermodel extraction both in the bi-neighbourhood and in the standard neighbourhood semantics, the latter obtained by Proposition 1.

Example 2 (Proof search for axiom M *in* $\mathcal{H}_E$ *and countermodels).*

$$
\cfrac{
  \cfrac{
    \text{closed} \quad \cfrac{
      \cfrac{\text{closed}}{\dots \mid p \Rightarrow p \land q, p} \qquad \langle p \land q \rangle, \Box(p \land q) \Rightarrow \Box p \mid p \Rightarrow p \land q, q
    }{\langle p \land q \rangle, \Box(p \land q) \Rightarrow \Box p \mid p \Rightarrow p \land q} \land_R
  }{\langle p \land q \rangle, \Box(p \land q) \Rightarrow \Box p} \Box_R
}{}
$$

closed

$\dots \mid p \land q \Rightarrow p$

closed

$\dots \mid p \Rightarrow p \land q, p$

saturated hypersequent H

$\langle p \land q \rangle, \Box(p \land q) \Rightarrow \Box p \mid p \Rightarrow p \land q, q$

$\cfrac{\langle p \land q \rangle, \Box(p \land q) \Rightarrow \Box p \mid p \Rightarrow p \land q}{}$ $\land_R$

$\cfrac{\langle p \land q \rangle, \Box(p \land q) \Rightarrow \Box p}{\Box(p \land q) \Rightarrow \Box p}$ $\Box_L$ $\Box_R$

Bi-neighbourhood Countermodel. Let us consider the enumeration for the compontents of $\mathcal{B}$ where $1 \mapsto \langle p \land q \rangle, \Box(p \land q) \Rightarrow \Box p$, and $2 \mapsto p \Rightarrow p \land q, q$. According to the construction in Definition 5, from H we obtain the following countermodel $\mathcal{M} = \langle \mathcal{W}, \mathcal{N}, \mathcal{V} \rangle$: $\mathcal{W} = \{1, 2\}$, $\mathcal{V}(p) = \{2\}$, $\mathcal{V}(q) = \emptyset$, $\mathcal{N}(2) = \emptyset$, and $\mathcal{N}(1) = \{(\emptyset, \{2\})\}$, as $\mathcal{N}(1) = \{(p \land q^+, p \land q^-)\}$ and $p \land q^+ = \emptyset$, $p \land q^- = \{2\}$. *Neighbourhood Countermodel.* Considering $\mathcal{S} = \{\Box(p \land q), \Box p, p \land q, p, q\}$, by Proposition 1 we obtain from $\mathcal{M}$ the standard model $\mathcal{M}'$ with $\mathcal{N}'(1) = \{\emptyset\}$, as $\mathcal{N}'(1) = \{[p \land q]_{\mathcal{M}}\}$ and $[p \land q]_{\mathcal{M}} = \emptyset$.

Example 3 (Proof search for axiom K *in* $\mathcal{H}_{EC}$ *and countermodels).* By bottom-up proof search for $\Box(p \to q) \to (\Box p \to \Box q)$ in $\mathcal{H}_{EC}$ we obtain the following branch ending with a saturated hypersequent H (for lack of space we do not show the whole tree).

$$
\cfrac{
  \cfrac{
    \cfrac{
      \cfrac{
        \cfrac{
          \Box(p \to q), \Box p, \langle p \to q \rangle, \langle p \rangle, \langle p \to q, p \rangle \Rightarrow \Box q \mid q \Rightarrow p \mid p \to q \Rightarrow q, p
        }{\Box(p \to q), \Box p, \langle p \to q \rangle, \langle p \rangle, \langle p \to q, p \rangle \Rightarrow \Box q \mid q \Rightarrow p \mid p \to q \Rightarrow q} \to_L
      }{\Box(p \to q), \Box p, \langle p \to q \rangle, \langle p \rangle, \langle p \to q, p \rangle \Rightarrow \Box q \mid q \Rightarrow p} \Box_R
    }{\Box(p \to q), \Box p, \langle p \to q \rangle, \langle p \rangle, \langle p \to q, p \rangle \Rightarrow \Box q} \Box_R
  }{\Box(p \to q), \Box p, \langle p \to q \rangle, \langle p \rangle \Rightarrow \Box q} C
}{\Box(p \to q), \Box p \Rightarrow \Box q} \Box_L{}^2
$$

Bi-neighbourhood Countermodel. We consider the following enumeration of the compontents of H: $1 \mapsto \Box(p \to q), \Box p, \langle p \to q \rangle, \langle p \rangle, \langle p \to q, p \rangle \Rightarrow \Box q$; $2 \mapsto q \Rightarrow p$; $3 \mapsto p \to q \Rightarrow q, p$. From H we obtain $\mathcal{M} = \langle \mathcal{W}, \mathcal{N}, \mathcal{V} \rangle$, where $\mathcal{W} = \{1, 2, 3\}$, $\mathcal{V}(p) = \emptyset$, $\mathcal{V}(q) = \{2\}$, $\mathcal{N}(2) = \mathcal{N}(3) = \emptyset$, and $\mathcal{N}(1) = \{(\emptyset, \{2, 3\}), (\{3\}, \emptyset)\}$, as $\mathcal{N}(1) = \{(p^+, p^-), (p \to q^+, p \to q^-), (p, p \to q^+, p, p \to q^-)\}$ and $p^+ = \emptyset$,

$p^- = \{2,3\}$, $p \to q^+ = \{3\}$, $p \to q^- = \emptyset$, $p, p \to q^+ = \emptyset$, $p, p \to q^- = \{2,3\}$. It is easy to verify that $\mathcal{M}$ is a C-model.

Neighbourhood Countermodel. By logical equivalence we can take $\mathcal{S} = \{\Box(p \to q), \Box p, \Box q, p \to q, p, q, \Box((p \to q) \wedge p), \Box(p \wedge q)\}$. We obtain the standard model $\mathcal{M}'$ with $\mathcal{N}'(1) = \{\emptyset, \mathcal{W}\}$.

6 Extensions with Axioms T, P, and D

We aim to extend our calculi to systems containing further modal axioms. As a starting point, we consider in this section extensions of non-normal modal logics with axioms T, P, and D:

$$\text{T} \quad \Box A \to A \qquad \text{P} \quad \neg \Box \bot \qquad \text{D} \quad \neg(\Box A \wedge \Box \neg A)$$

T is a standard axiom, and P and D are of specific interest in deontic logic: If $\Box A$ is read as 'A is obligatory', then axiom D means that there cannot be two contradicting obligations, whereas axiom P means that there cannot be inconsistent obligations. It is worth noticing that axioms D and P are equivalent in normal modal logics, but are not necessarily equivalent in non-normal ones. The following dependencies hold: $\vdash_{\mathbf{ET}}$ D; $\vdash_{\mathbf{ET}}$ P; $\vdash_{\mathbf{MD}}$ P; $\vdash_{\mathbf{END}}$ P; $\vdash_{\mathbf{ECP}}$ D.

Like the systems of the classical cube, their extensions with axioms T, P, D can be characterised by certain classes of bi-neighbourhood models ([4,5]). The corresponding conditions are (T) if $(\alpha, \beta) \in \mathcal{N}(w)$, then $w \in \alpha$; (P) if $(\alpha, \beta) \in \mathcal{N}(w)$, then $\alpha \neq \emptyset$; and (D) if $(\alpha, \beta), (\gamma, \delta) \in \mathcal{N}(w)$, then $\alpha \cap \gamma \neq \emptyset$ or $\beta \cap \delta \neq \emptyset$.

Sequent calculi for NNMLs containing axioms T, P, or D have been studied in [9,10,15], although in none of them semantic completeness and countermodel extraction are considered. Here we define hypersequent calculi by the rules below.

$$\text{T} \; \frac{G \mid \Sigma, \langle \Sigma \rangle, \Gamma \Rightarrow \Delta}{G \mid \langle \Sigma \rangle, \Gamma \Rightarrow \Delta} \qquad \text{P} \; \frac{G \mid \langle \Sigma \rangle, \Gamma \Rightarrow \Delta \mid \Sigma \Rightarrow}{G \mid \langle \Sigma \rangle, \Gamma \Rightarrow \Delta} \qquad \text{D}_\mathsf{M} \; \frac{G \mid \langle \Sigma \rangle, \langle \Pi \rangle, \Gamma \Rightarrow \Delta \mid \Sigma, \Pi \Rightarrow}{G \mid \langle \Sigma \rangle, \langle \Pi \rangle, \Gamma \Rightarrow \Delta}$$

$$\text{D} \; \frac{G \mid \langle \Sigma \rangle, \langle \Pi \rangle, \Gamma \Rightarrow \Delta \mid \Sigma, \Pi \Rightarrow \qquad \{G \mid \langle \Sigma \rangle, \langle \Pi \rangle, \Gamma \Rightarrow \Delta \mid \Rightarrow A, B\}_{A \in \Sigma, B \in \Pi}}{G \mid \langle \Sigma \rangle, \langle \Pi \rangle, \Gamma \Rightarrow \Delta}$$

The above rules allow us to derive the corresponding axioms as follows:

$$\frac{\dfrac{A, \langle A \rangle, \Box A \Rightarrow A}{\langle A \rangle, \Box A \Rightarrow A} \; \text{T}}{\Box A \Rightarrow A} \; \Box_\mathsf{L} \qquad \frac{\dfrac{\langle \bot \rangle, \Box \bot \Rightarrow \mid \bot \Rightarrow}{\langle \bot \rangle, \Box \bot \Rightarrow} \; \text{P}}{\Box \bot \Rightarrow} \; \Box_\mathsf{L}$$

$$\frac{\dfrac{\langle A \rangle, \langle \neg A \rangle, \Box A, \Box \neg A \Rightarrow \mid A, \neg A \Rightarrow \qquad \langle A \rangle, \langle \neg A \rangle, \Box A, \Box \neg A \Rightarrow \mid \Rightarrow A, \neg A}{\langle A \rangle, \langle \neg A \rangle, \Box A, \Box \neg A \Rightarrow}}{\Box A, \Box \neg A \Rightarrow} \; \Box_\mathsf{L} \times 2 \quad \text{D}$$

Cut-free calculi for systems *containing* axioms T or P are obtained just by adding T rule or P rule. For non-monotonic logics containing D (but neither T nor P) it *seems necessary* (at present) to add $\mathsf{ctr}_{\langle \rangle}$ or the contracted version of D:

$$D_{aux} \frac{G \mid \langle \Sigma \rangle, \Gamma \Rightarrow \Delta \mid \Sigma \Rightarrow \quad \{G \mid \langle \Sigma \rangle, \Gamma \Rightarrow \Delta \mid \Rightarrow A\}_{A \in \Sigma}}{G \mid \langle \Sigma \rangle, \Gamma \Rightarrow \Delta}$$

In constrast, for monotonic logics containing D we can add rule P (notice that axiom P is derivable in **MD**). As before, one can prove soundness, syntactic completeness, and semantic completeness of hypersequent calculi. Since the calculi are defined modularly, it suffices to extend the proofs in previous sections.

Theorem 7. *If H is derivable in $\mathcal{H}_{E(T/P/D)^*}$, then it is valid in the corresponding bi-neighbourhood models.*

Proof. T) Assume $G \mid \Sigma, \langle \Sigma \rangle, \Gamma \Rightarrow \Delta$ valid. Then for all $\mathcal{M}$, $\mathcal{M} \models G$ or $\mathcal{M} \models \Sigma, \langle \Sigma \rangle, \Gamma \Rightarrow \Delta$. In the second case, for all wolds w of $\mathcal{M}$, $w \Vdash i(\Sigma, \langle \Sigma \rangle, \Gamma \Rightarrow \Delta)$, which is equivalent to $\bigwedge \Sigma \wedge \Box \bigwedge \Sigma \to i(\Gamma \Rightarrow \Delta)$. Since $\mathcal{M}$ is a T-model, this is equivalent to $\Box \bigwedge \Sigma \to i(\Gamma \Rightarrow \Delta)$. Thus $w \Vdash i(\langle \Sigma \rangle, \Gamma \Rightarrow \Delta)$.

P) Assume $G \mid \langle \Sigma \rangle, \Gamma \Rightarrow \Delta \mid \Sigma \Rightarrow$ valid. If $\mathcal{M} \models G \mid \langle \Sigma \rangle, \Gamma \Rightarrow \Delta$ we are done. Otherwise $\mathcal{M} \models \Sigma \Rightarrow$, that is $[\bigwedge \Sigma] = \emptyset$. Since $\mathcal{M}$ is a P-model, $w \not\Vdash \Box \bigwedge \Sigma$ for all worlds w of $\mathcal{M}$, which implies $\mathcal{M} \models \langle \Sigma \rangle, \Gamma \Rightarrow \Delta$.

The cases D, D_M and D_{aux} are analogous. □

It is possible to prove the admissibility of structural rules in Proposition 2. In particular, in case of calculi containing rules for D one may need to use the auxiliary rules D_{aux} and P in order to show admissibility of contraction. As before, this allows us to prove cut elimination, whence syntactic completeness of hypersequent calculi.

Theorem 8. *If $\mathcal{H}_{E(T/P/D)^*}$ contains $\Box_R$, then the rules* cut *and* sub *are admissible in $\mathcal{H}_{E(T/P/D)^*}$, otherwise* cut *and* sub_M *are admissible in $\mathcal{H}_{E(T/P/D)^*}$.*

Sketch of Proof. By extending the proof of Theorem 2. In particular we need to extend point **(C)** *(ii)* to the cases where the last rule applied in the derivation of $G \mid \langle A^{n_1}, \Pi_1 \rangle, ..., \langle A^{n_k}, \Pi_k \rangle, \Gamma \Rightarrow \Delta$ is T, P, D, or D_{aux} (resp. T, P, or D_M in monotonic case). □

Theorem 9 (Syntactic completeness). *If $i(\Gamma \Rightarrow \Delta)$ is derivable in* **E(T/P/D)***, then $\Gamma \Rightarrow \Delta$ is derivable in $\mathcal{H}_{E(T/P/D)^*}$.*

In order to get semantic completeness we need to extend the notion of saturation to the rules for T, P, D. The saturation conditions are the obvious ones, for instance the condition corresponding to T is: If $\langle \Sigma \rangle \in \Gamma_i$, then $\Sigma \subseteq \Gamma_i$. Then, given a saturated hypersequent one can define a countermodel $\mathcal{M}$ by means of the same construction as in Definition 5. We can prove the following lemma.

Lemma 2. *Let H be a saturated hypersequent occurring in a proof for H', and $\mathcal{M}$ be defined as in Definition 5. Then (a) $\mathcal{M}$ is a T-model if $\mathcal{H}_{E^*}$ contains rule T; (b) $\mathcal{M}$ is a P-model if $\mathcal{H}_{E^*}$ contains rule P; and (c) $\mathcal{M}$ is a D-model if $\mathcal{H}_{E^*}$ contains rules D and D_{aux}, or it contains rules D_M and P.*

Proof. (a) Assume $(\Sigma^+, \Sigma^-) \in \mathcal{N}(n)$. Then there is $\langle \Lambda \rangle \in \Gamma_n$ s.t. $\Sigma^+ = \Lambda^+, \Sigma^- = \Lambda^-$. By saturation of T, $\mathsf{set}(\Lambda) \subseteq \Gamma_n$, then by definition $n \in \Lambda^+ = \Sigma^+$.

(b) Assume $(\Sigma^+, \Sigma^-) \in \mathcal{N}(n)$. Then there is $\langle \Lambda \rangle \in \Gamma_n$ s.t. $\Sigma^+ = \Lambda^+, \Sigma^- = \Lambda^-$. By saturation of P, there is $m \in \mathcal{W}$ such that $\mathsf{set}(\Lambda) \subseteq \Gamma_m$, then by definition $m \in \Lambda^+ = \Sigma^+$, that is $\Sigma^+ \neq \emptyset$.

(c) Assume $(\Sigma^+, \Sigma^-), (\Pi^+, \Pi^-) \in \mathcal{N}(n)$. If $(\Sigma^+, \Sigma^-) \neq (\Pi^+, \Pi^-)$, then there are $\langle \Lambda \rangle, \langle \Theta \rangle \in \Gamma_n$ s.t. $\Sigma^+ = \Lambda^+, \Sigma^- = \Lambda^-, \Pi^+ = \Theta^+$ and $\Pi^- = \Theta^-$. If $\mathcal{H}_{\mathsf{E}*}$ is non-monotonic, by saturation of D there is $m \in \mathcal{W}$ such that $\mathsf{set}(\Lambda, \Theta) \subseteq \Gamma_m$ or there is $m \in \mathcal{W}$ such that $A, B \in \Delta_m$ for $A \in \Sigma$ and $B \in \Pi$. In the first case, $\mathsf{set}(\Lambda) \subseteq \Gamma_m$ and $\mathsf{set}(\Theta) \subseteq \Gamma_m$, thus by definition $m \in \Lambda^+ = \Sigma^+$ and $m \in \Theta^+ = \Pi^+$, that is $\Sigma^+ \cap \Pi^+ \neq \emptyset$. In the second case, $m \in \Lambda^- = \Sigma^-$ and $m \in \Theta^- = \Pi^-$, that is $\Sigma^- \cap \Pi^- \neq \emptyset$. If instead $\mathcal{H}_{\mathsf{E}*}$ is monotonic, by saturation of $\mathsf{D_M}$ there is $m \in \mathcal{W}$ such that $\mathsf{set}(\Lambda, \Theta) \subseteq \Gamma_m$. Then $\mathsf{set}(\Lambda) \subseteq \Gamma_m$ and $\mathsf{set}(\Theta) \subseteq \Gamma_m$, thus by definition $m \in \Lambda^+ = \Sigma^+$ and $m \in \Theta^+ = \Pi^+$, that is $\Sigma^+ \cap \Pi^+ \neq \emptyset$. The other possibility is that $(\Sigma^+, \Sigma^-) = (\Pi^+, \Pi^-)$. Then there is $\langle \Lambda \rangle \in \Gamma_n$ s.t. $\Sigma^+ = \Lambda^+$ and $\Sigma^- = \Lambda^-$. In the non-monotonic case, by saturation of $\mathsf{D_{aux}}$ there is $m \in \mathcal{W}$ such that $\mathsf{set}(\Lambda) \subseteq \Gamma_m$ or there is $m \in \mathcal{W}$ such that $A \in \Delta_m$ for $A \in \Sigma$. In the first case, by definition $m \in \Lambda^+ = \Sigma^+$. In the second case, $m \in \Lambda^- = \Sigma^-$. Thus $\Sigma^+ \neq \emptyset$ or $\Sigma^- \neq \emptyset$. In the monotonic case we can consider saturation of P and conclude that $\Sigma^+ \neq \emptyset$. □

Since every failed proof returns a saturated hypersequent, this implies semantic completeness of hypersequent calculi.

Theorem 10. *If H is valid in (M/C/N/T/P/D-)models, then it is derivable in* $\mathcal{H}_{\mathsf{E(M/C/N/T/P/D)}}$.

7   Conclusion

In this paper we have provided hypersequent calculi for the cube of classical Non-Normal Modal logics and some deontic extensions. The Hypersequent formulation is possibly the most adequate, in particular for non-monotone non-normal modal logics, as it ensures good semantic, computational, as well as structural properties. First of all, from a failed proof we can easily extract a countermodel (of polynomial size for logics without C) in the bi-neighbourhood semantics, whence in the standard one. The calculi provide a decision procedure of optimal complexity and enjoy syntactic cut elimination. Finally, they have a natural "almost internal" interpretation, as each component of a hypersequent can be read as a formula of the language. In future research, we intend to extend the calculi to further non-normal modal logics obtained by adding standard modal axioms, possibly including also regular logics which have a non standard relational semantics. Moreover, we intend to use the calculi also for metalogical investigation, *e.g.*, for obtaining proof-theoretic constructive proofs of interpolation complementing the general result in [17] and completing [15]. Finally we wish to study the formal relation with other recent calculi in the literature such as [4,13] in the form of mutual simulation. We also think of implementing our calculi and comparing them with the theorem prover proposed recently in [6].

Appendix

Theorem 2. *If $\mathcal{H}_{\mathsf{E}^\star}$ contains $\Box_\mathsf{R}$, then the rules* cut *and* sub *are admissible in $\mathcal{H}_{\mathsf{E}^\star}$, otherwise* cut *and* sub$_\mathsf{M}$ *are admissible in $\mathcal{H}_{\mathsf{E}^\star}$.*

Proof. We prove that cut and sub are admissible in non-monotonic $\mathcal{H}_{\mathsf{E}^\star}$; the proof in the monotonic cases is analogous. Recall that, for an application of cut, the *cut formula* is the formula which is deleted by that application, while the *cut height* is the sum of the heights of the derivations of the premises of cut.

The theorem is a consequence of the following claims, where $Cut(c, h)$ means that all applications of cut of height h on a cut formula of weight c are admissible, and $Sub(c)$ means that all applications of sub where A has weight c are admissible (for all $\Sigma, \Pi_1, ..., \Pi_k$): **(A)** $\forall c.Cut(c, 0)$. **(B)** $\forall h.Cut(0, h)$. **(C)** $\forall c.(\forall h.Cut(c, h) \rightarrow Sub(c))$. **(D)** $\forall c.\forall h.((\forall c' < c.(Sub(c') \wedge \forall h'.Cut(c', h')) \wedge \forall h'' < h.Cut(c, h'')) \rightarrow Cut(c, h))$.

(A) deals with applications of cut to initial sequents and is trivial.

(B) If the cut formula has weight 0, then it is $\bot$, $\top$, or a propositional variable p. In both situations the proof is by complete induction on h. The basic case $h = 0$ is a particular case of **(A)**. For the inductive step, we distinguish three cases.

(i) The cut formula $\bot$, $\top$, or p is not principal in the last rule applied in the derivation of the left premise. By examining all possible rule applications, we show that the application of cut can be replaced by one o more applications of cut at a smaller height. For instance, assume that the last rule applied is $\Box_\mathsf{L}$.

$$\Box_\mathsf{L} \cfrac{\cfrac{G \mid \langle A \rangle, \Box A, \Gamma \Rightarrow \Delta, \bot}{G \mid \Box A, \Gamma \Rightarrow \Delta, \bot} \qquad G \mid \bot, \Box A, \Gamma \Rightarrow \Delta}{G \mid \Box A, \Gamma \Rightarrow \Delta} \; \text{cut}$$

The derivation is transformed as follows, with a hp-application of wk and an application of cut of smaller height.

$$\Box_\mathsf{L} \cfrac{\cfrac{G \mid \langle A \rangle, \Box A, \Gamma \Rightarrow \Delta, \bot \qquad \cfrac{G \mid \bot, \Box A, \Gamma \Rightarrow \Delta}{G \mid \bot, \langle A \rangle, \Box A, \Gamma \Rightarrow \Delta} \; \text{wk}_\mathsf{L}}{G \mid \langle A \rangle, \Box A, \Gamma \Rightarrow \Delta} \; \text{cut}}{G \mid \Box A, \Gamma \Rightarrow \Delta}$$

The situation is similar if the last rule in the derivation of the left premise is applied to some sequent in G.

(ii) The cut formula $\bot$, $\top$, or p is not principal in the last rule applied in the derivation of the right premise. The case is analogous to (i). As an example, suppose that the last rule applied is $_\mathsf{M}\Box_\mathsf{R}$.

$$\cfrac{G \mid \langle \Sigma \rangle, \Gamma \Rightarrow \Delta, \Box B, \bot \qquad {}_\mathsf{M}\Box_\mathsf{R}\cfrac{G \mid \bot, \langle \Sigma \rangle, \Gamma \Rightarrow \Delta, \Box B \mid \Sigma \Rightarrow B}{G \mid \bot, \langle \Sigma \rangle, \Gamma \Rightarrow \Delta, \Box B}}{G \mid \langle \Sigma \rangle, \Gamma \Rightarrow \Delta, \Box B} \; \text{cut}$$

The derivation is converted into

$$
\text{ew} \frac{G \mid \langle \Sigma \rangle, \Gamma \Rightarrow \Delta, \Box B, \bot}{\dfrac{\dfrac{G \mid \langle \Sigma \rangle, \Gamma \Rightarrow \Delta, \Box B, \bot \mid \Sigma \Rightarrow B \qquad G \mid \bot, \langle \Sigma \rangle, \Gamma \Rightarrow \Delta, \Box B \mid \Sigma \Rightarrow B}{G \mid \langle \Sigma \rangle, \Gamma \Rightarrow \Delta, \Box B \mid \Sigma \Rightarrow B} \text{cut}}{G \mid \langle \Sigma \rangle, \Gamma \Rightarrow \Delta, \Box B} \, \text{M}\Box\text{R}}
$$

where cut is applied at a smaller height.

(iii) The cut formula $\bot$, $\top$, or p is principal in the last rule applied in the derivation of both premises. Then the cut formula is p, as $\bot$ (resp. $\top$) is never principal on the right-hand side (resp. left-hand side) of the conclusion of any rule application. This means that both premises are derived by init, which implies $h = 0$. Then we are back to case **(A)**.

(C) Assume $\forall h Cut(c, h)$. The proof is by induction on the height m of the derivation of $G \mid \langle A^{n_1}, \Pi_1 \rangle, ..., \langle A^{n_k}, \Pi_k \rangle, \Gamma \Rightarrow \Delta$. Here we only consider the case where $m > 0$ and the last rule applied in the derivation is $\Box$R, with one block among $\langle A, \Pi_1 \rangle, ..., \langle A, \Pi_k \rangle$ principal in the rule application:

$$
\frac{G \mid \langle A^{n_i}, \Pi_i \rangle, \Gamma' \Rightarrow \Delta', \Box D \mid A^{n_i}, \Pi_i \Rightarrow D \qquad \{G \mid \langle A^{n_i}, \Pi_i \rangle, \Gamma' \Rightarrow \Delta', \Box D \mid D \Rightarrow A\}_1^{n_i} \quad \begin{matrix} \text{①} \\ \{G \mid \langle A^{n_i}, \Pi_i \rangle, \Gamma' \Rightarrow \Delta', \Box D \mid D \Rightarrow C\}_{C \in \Pi_i} \\ \vdots \end{matrix}}{G \mid \langle A^{n_i}, \Pi_i \rangle, \Gamma' \Rightarrow \Delta', \Box D} \, \Box\text{R}
$$

The derivation is converted as follows. First we derive:

$$
\frac{\dfrac{G \mid \Sigma \Rightarrow A}{G \mid \Sigma \Rightarrow A \mid A^{n_i}, \Pi_i \Rightarrow D} \, \text{ew} \quad \left\{ \dfrac{G \mid A \Rightarrow B}{G \mid A \Rightarrow B \mid A^{n_i}, \Pi_i \Rightarrow D} \, \text{ew} \right\}_{B \in \Sigma} \qquad \text{①}}{G \mid \langle \Sigma^{n_i}, \Pi_i \rangle, \Gamma' \Rightarrow \Delta', \Box D \mid A^{n_i}, \Pi_i \Rightarrow D} \, \text{sub}
$$

Moreover, by applying ew to $G \mid \Sigma \Rightarrow A$ we obtain $G \mid \langle \Sigma^{n_i}, \Pi_i \rangle, \Gamma' \Rightarrow \Delta', \Box D \mid \Sigma \Rightarrow A$. By auxiliary applications of wk we can cut A and get $G \mid \langle \Sigma^{n_i}, \Pi_i \rangle, \Gamma' \Rightarrow \Delta', \Box D \mid \Sigma, A^{n_i-1}, \Pi_i \Rightarrow D$. Then with further applications of cut (each time with auxiliary applications of wk) we obtain $G \mid \langle \Sigma^{n_i}, \Pi_i \rangle, \Gamma' \Rightarrow \Delta', \Box D \mid \Sigma^{n_i}, \Pi_i \Rightarrow D$. By doing the same with the other premises of $\Box$R in the initial derivation we obtain also $\{G \mid \langle \Sigma^{n_i}, \Pi_i \rangle, \Gamma' \Rightarrow \Delta', \Box D \mid D \Rightarrow B\}_{B \in \Sigma \, (1...n_1)}$ and $\{G \mid \langle \Sigma^{n_i}, \Pi_i \rangle, \Gamma' \Rightarrow \Delta', \Box D \mid D \Rightarrow C\}_{C \in \Pi_i}$. Then by $\Box$R we derive the conclusion of sub $G \mid \langle \Sigma^{n_i}, \Pi_i \rangle, \Gamma' \Rightarrow \Delta', \Box D$.

(D) Assume $\forall c' < c. (Sub(c') \wedge \forall h'. Cut(c', h'))$ and $\forall h'' < h. Cut(c, h'')$. We show that all applications of cut of height h on a cut formula of weight c can be replaced by different applications of cut, either of smaller height or on a cut formula of smaller weight. We can assume $c, h > 0$ as the cases $c = 0$ and $h = 0$ have been considered already in **(B)** and **(A)**. We distinguish two cases.

(i) The cut formula is not principal in the last rule application in the derivation of at least one of the two premises of cut. This case is analogous to (i) or (ii) in **(B)**.

(ii) The cut formula is principal in the last rule application in the derivation of both premises. Then the cut formula is either $B \to C$, or $B \wedge C$, or $\Box B$.

— If the cut formula is $B \to C$ we have

$$
\frac{\dfrac{\text{①} \, G \mid B, \Gamma \Rightarrow \Delta, B \to C, C}{\text{②} \, G \mid \Gamma \Rightarrow \Delta, B \to C} \to\text{R} \qquad \dfrac{\text{③} \, G \mid B \to C, \Gamma \Rightarrow \Delta, B \qquad \text{④} \, G \mid C, B \to C, \Gamma \Rightarrow \Delta}{\text{⑤} \, G \mid B \to C, \Gamma \Rightarrow \Delta} \to\text{L}}{G \mid \Gamma \Rightarrow \Delta} \, \text{cut}
$$

The derivation is converted into the following one:

$$
\mathsf{cut}\cfrac{
\mathsf{wk_R}\cfrac{②}{\mathsf{wk_R}\cfrac{G \mid \varGamma \Rightarrow \varDelta, B \rightarrow C, B}{G \mid \varGamma \Rightarrow \varDelta, B}}\quad ③
}{
\mathsf{cut}\cfrac{\mathsf{wk_R}\cfrac{}{G \mid \varGamma \Rightarrow \varDelta, B, C}\quad \mathsf{cut}\cfrac{①\quad \mathsf{wk_R}\cfrac{\mathsf{wk_L}\cfrac{⑤}{G \mid B, B \rightarrow C, \varGamma \Rightarrow \varDelta}}{G \mid B, B \rightarrow C, \varGamma \Rightarrow \varDelta, C}}{G \mid B, \varGamma \Rightarrow \varDelta, C}\quad \mathsf{wk_L}\cfrac{\mathsf{cut}\cfrac{②}{G \mid C, \varGamma \Rightarrow \varDelta, B \rightarrow C}\quad ④}{G \mid C, \varGamma \Rightarrow \varDelta}}{G \mid \varGamma \Rightarrow \varDelta, C}
}{G \mid \varGamma \Rightarrow \varDelta}
$$

— If the cut formula is $B \wedge C$ the situation is similar.
— If the cut formula is $\Box B$ we have

① $G \mid \langle \varSigma \rangle, \varGamma \Rightarrow \varDelta, \Box B \mid \varSigma \Rightarrow B$

$$
\mathsf{cut}\cfrac{
\Box_\mathsf{R}\cfrac{\vdots\quad \{②_C\ G \mid \langle \varSigma \rangle, \varGamma \Rightarrow \varDelta, \Box B \mid B \Rightarrow C\}_{C \in \varSigma}}{③\ G \mid \langle \varSigma \rangle, \varGamma \Rightarrow \varDelta, \Box B}
\quad
\Box_\mathsf{L}\cfrac{④\ G \mid \langle B \rangle, \Box B, \langle \varSigma \rangle, \varGamma \Rightarrow \varDelta}{⑤\ G \mid \Box B, \langle \varSigma \rangle, \varGamma \Rightarrow \varDelta}
}{G \mid \langle \varSigma \rangle, \varGamma \Rightarrow \varDelta}
$$

The derivation is converted as follows, with several applications of cut of smaller height.

$$
\mathsf{ew}\cfrac{\mathsf{cut}\cfrac{\mathsf{wk_{\langle\rangle}}\cfrac{③}{G \mid \langle B \rangle, \langle \varSigma \rangle, \varGamma \Rightarrow \varDelta, \Box B}\quad ④}{G \mid \langle B \rangle, \langle \varSigma \rangle, \varGamma \Rightarrow \varDelta}}{⊛\ G \mid \langle \varSigma \rangle, \varGamma \Rightarrow \varDelta \mid \langle B \rangle, \langle \varSigma \rangle, \varGamma \Rightarrow \varDelta}
$$

$$
\mathsf{ec}\cfrac{\mathsf{ctr_{\langle\rangle}}\cfrac{\mathsf{sub}\cfrac{\mathsf{cut}\cfrac{①\quad \mathsf{ew}\cfrac{⑤}{G \mid \Box B, \langle \varSigma \rangle, \varGamma \Rightarrow \varDelta \mid \varSigma \Rightarrow B}}{G \mid \langle \varSigma \rangle, \varGamma \Rightarrow \varDelta \mid \varSigma \Rightarrow B}\quad \left(②_C\ \mathsf{cut}\cfrac{\mathsf{ew}\cfrac{⑤}{G \mid \Box B, \langle \varSigma \rangle, \varGamma \Rightarrow \varDelta \mid B \Rightarrow C}}{G \mid \langle \varSigma \rangle, \varGamma \Rightarrow \varDelta \mid B \Rightarrow C}\right)_{C \in \varSigma}\quad ⊛}{G \mid \langle \varSigma \rangle, \varGamma \Rightarrow \varDelta \mid \langle \varSigma \rangle, \langle \varSigma \rangle, \varGamma \Rightarrow \varDelta}}{G \mid \langle \varSigma \rangle, \varGamma \Rightarrow \varDelta \mid \langle \varSigma \rangle, \varGamma \Rightarrow \varDelta}}{G \mid \langle \varSigma \rangle, \varGamma \Rightarrow \varDelta}
$$

$\square$

Theorem 8. *If $\mathcal{H}_{\mathsf{E(T/P/D)^\star}}$ contains $\Box_\mathsf{R}$, then the rules cut and sub are admissible in $\mathcal{H}_{\mathsf{E(T/P/D)^\star}}$, otherwise cut and $\mathsf{sub_M}$ are admissible in $\mathcal{H}_{\mathsf{E(T/P/D)^\star}}$.*

Proof. We extend point **(C)** *(ii)* in the proof of Theorem 2 to the cases where the last rule applied in the derivation of $G \mid \langle A^{n_1}, \varPi_1 \rangle, ..., \langle A^{n_k}, \varPi_k \rangle, \varGamma \Rightarrow \varDelta$ is T, P, D, or $\mathsf{D_{aux}}$ (resp. T, P, or $\mathsf{D_M}$ in monotonic case). We consider as examples the following two cases.

— The last rule is T:

$$
\mathsf{T}\cfrac{G \mid A^{n_i}, \varPi_i, \langle A^{n_i}, \varPi_i \rangle, \varGamma \Rightarrow \varDelta}{G \mid \langle A^{n_i}, \varPi_i \rangle, \varGamma \Rightarrow \varDelta}
$$

By applying the inductive hypothesis to the premiss we obtain $G \mid A^{n_i}, \varPi_i, \langle \varSigma^{n_i}, \varPi_i \rangle, \varGamma \Rightarrow \varDelta$. Then, from this and $G \mid \varSigma \Rightarrow A$, by several applications of cut (each time with auxiliary applications of wk) we obtain $G \mid \varSigma^{n_i}, \varPi_i, \langle \varSigma^{n_i}, \varPi_i \rangle, \varGamma \Rightarrow \varDelta$. Finally, by T we derive $G \mid \langle \varSigma^{n_i}, \varPi_i \rangle, \varGamma \Rightarrow \varDelta$.

— The last rule is P:

$$
\mathsf{P}\cfrac{G \mid \langle A^{n_i}, \varPi_i \rangle, \varGamma \Rightarrow \varDelta \mid A^{n_i}, \varPi_i \Rightarrow}{G \mid \langle A^{n_i}, \varPi_i \rangle, \varGamma \Rightarrow \varDelta}
$$

By applying the inductive hypothesis to the premiss (aftar auxiliary applications of ew to the other premisses of sub) we obtain $G \mid \langle \Sigma^{n_i}, \Pi_i \rangle, \Gamma \Rightarrow \Delta \mid A^{n_i}, \Pi_i \Rightarrow$. Then, from this and $G \mid \Sigma \Rightarrow A$, by several applications of cut (each time with auxiliary applications of wk) we obtain $G \mid \langle \Sigma^{n_i}, \Pi_i \rangle, \Gamma \Rightarrow \Delta \mid \Sigma^{n_i}, \Pi_i \Rightarrow$. Finally, by P we derive $G \mid \langle \Sigma^{n_i}, \Pi_i \rangle, \Gamma \Rightarrow \Delta$. □

References

1. Askounis, D., Koutras, C.D., Zikos, Y.: Knowledge means "all", belief means "most". J. Appl. Non-Class. Logics **26**(3), 173–192 (2016)
2. Avron, A.: The method of hypersequents in the proof theory of propositional non-classical logics. In: Logic: From Foundations to Applications. Clarendon P. (1996)
3. Chellas, B.F.: Modal Logic. Cambridge University Press, Cambridge (1980)
4. Dalmonte, T., Olivetti, N., Negri, S.: Non-normal modal logics: bi-neighbourhood semantics and its labelled calculi. In: Proceedings of AiML (2018)
5. Dalmonte, T., Olivetti, N., Negri, S.: Bi-neighbourhood semantics and labelled sequent calculi for non-normal modal logics. Part 1 (draft) (2019). http://www.lsis.org/olivetti/TR2019/TR-BINS.pdf
6. Dalmonte, T., Olivetti, N., Negri, S., Pozzato, G.L.: PRONOM: proof-search and countermodel generation for non-normal modal logics. In: Proceedings of AIIA (2019, to appear)
7. Gilbert, D., Maffezioli, P.: Modular sequent calculi for classical modal logics. Studia Logica **103**(1), 175–217 (2015)
8. Goble, L.: Prima facie norms, normative conflicts, and dilemmas. Handb. Deontic Logic Normative Syst. **1**, 241–352 (2013)
9. Indrzejczak, A.: Sequent calculi for monotonic modal logics. Bull. Section Logic **34**(3), 151–164 (2005)
10. Indrzejczak, A.: Admissibility of cut in congruent modal logics. Logic Logical Philos. **21**, 189–203 (2011)
11. Lavendhomme, R., Lucas, T.: Sequent calculi and decision procedures for weak modal systems. Studia Logica **65**, 121–145 (2000)
12. Lellmann, B.: Combining monotone and normal modal logic in nested sequents – with countermodels. In: Cerrito, S., Popescu, A. (eds.) TABLEAUX 2019. LNCS (LNAI), vol. 11714, pp. 203–220. Springer, Cham (2019). https://doi.org/10.1007/978-3-030-29026-9_12
13. Lellmann, B., Pimentel, E.: Modularisation of sequent calculi for normal and non-normal modalities. ACM Trans. Comput. Logic **20**(2), 7:1–7:46 (2019)
14. Negri, S.: Proof theory for non-normal modal logics: the neighbourhood formalism and basic results. IfCoLog J. Log. Appl **4**(4), 1241–1286 (2017)
15. Orlandelli, E.: Sequent calculi and interpolation for non-normal logics. arXiv preprint arXiv:1903.11342 (2019)
16. Pacuit, E.: Neighborhood Semantics for Modal Logic. Springer, Heidelberg (2017)
17. Pattinson, D.: The logic of exact covers: completeness and uniform interpolation. In: LICS 2013 (2013)
18. Pauly, M.: A modal logic for coalitional power in games. J. Log. Comput. **12**(1), 149–166 (2002)
19. Vardi, M.Y.: On epistemic logic and logical omniscience. In: Theoretical Aspects of Reasoning About Knowledge, pp. 293–305. Elsevier (1986)
20. Vardi, M.Y.: On the complexity of epistemic reasoning. In: Proceedings of 4th IEEE Symposium on Logic in Computer Science, pp. 243–252 (1989)

Completeness Theorems for First-Order Logic Analysed in Constructive Type Theory

Yannick Forster◉, Dominik Kirst$^{(\boxtimes)}$◉, and Dominik Wehr◉

Saarland University, Saarland Informatics Campus, Saarbrücken, Germany
{forster,kirst}@ps.uni-saarland.de, dwehr@dortselb.st

Abstract. We study various formulations of the completeness of first-order logic phrased in constructive type theory and mechanised in the Coq proof assistant. Specifically, we examine the completeness of variants of classical and intuitionistic natural deduction and sequent calculi with respect to model-theoretic, algebraic, and game semantics. As completeness with respect to standard model-theoretic semantics is not readily constructive, we analyse the assumptions necessary for particular syntax fragments and discuss non-standard semantics admitting assumption-free completeness. We contribute a reusable Coq library for first-order logic containing all results covered in this paper.

1 Introduction

Completeness theorems are central to the field of mathematical logic. Once completeness of a sound deduction system with respect to a semantic account of the syntax is established, the typically infinitary notion of semantic validity is reduced to the finitary, and hence algorithmically more tractable, notion of syntactic deduction. In the case of first-order logic, being the formalism underlying traditional mathematics based on a set-theoretic foundation, completeness enables the use of semantic techniques to study the deductive consequence of axiomatic systems.

The seminal completeness theorem for first-order logic proven by Gödel [18] and later refined by Henkin [21,20] yields a syntactic deduction of every formula valid in the canonical Tarski semantics based on interpreting the non-logical function and relation symbols in models providing the corresponding structure. However, this result may not be understood as an effective procedure in the sense that a formal deduction for a formula satisfied by all models can be computed by an algorithm, since even for finite signatures the proof relies on non-constructive assumptions. Specifically, when admitting all logical connectives, completeness is equivalent to a weak form of König's lemma [32]. Even restricted to the classically sufficient $\to, \forall, \bot$-fragment, the classically vacuous but constructively contested[1] assumption of Markov's principle, asserting that every non-diverging computation terminates, is necessary [28]. We defer a more detailed overview of known dependencies to the discussion of related work in Sect. 7.1.

---

[1] Accepted in Russian constructivism while in conflict with Brouwer's intuitionism.

© Springer Nature Switzerland AG 2020
S. Artemov and A. Nerode (Eds.): LFCS 2020, LNCS 11972, pp. 47–74, 2020.
https://doi.org/10.1007/978-3-030-36755-8_4

The aim of this paper is to coherently analyse the computational content of completeness theorems concerning various semantics and deduction systems. Naturally, such matters of *constructive reverse mathematics* [25] need to be addressed in an intuitionistic meta-logic such as constructive type theory. In fact, the results in this paper are formalised in the Coq proof assistant [50] that implements the *predicative calculus of cumulative inductive constructions* (pCuIC) [51], yielding executable programs for all constructively given completeness proofs. For ease of language, we reserve the term "constructive" for statements provable in this specific system, hence excluding Markov's principle [7,41]. In fact, coming with an internal notion of computation, constructive type theory allows us to state Markov's principle both internally (MP) as well as for any concrete model of computation (MP_L), whereby the former implies the latter and both can be related to completeness statements. The two main questions in focus are which specific assumptions are necessary for particular formulations of completeness and how the statements can be modified such that they hold constructively.

Applying this strategy to Tarski semantics, a first observation is that the model existence theorem, central to Henkin's completeness proof, holds constructively [22]. Model existence directly implies that valid formulas cannot be unprovable. Thus, a single application of MP, rendering enumerable predicates such as deduction stable under double negation, yields completeness. Similarly, MP_L yields the stability of deduction from finite contexts and hence the corresponding form of completeness. Because MP is admissible in pCuIC [41], so are MP_L and the two completeness statements. Finally, we illustrate that completeness for the minimal $\rightarrow, \forall$-fragment does not depend on additional assumptions and, consequently, how the interpretation of $\bot$ can be relaxed to *exploding models* [53,30] admitting a constructive completeness proof for the $\rightarrow, \forall, \bot$-fragment.

Turning to intuitionistic logic, we discuss analogous relationships for Kripke semantics and a cut-free intuitionistic sequent calculus [23]. Again, completeness for the $\rightarrow, \forall, \bot$-fragment is equivalent to Markov's principle while being constructive if restricted to the minimal $\rightarrow, \forall$-fragment or employing a relaxed treatment of $\bot$. The intuitionistically undefinable connectives $\vee$ and $\exists$ add further complexity [24] and need to remain untreated in this paper. As a side note, we explain how the constructivised completeness theorem for intuitionistic logic can be used to implement a semantic cut-elimination procedure.

After considering such model-theoretic semantics, mainly based on embedding the object-logic into the meta-logic, we exemplify two rather different approaches to assigning meaning to formulas, namely algebraic semantics and game semantics. Differing fundamentally from model-theoretic semantics, both share a constructive rendering of completeness for the full syntax of first-order logic, agnostic to the intuitionistic or classical flavour of the deduction system.

In algebraic semantics, the embedding of formulas into the meta-logic is generalised to an evaluation in algebras providing the structure of the logical connectives. In this setting, completeness follows from the observation that provability induces such an algebra on formulas. We discuss intuitionistic and classical logic evaluated in complete Heyting and complete Boolean algebras (cf. [46]).

Dialogue game semantics as introduced by Lorenzen [34,35], on the other hand, completely disposes of interpreting logical connectives as operations and instead understand logic as a dialectic game of assertion and argument. An assertion is considered valid if every sceptic can be convinced through substantive reasoning, i.e. if there is a strategy such that every argument about the assertion can be won. Hence, game semantics are inherently closer to deduction systems than the previous semantic accounts and in fact a very general isomorphism of winning strategies and formal deductions has been established [47]. We instantiate this isomorphism to a first-order intuitionistic sequent calculus.

Contributions. We present a comprehensive analysis of the computational content of completeness theorems for first-order logic considering various semantics and deduction systems. Concerning model-theoretic semantics, we refine the well-known relation of completeness for $\to, \forall, \bot$-formulas to Markov's principle to constructive completeness up to double negation, hence entailing the admissibility of completeness in pCuIC. Our elaboration of game semantics introduces a streamlined representation of dialogues as state transition systems suitable for mechanisation and translates the generic completeness result for classical logic from [47] to the case of intuitionistic first-order logic. Finally, we provide a reusable Coq library[2] for first-order logic including all results covered in this paper. Notably, the development is based on a de Bruijn encoding of binders [8,49] and is parametric in the signature of non-logical symbols and thus adjustable to any particular first-order theory (see Appendix B for more formalisation details).

Outline. In Sect. 2, we begin with some preliminary definitions concerning the syntax of first-order logic, deduction systems, and synthetic computability. We then analyse completeness for model-theoretic semantics (Sect. 3) and its connection to Markov's principle (Sect. 4). Subsequently, we give constructive completeness proofs for algebraic semantics (Sect. 5) and game semantics (Sect. 6). We end with a discussion of related and future work in Sect. 7.

2   Syntax, Deduction, Computability

We work in a constructive type theory with a predicative hierarchy of type universes above a single impredicative universe $\mathbb{P}$ of propositions. Assumed type formers are function spaces $X \to Y$, products $X \times Y$, sums $X + Y$, dependent products $\forall x : X. F\,x$, and dependent sums $\Sigma\,x : X. F\,x$. The propositional versions of these connectives are denoted by the usual logical symbols ($\to, \wedge, \vee, \forall$, and $\exists$) in addition to $\top : \mathbb{P}$ and $\bot : \mathbb{P}$ denoting truth and falsity.

Basic inductive types are the Booleans $\mathbb{B} ::= \mathsf{tt} \mid \mathsf{ff}$ and the natural numbers $\mathbb{N} ::= 0 \mid \mathsf{S}\,n$ for $n : \mathbb{N}$. Given a type X, we further define options $\mathcal{O}(X) ::= \emptyset \mid \ulcorner x \urcorner$ and lists $\mathcal{L}(X) ::= [] \mid x :: A$ for $x : X$ and $A : \mathcal{L}(X)$. On lists we employ the standard notation for membership $x \in A$, inclusion $A \subseteq B$, concatenation $A \mathbin{+\!\!+} B$, and map $f\,@\,A$. These notations are shared with vectors $\boldsymbol{x} : X^n$ of fixed length $n : \mathbb{N}$. Possibly infinite collections are expressed by sets $p : X \to \mathbb{P}$ with set-theoretic notations like $x \in p$, $p \subseteq q$, and $p \cap q$.

---

[2] On www.ps.uni-saarland.de/extras/fol-completeness and hyperlinked with this pdf.

2.1 Syntax of First-Order Logic

We represent the terms and formulas of first-order logic as inductive types over a fixed signature $\Sigma = (\mathcal{F}_\Sigma, \mathcal{P}_\Sigma)$ specialising function symbols $f : \mathcal{F}_\Sigma$ and predicate symbols $P : \mathcal{P}_\Sigma$ together with their arities $|f| : \mathbb{N}$ and $|P| : \mathbb{N}$. Variable binding is implemented using de Bruijn indices [8] well-suited for formalisation [49].

Definition 1. We define the terms and formulas of first-order logic by

$$t : \mathbb{T} ::= x \mid f\,\boldsymbol{t} \qquad \varphi, \psi : \mathbb{F} ::= \dot{\bot} \mid P\,\boldsymbol{t} \mid \varphi\,\dot{\rightarrow}\,\psi \mid \varphi\,\dot{\wedge}\,\psi \mid \varphi\,\dot{\vee}\,\psi \mid \dot{\forall}\varphi \mid \dot{\exists}\varphi \qquad x : \mathbb{N}, f : \mathcal{F}_\Sigma, P : \mathcal{P}_\Sigma$$

where the vectors $\boldsymbol{t}$ are of the expected lengths $|f|$ and $|P|$, respectively. We set $\dot{\neg}\varphi := \varphi\,\dot{\rightarrow}\,\dot{\bot}$ and isolate the type $\mathbb{F}^*$ of formulas in the $\rightarrow, \forall, \bot$-fragment.

A bound variable is encoded as the number of quantifiers shadowing its relevant binder, e.g. $P\,x\,y \rightarrow \forall x.\,\exists y.\,P\,x\,y$ may be represented by $P\,7\,4\,\dot{\rightarrow}\,\dot{\forall}\,\dot{\exists}\,P\,1\,0$. The variables 7 and 4 in this example are called *free* and variables that do not occur freely are called *fresh*. A formula with no free variables is called *closed*.

Definition 2. Instantiating with a substitution $\sigma : \mathbb{N} \rightarrow \mathbb{T}$ is defined by

$$
\begin{aligned}
x[\sigma] &:= \sigma\,x & \dot{\bot}[\sigma] &:= \dot{\bot} & (\varphi\,\square\,\psi)[\sigma] &:= \varphi[\sigma]\,\square\,\psi[\sigma] \\
(f\,\boldsymbol{t})[\sigma] &:= f\,(\boldsymbol{t}\,[\sigma]) & (P\,\boldsymbol{t})[\sigma] &:= P\,(\boldsymbol{t}\,[\sigma]) & (\square\,\varphi)[\sigma] &:= \square\,\varphi[\Uparrow\sigma]
\end{aligned}
$$

where $\boldsymbol{t}\,[\sigma]$ is short for $(\lambda t.\,t[\sigma])\,@\,\boldsymbol{t}$, $\Uparrow\sigma$ denotes the substitution $\lambda n.\,\sigma\,(\mathsf{S}\,n)$, and $\square$ is used as placeholder for the logical connectives and quantifiers, respectively.

Useful shorthands are $\varphi[t; \sigma]$ for instantiating 0 with t and $\mathsf{S}\,x$ with $\sigma\,x$, $\varphi[t]$ for $\varphi[t; \lambda x.\,x]$, and $\Uparrow\varphi$ for the shift $\varphi[\lambda x.\,\mathsf{S}\,x]$. All terminology and notation concerning formulas carries over to *contexts* $\Gamma : \mathcal{L}(\mathbb{F})$ and *theories* $\mathcal{T} : \mathbb{F} \rightarrow \mathbb{P}$. For ease of notation we freely identify contexts Γ with their theory $\lambda\varphi.\,\varphi \in \Gamma$.

2.2 Deduction Systems

We represent deduction systems as inductive predicates of type $\mathcal{L}(\mathbb{F}) \rightarrow \mathbb{F} \rightarrow \mathbb{P}$ or similar. The archetypal system is natural deduction (ND), exemplified by an intuitionistic version $\Gamma \vdash \varphi$ as defined in Definition 55 of Appendix A. Since most rules are standard, we only discuss the quantifier rules in more detail as they rely on the de Bruijn representation of formulas:

$$
\frac{\Uparrow\Gamma \vdash \varphi}{\Gamma \vdash \dot{\forall}\varphi}\,\text{AI} \qquad \frac{\Gamma \vdash \dot{\forall}\varphi}{\Gamma \vdash \varphi[t]}\,\text{AE} \qquad \frac{\Gamma \vdash \varphi[t]}{\Gamma \vdash \dot{\exists}\varphi}\,\text{EI} \qquad \frac{\Gamma \vdash \dot{\exists}\varphi \quad \Uparrow\Gamma, \varphi \vdash \Uparrow\psi}{\Gamma \vdash \psi}\,\text{EE}
$$

Note that $\Uparrow\Gamma, \varphi$ is notation for $\varphi :: \Uparrow\Gamma$. In a shifted context $\Uparrow\Gamma$ there is no reference to the variable 0 which hence plays the role of an arbitrary but fixed individual. So if $\Uparrow\Gamma \vdash \varphi$ then we can conclude $\Gamma \vdash \dot{\forall}\varphi$ as expressed by the rule (AI) for $\forall$-introduction. Similarly, the shifts in the rule (EE) for $\exists$-elimination simulate that Γ together with φ instantiated to the witness provided by $\Gamma \vdash \dot{\exists}\varphi$ proves ψ and hence admits the conclusion that already $\Gamma \vdash \psi$. For many proofs it will be helpful to employ fresh variables explicitly as justified by Lemma 4, which we state after observing *weakening* and *substitutivity*:

Lemma 3. *If $\Gamma \vdash \varphi$, then $\Delta \vdash \varphi$ for all $\Delta \supseteq \Gamma$ and $\Gamma[\sigma] \vdash \varphi[\sigma]$ for all σ.*

Lemma 4. *Given Γ, φ, and ψ one can compute a fresh variable x such that*

1. *$\uparrow\!\Gamma \vdash \varphi$ iff $\Gamma \vdash \varphi[x]$ and*   2. *$\uparrow\!\Gamma, \varphi \vdash \uparrow\!\psi$ iff $\Gamma, \varphi[x] \vdash \psi$.*

A classical variant $\Gamma \vdash_c \varphi$ of the ND system can be obtained without referring to $\bot$ by adding the axiom $\Gamma \vdash_c ((\varphi \dot\to \psi) \dot\to \varphi) \dot\to \varphi$ expressing Peirce's law (Definition 56). Then the structural properties stated in the two lemmas above are maintained while the typical classical proof rules become available.

Deduction systems such as intuitionistic ND introduced above naturally extend to theories by writing $\mathcal{T} \vdash \varphi$ if there is a finite context $\Gamma \subseteq \mathcal{T}$ with $\Gamma \vdash \varphi$. Then $\mathcal{T} \vdash \varphi$ satisfies proof rules analogous to $\Gamma \vdash \varphi$.

2.3 Synthetic Computability

Since every function definable in constructive type theory is computable, the standard notions of computability theory can be synthesised by type-level operations [1,14], eliminating references to a concrete model of computation such as Turing machines, μ-recursive functions, or the untyped lambda calculus.

Definition 5. *Let X be a type and $p : X \to \mathbb{P}$ be a predicate.*

- *p is decidable if there is $f : X \to \mathbb{B}$ with $\forall x.\, p\,x \leftrightarrow f\,x = \mathsf{tt}$.*
- *p is enumerable if there is $f : \mathbb{N} \to \mathcal{O}(X)$ with $\forall x.\, p\,x \leftrightarrow \exists n.\, f\,n = \ulcorner x \urcorner$.*

These two notions generalise to predicates of higher arity as expected.

- *X is enumerable if there is $f : \mathbb{N} \to \mathcal{O}(X)$ with $\forall x. \exists n.\, f = \ulcorner x \urcorner$.*
- *X is discrete if equality $\lambda xy.x = y$ on X is decidable.*
- *X is a data type if it is both enumerable and discrete.*

We assume that the components $\mathcal{F}_\Sigma$ and $\mathcal{P}_\Sigma$ of our fixed signature Σ are data types. Then applying the terminology to the syntax and deductions systems introduced in the previous sections leads to the following observations.

Fact 6. *$\mathbb{T}$ and $\mathbb{F}$ are data types and $\Gamma \vdash \varphi$ and $\Gamma \vdash_c \varphi$ are enumerable.*

Proof. By the techniques discussed in [14], e.g. Fact 3.19. $\square$

The standard model-theoretic completeness proofs analysed in Sect. 3 require the assumption of Markov's principle. A proposition $P : \mathbb{P}$ is called *stable* if $\neg\neg P \to P$ and, analogously, a predicate $p : X \to \mathbb{P}$ is called stable if $p\,x$ is stable for all x. A synthetic version of Markov's principle states that satisfiability of Boolean sequences is stable (cf. [37]):

$$\mathsf{MP} := \forall f : \mathbb{N} \to \mathbb{B}.\, \neg\neg(\exists n.\, f\,n = \mathsf{tt}) \to \exists n.\, f\,n = \mathsf{tt}$$

Note that MP is trivially implied by excluded middle $\mathsf{EM} := \forall P : \mathbb{P}.\, P \vee \neg P$. Moreover, MP regulates the behaviour of computationally tractable predicates:

Fact 7. MP *implies that enumerable predicates on data types are stable.*

Proof. This is Fact 2.18 in [14]. □

As a consequence of Facts 6 and 7, MP implies that the deduction systems $\Gamma \vdash \varphi$ and $\Gamma \vdash_c \varphi$ are stable. In fact, only these stabilities are required for the standard model-theoretic completeness proofs discussed in the next section and they are equivalent to $\mathsf{MP_L}$, a version of Markov's principle stated for the call-by-value λ-calculus L [42,17] and its halting problem $\mathcal{E}$:

$$\mathsf{MP_L} := \forall s.\ \neg\neg\mathcal{E}s \to \mathcal{E}s$$

We will prove the following in Sect. 4:

Lemma 8. $\mathsf{MP_L}$*, stability of* $\Gamma \vdash \varphi$ *and stability of* $\Gamma \vdash_c \varphi$ *are all equivalent.*

3   Model-Theoretic Semantics

The first variant of semantics we consider is based on the idea of interpreting terms as objects in a model and embedding the logical connectives into the meta-logic. A formula is considered valid if it is satisfied by all models. The simplest case is Tarski semantics, coinciding with classical deduction via Henkin's completeness proof factoring through a (constructive) model-existence theorem [21]. Kripke semantics, coinciding with intuitionistic deduction, add more structure by connecting several models through an accessibility relation and admit a simpler completeness proof using a universal model. In this section, we only consider formulas $\varphi : \mathbb{F}^*$ in the $\to, \forall, \bot$-fragment.

3.1   Tarski Semantics

Definition 9. A *(Tarski) model* $\mathcal{M}$ over a domain D is a pair of functions

$$\_^{\mathcal{M}} \ :\ \forall f : \mathcal{F}_\Sigma.\ D^{|f|} \to D \qquad\qquad \_^{\mathcal{M}} \ :\ \forall P : \mathcal{P}_\Sigma.\ D^{|P|} \to \mathbb{P}.$$

Assignments $\rho : \mathbb{N} \to D$ are extended to *evaluations* $\hat\rho : \mathbb{T} \to D$ by $\hat\rho\, x := \rho\, x$ and $\hat\rho\,(f\,t) := f^{\mathcal{M}}\,(\hat\rho\,@\,t)$ and to formulas via the relation $\mathcal{M} \vDash_\rho \varphi$ defined by

$$\mathcal{M} \vDash_\rho \bot := \bot \qquad\qquad \mathcal{M} \vDash_\rho \varphi \dot\to \psi := \mathcal{M} \vDash_\rho \varphi \to \mathcal{M} \vDash_\rho \psi$$
$$\mathcal{M} \vDash_\rho P\,t := P^{\mathcal{M}}\,(\hat\rho\,@\,t) \qquad\qquad \mathcal{M} \vDash_\rho \dot\forall\varphi := \forall a : D.\,\mathcal{M} \vDash_{a;\rho} \varphi$$

where the assignment $a;\rho$ maps 0 to a and $\mathsf{S}\,x$ to $\rho\,x$. We write $\mathcal{M} \vDash \varphi$ if $\mathcal{M} \vDash_\rho \varphi$ for all ρ. $\mathcal{M}$ is called *classical* if it validates all instances of Peirce's law, i.e. $\mathcal{M} \vDash ((\varphi \dot\to \psi) \dot\to \varphi) \dot\to \varphi$ for all $\varphi, \psi : \mathbb{F}^*$. We write $\mathcal{M} \vDash_\rho \mathcal{T}$ if $\mathcal{M}_\rho \vDash \varphi$ for all $\varphi \in \mathcal{T}$ and $\mathcal{T} \vDash \varphi$ if $\mathcal{M} \vDash_\rho \varphi$ for every classical $\mathcal{M}$ and ρ with $\mathcal{M} \vDash_\rho \mathcal{T}$.

We first show that the classical deduction system $\Gamma \vdash_c \varphi$ (restricted to the considered $\to, \forall, \bot$-fragment) is *sound* for Tarski semantics.

Fact 10. $\Gamma \vdash_c \varphi$ *implies* $\Gamma \vDash \varphi$.

Proof. By induction on $\Gamma \vdash_c \varphi$ similar to the soundness proof in [14; Fact 3.14]. The classical Peirce axioms $\Gamma \vdash_c ((\varphi \dot\to \psi) \dot\to \varphi) \dot\to \varphi$ are sound given that we only consider classical models. □

Formally, *completeness* denotes the converse property, i.e. that $\Gamma \vDash \varphi$ implies $\Gamma \vdash_c \varphi$. We now outline a Henkin-style completeness proof for $\Gamma \vdash_c \varphi$ based on the presentation by Herbelin and Ilik [22]. The main idea is to factor through a model existence theorem, stating that every consistent context is satisfied by a syntactic model. The model existence theorem in turn is based on a theory extension lemma generalising the role of $\dot\bot$ to an arbitrary substitute $\varphi_\bot$:

Lemma 11. *For every closed formula* $\varphi_\bot$ *and closed* T *there is* $T' \supseteq T$ *with:*

1. T' *maintains* $\varphi_\bot$*-consistency, i.e.* $T \vdash_c \varphi_\bot$ *whenever* $T' \vdash_c \varphi_\bot$.
2. T' *is deductively closed, i.e.* $\varphi \in T'$ *whenever* $T' \vdash_c \varphi$.
3. T' *respects implication, i.e.* $\varphi \dot\to \psi \in T'$ *iff* $\varphi \in T' \to \psi \in T'$.
4. T' *respects universal quantification, i.e.* $\dot\forall \varphi \in T'$ *iff* $\forall t. \varphi[t] \in T'$.

Proof. We fix an enumeration φ_n of $\mathbb{F}^*$ such that x is fresh for φ_n if $x \geq n$. The extension can be separated into three steps, all maintaining $\varphi_\bot$-consistency:

a. $\mathcal{E} \supseteq T$ which is *exploding*, i.e. $(\varphi_\bot \dot\to \varphi) \in \mathcal{E}$ for all closed φ.
b. $\mathcal{H} \supseteq \mathcal{E}$ which is *Henkin*, i.e. $(\varphi_n[n] \dot\to \dot\forall \varphi_n) \in \mathcal{H}$ for all n.
c. $\Omega \supseteq \mathcal{H}$ which is *maximal*, i.e. $\varphi \in \Omega$ whenever $\Omega, \varphi \vdash_c \varphi_\bot$ implies $\Omega \vdash_c \varphi_\bot$.

Note that being exploding allows to use $\varphi_\bot$ analogously to $\dot\bot$ and that being Henkin ensures that there is no mismatch between the provability of a universal formula and all its instances. We first argue why Ω satisfies the claims (1)–(4) of the extension lemma.

1. Ω is a $\varphi_\bot$-consistent extension of T since all steps maintain $\varphi_\bot$-consistency.
2. Let $\Omega \vdash_c \varphi$ and assume $\Omega, \varphi \vdash_c \varphi_\bot$, so $\Omega \vdash_c \varphi_\bot$. Thus $\varphi \in \Omega$ per maximality.
3. The first direction is immediate as Ω is deductively closed. We prove the converse using maximality, so assume $\Omega, \varphi \dot\to \psi \vdash_c \varphi_\bot$. It suffices to show that $\Omega \vdash_c \varphi$ since then $\varphi \in \Omega$, $\psi \in \Omega$, and ultimately $\Omega \vdash_c \varphi_\bot$ follow. $\Omega \vdash_c \varphi$ can be derived by proof rules for $\varphi_\bot$ analogous to the ones for $\dot\bot$.
4. The first direction is again immediate by Ω being deductively closed and the converse exploits that Ω is Henkin as follows. Suppose $\forall t. \varphi[t] \in \Omega$ and let φ be φ_n in the given enumeration. Then in particular $\varphi_n[n] \in \Omega$ and since Ω is Henkin also $\varphi_n[n] \dot\to \dot\forall \varphi_n \in \Omega$ which is enough to derive $\dot\forall \varphi \in \Omega$.

We now discuss the three extension steps separately:

a. Since the requirement is unconditional, we just add all needed formulas:

$$\mathcal{E} := T \cup \{\varphi_\bot \dot\to \varphi \mid \varphi \text{ closed}\}$$

We only have to argue that $\mathcal{E}$ maintains $\varphi_\perp$-consistency over $\mathcal{T}$. So suppose $\mathcal{E} \vdash_c \varphi_\perp$, meaning that $\Gamma \vdash_c \varphi_\perp$ for some $\Gamma \subseteq \mathcal{E}$. We show that all added instances of explosion for $\varphi_\perp$ in Γ can be eliminated. Indeed, for $\Gamma = \Delta, \varphi_\perp \dot{\to} \varphi$ we have $\Delta \vdash_c (\varphi_\perp \dot{\to} \varphi) \dot{\to} \varphi_\perp$ and hence $\Delta \vdash_c \varphi_\perp$ by the Peirce rule. Thus by iteration there is $\Gamma' \subseteq \mathcal{T}$ with $\Gamma' \vdash_c \varphi_\perp$, justifying $\mathcal{T} \vdash_c \varphi_\perp$.

b. As above, to make $\mathcal{E}$ Henkin we just add all necessary Henkin-axioms

$$\mathcal{H} := \mathcal{E} \cup \{\varphi_n[n] \dot{\to} \dot{\forall} \varphi_n \mid n : \mathbb{N}\}$$

and justify that the extension maintains $\varphi_\perp$-consistency. So let $\Gamma \vdash_c \varphi_\perp$ for some $\Gamma \subseteq \mathcal{H}$, we again show that all added instances can be eliminated. Hence suppose $\Gamma = \Delta, \varphi_n[n] \dot{\to} \dot{\forall} \varphi_n$. Once can show that in a context Δ' extending Δ by suitable instances of $\varphi_\perp$-explosion one can derive $\Delta' \vdash_c \varphi_\perp$. In this derivation one exploits that n is fresh for φ_n and that the input theory $\mathcal{E}$ is closed. Thus ultimately $\mathcal{E} \vdash_c \varphi_\perp$.

c. The last step maximises $\mathcal{H}$ by adding all formulas maintaining $\varphi_\perp$-consistency:

$$\Omega_0 := \mathcal{H} \quad \Omega_{n+1} := \Omega_n \cup \{\varphi_n \mid \Omega_n, \varphi_n \vdash_c \varphi_\perp \text{ implies } \Omega_n \vdash_c \varphi_\perp\} \quad \Omega := \bigcup_{n:\mathbb{N}} \Omega_n$$

Note that Ω maintains $\varphi_\perp$-consistency over all Ω_n and hence $\mathcal{H}$ by construction so it remains to justify that Ω is maximal. So suppose $\Omega, \varphi_n \vdash_c \varphi_\perp$ implies $\Omega \vdash_c \varphi_\perp$, we have to show that $\varphi_n \in \Omega$. This is the case if the condition in the definition of Ω_{n+1} is satisfied, so let $\Omega_n, \varphi_n \vdash_c \varphi_\perp$. Then by the assumed implication $\Omega \vdash_c \varphi_\perp$ and since Ω maintains $\varphi_\perp$-consistency over Ω_n also $\Omega_n \vdash_c \varphi_\perp$ as required. □

Since the proof of this lemma relies on the input theory $\mathcal{T}$ to be closed, we only consider completeness for closed formulas. This is in fact enough for usual applications but we refer to the Coq development and [54] for a technically more involved generalisation incorporating formulas with free variables.

The generalisation via the falsity substitute $\varphi_\perp$ will become important later, for now the instance $\varphi_\perp := \dot{\perp}$ suffices. Also note that in usual jargon the extension $\mathcal{T}'$ of a consistent theory $\mathcal{T}$ is called *maximal consistent*, as no further formulas can be added to $\mathcal{T}'$ without breaking consistency.

Maximal consistent theories $\mathcal{T}$ give rise to equivalent *syntactic models* $\mathcal{M}_\mathcal{T}$ over the domain $\mathbb{T}$ of terms by setting $f^\mathcal{T} \, t := f\,t$ and $P^\mathcal{T} \, t := (P\,t \in \mathcal{T})$. We then observe that $\mathcal{M}_\mathcal{T} \vDash_\sigma \varphi$ iff $\varphi[\sigma] \in \mathcal{T}$ for all substitutions σ by a straighforward induction on φ using the properties stated in Lemma 11. Hence in particular $\mathcal{M}_\mathcal{T} \vDash_{\mathrm{id}} \varphi$ iff $\varphi \in \mathcal{T}$ for the identity substitution $\mathrm{id}\,x := x$. From this observation we directly conclude the model existence theorem:

Theorem 12. *Every closed consistent theory is satisfied in a classical model.*

Proof. Let $\mathcal{T}$ be closed and consistent and let $\mathcal{T}'$ be its extension per Lemma 11 for $\varphi_\perp := \dot{\perp}$. To show $\mathcal{M}_{\mathcal{T}'} \vDash_{\mathrm{id}} \mathcal{T}$, let $\varphi \in \mathcal{T}$, hence $\varphi \in \mathcal{T}'$. Then since $\mathcal{M}_{\mathcal{T}'}$ is equivalent to $\mathcal{T}'$ we conclude $\mathcal{M}_{\mathcal{T}'} \vDash_{\mathrm{id}} \varphi$ as desired. Finally, $\mathcal{M}_{\mathcal{T}'}$ is classical due to (2) of Lemma 11. □

The model existence theorem yields completeness up to double negation:

Fact 13. $T \vDash \varphi$ implies $\neg\neg(T \vdash_c \varphi)$ for closed T and φ.

Proof. Suppose that $T \vDash \varphi$ for closed T and φ and assume $T \nvdash_c \varphi$ which is equivalent to $T, \dot\neg\varphi$ being consistent. But then there must be a model of $T, \dot\neg\varphi$ in conflict to the assumption $T \vDash \varphi$. $\square$

In fact, the remaining double negation elimination turns out to be necessary:

Fact 14. *Completeness of* $\Gamma \vdash_c \varphi$ *is equivalent to stability of* $\Gamma \vdash_c \varphi$.

Proof. Assuming stability, Fact 13 directly yields the completeness of $\Gamma \vdash_c \varphi$. Conversely, assume completeness and let $\neg\neg(\Gamma \vdash_c \varphi)$. Employing completeness, to get $\Gamma \vdash_c \varphi$ it suffices to show $\Gamma, \dot\neg\varphi \vDash \bot$, so suppose $\mathcal{M} \vDash_\rho \Gamma, \dot\neg\varphi$ for some $\mathcal{M}$ and ρ. As we now aim at a contradiction, we can turn $\neg\neg(\Gamma \vdash_c \varphi)$ into $\Gamma \vdash_c \varphi$ and therefore obtain $\Gamma \vDash_c \varphi$ by soundness, a conflict to $\mathcal{M} \vDash_\rho \Gamma, \dot\neg\varphi$. $\square$

Hence, we can characterise completeness of classical ND as follows.

Theorem 15. *1. Completeness of* $\Gamma \vdash_c \varphi$ *is equivalent to* $\mathsf{MP_L}$.
2. Completeness of $T \vdash_c \varphi$ *for enumerable* T *is equivalent to* MP.
3. Completeness of $T \vdash_c \varphi$ *for arbitrary* T *is equivalent to* EM.

Proof. 1. By Fact 14 completeness is equivalent to the stability of $\Gamma \vdash_c \varphi$ which is shown equivalent to $\mathsf{MP_L}$ in Sect. 4.
2. $T \vdash_c \varphi$ for enumerable T is enumerable, hence stable under MP and thus complete per Fact 13. For the converse, assume a function $f : \mathbb{N} \to \mathbb{B}$ and consider $T := (\lambda\varphi. \varphi = \dot\bot \wedge \exists n. f\, n = \mathsf{tt})$. Since T is enumerable, completeness yields that $T \vDash \bot$ is equivalent to $T \vdash_c \bot$ which in turn is equivalent to $\exists n. f\, n = \mathsf{tt}$. Then since $T \vDash \bot$ is stable so must be $\exists n. f\, n = \mathsf{tt}$.
3. EM particularly implies that $T \vdash_c \varphi$ is stable and hence complete. Conversely given a proposition $P : \mathbb{P}$, completeness for $T := (\lambda\varphi. \varphi = \dot\bot \wedge P)$ yields the stability of P with an argument as in (2). $\square$

Having analysed the usual Henkin-style completeness proof, we now turn to its constructivisation. The central observation is that completeness already holds constructively for the minimal $\to, \forall$-fragment, by an elaboration of the classical proof for the minimal fragment given in [45]. To this end, we further restrict the deduction system and semantics to the minimal fragment and prove completeness via a suitable form of model existence.

Lemma 16. *In the* $\to, \forall$-fragment, for closed T and φ there is a classical model $\mathcal{M}$ and an assignment ρ s.t. (1) $\mathcal{M} \vDash_\rho T$ and (2) $\mathcal{M} \vDash_\rho \varphi$ implies $T \vdash_c \varphi$.

Proof. Let T' be the extension of T for $\varphi_\bot := \varphi$. As before, we have $\mathcal{M}_{T'} \vDash_{id} T'$. So now let $\mathcal{M}_{T'} \vDash_{id} \varphi$, then $\varphi \in T'$ and $T \vdash_c \varphi$ by (1) of Lemma 11. $\square$

Corollary 17. *In the* $\to, \forall$-fragment, $\Gamma \vDash \varphi$ implies $\Gamma \vdash_c \varphi$ for closed Γ and φ.

As opposed to completeness for fomulas incorporating $\dot{\perp}$, completeness in the minimal fragment does not rely on consistency requirements. Consequently, if these requirements are eliminated by allowing models treating inconsistency more liberal, completeness for formulas with $\perp$ can be established constructively (cf. [53,30]).

So we now turn back to the $\rightarrow, \forall, \perp$-fragment and define a satisfaction relation $\mathcal{M} \vDash_\rho^A \varphi$ for arbitrary propositions A with the relaxed rule $(\mathcal{M} \vDash_\rho^A \dot{\perp}) := A$. A model $\mathcal{M}$ is A-exploding if $\mathcal{M} \vDash^A \dot{\perp} \rightarrow \varphi$ for all φ and exploding if it is A-exploding for some choice of A. Note that $A := \top$ and $P^{\mathcal{M}} t := \top$ in particular yields an exploding model satisfying all formulas, hence accommodating inconsistent theories. This leads to the following formulation of model existence.

Lemma 18. *For every closed theory $\mathcal{T}$ there is an exploding classical model $\mathcal{M}$ and an assignment ρ such that (1) $\mathcal{M} \vDash_\rho^A \mathcal{T}$ and (2) $\mathcal{M} \vDash_\rho^A \dot{\perp}$ implies $\mathcal{T} \vdash_c \dot{\perp}$.*

Proof. Let $\mathcal{T}$ be closed and let $\mathcal{T}'$ be its extension for $\varphi_\perp := \dot{\perp}$. We set $A := \dot{\perp} \in \mathcal{T}'$ and observe that the syntactic model $\mathcal{M}_{\mathcal{T}'}$ still coincides with $\mathcal{T}'$, i.e. $\mathcal{M}_{\mathcal{T}'} \vDash_\sigma^A \varphi$ iff $\varphi[\sigma] \in \mathcal{T}'$. Hence we have (1) $\mathcal{M}_{\mathcal{T}'} \vDash_{\mathrm{id}}^A \mathcal{T}$. Moreover, $\mathcal{M}_{\mathcal{T}'}$ is A-exploding since proving $\mathcal{M}_{\mathcal{T}'} \vDash_\sigma^A \dot{\perp} \rightarrow \varphi$ in this case means to prove that $\dot{\perp} \dot{\rightarrow} \varphi[\sigma] \in \mathcal{T}'$, a straightforward consequence of $\mathcal{T}'$ being deductively closed. Finally, (2) follows from (1) of Lemma 11 as seen before.   □

We write $\Gamma \vDash_e \varphi$ if $\mathcal{M} \vDash_\rho^A \varphi$ for all $A : \mathbb{P}$ and A-exploding $\mathcal{M}$ and ρ with $\mathcal{M} \vDash_\rho^A \Gamma$ and finally establish completeness with respect to exploding models:

Fact 19. $\Gamma \vDash_e \varphi$ *implies* $\Gamma \vdash_c \varphi$ *for closed Γ and φ.*

Proof. Let $\Gamma \vDash_e \varphi$, then $\Gamma, \dot{\neg}\varphi \vdash_c \dot{\perp}$ follows by Lemma 18 for $\mathcal{T} := \Gamma, \dot{\neg}\varphi$.   □

3.2   Kripke Semantics

Turning to intuitionistic logic, we present Kripke semantics immediately generalised to arbitrary interpretations of falsity.

Definition 20. *A Kripke model $\mathcal{K}$ over a domain D is a preorder $(\mathcal{W}, \preceq)$ with*

$$\_^{\mathcal{K}} : \forall f : \mathcal{F}_\Sigma. D^{|f|} \rightarrow D \qquad \_^{\mathcal{K}} : \forall P : \mathcal{P}_\Sigma. \mathcal{W} \rightarrow D^{|P|} \rightarrow \mathbb{P} \qquad \perp^{\mathcal{K}} : \mathcal{W} \rightarrow \mathbb{P}.$$

The interpretations of predicates and falsity are required to be monotone, i.e. $P_v^{\mathcal{K}} a \rightarrow P_w^{\mathcal{K}} a$ and $\perp_v^{\mathcal{K}} \rightarrow \perp_w^{\mathcal{K}}$ whenever $v \preceq w$. Assignments ρ and their term evaluations $\hat{\rho}$ are extended to formulas via the relation $w \Vdash_\rho \varphi$ defined by

$$w \Vdash_\rho \dot{\perp} := \perp_w^{\mathcal{K}} \qquad\qquad w \Vdash_\rho \varphi \dot{\rightarrow} \psi := \forall v \succeq w. v \Vdash_\rho \varphi \rightarrow v \Vdash_\rho \psi$$

$$w \Vdash_\rho P t := P_w^{\mathcal{K}} (\hat{\rho} @ t) \qquad w \Vdash_\rho \dot{\forall} \varphi := \forall a : D. w \Vdash_{a;\rho} \varphi$$

We write $\mathcal{K} \Vdash \varphi$ if $w \Vdash_\rho \varphi$ for all ρ and w. $\mathcal{K}$ is standard if $\perp_w^{\mathcal{K}}$ implies $\perp$ for all w and exploding if $\mathcal{K} \Vdash \dot{\perp} \dot{\rightarrow} \varphi$ for all φ. We write $\mathcal{T} \Vdash \varphi$ if $\mathcal{K} \Vdash_\rho \varphi$ for all standard $\mathcal{K}$ and ρ with $\mathcal{K} \Vdash_\rho \mathcal{T}$, and $\mathcal{T} \Vdash_e \varphi$ when relaxing to exploding models.

Note that standard models are exploding, hence $\mathcal{T} \Vdash_e \varphi$ implies $\mathcal{T} \Vdash \varphi$. Moreover, the monotonicity required for the predicate and falsity interpretations lifts to all formulas, i.e. $w \Vdash_\rho \varphi$ implies $v \Vdash_\rho \varphi$ whenever $w \preceq v$. This property together with the usual facts about the interaction of assignments and substitutions yields soundness:

Fact 21. $\Gamma \vdash \varphi$ implies $\Gamma \Vdash_e \varphi$.

Proof. By induction on $\Gamma \vdash \varphi$ and analogous to [14; Fact 3.34]. □

Turning to completeness, instead of showing that $\Gamma \Vdash_e \varphi$ implies $\Gamma \vdash \varphi$ directly, we follow Herbelin and Lee [23] and reconstruct a formal derivation in the normal sequent calculus LJT, hence implementing a cut-elimination procedure. LJT is defined by judgements $\Gamma \Rightarrow \varphi$ and $\Gamma; \psi \Rightarrow \varphi$ for a focused formula ψ:

$$\frac{}{\Gamma; \varphi \Rightarrow \varphi}\ A \qquad \frac{\Gamma; \varphi \Rightarrow \psi \quad \varphi \in \Gamma}{\Gamma \Rightarrow \psi}\ C \qquad \frac{\Gamma \Rightarrow \varphi \quad \Gamma; \psi \Rightarrow \theta}{\Gamma; \varphi \dot\rightarrow \psi \Rightarrow \theta}\ IL$$

$$\frac{\Gamma, \varphi \Rightarrow \psi}{\Gamma \Rightarrow \varphi \dot\rightarrow \psi}\ IR \qquad \frac{\Gamma; \varphi[t] \Rightarrow \psi}{\Gamma; \dot\forall \varphi \Rightarrow \psi}\ AL \qquad \frac{\uparrow\Gamma \Rightarrow \varphi}{\Gamma \Rightarrow \dot\forall \varphi}\ AR \qquad \frac{\Gamma \Rightarrow \dot\bot}{\Gamma \Rightarrow \varphi}\ E$$

Fact 22. *Every sequent $\Gamma \Rightarrow \varphi$ can be translated into a normal derivation $\Gamma \vdash \varphi$.*

Proof. By simultaneous induction on both forms of judgements, where every sequent $\Gamma; \psi \Rightarrow \varphi$ is translated to an implication from $\Gamma \vdash \psi$ to $\Gamma \vdash \varphi$. □

By the previous fact, completeness for LJT implies completeness for intuitionistic ND. The technique to establish completeness for Kripke semantics is based on universal models coinciding with intuitionistic provability. We in fact construct two syntactic Kripke models over the domain $\mathbb{T}$.

- An exploding model $\mathcal{U}$ on contexts s.t. $\Gamma \Vdash_\sigma^{\mathcal{U}} \varphi$ iff $\Gamma \Rightarrow \varphi[\sigma]$.
- A standard model $\mathcal{C}$ on consistent contexts s.t. $\Gamma \Vdash_\sigma^{\mathcal{C}} \varphi$ iff $\neg\neg(\Gamma \Rightarrow \varphi[\sigma])$.

These constructions are adaptions of those in [54], which in turn are based on the proof and comments in [23]. We begin with the exploding model $\mathcal{U}$.

Definition 23. The model $\mathcal{U}$ over the domain $\mathbb{T}$ of terms is defined on the contexts Γ preordered by inclusion $\subseteq$. Further, we set:

$$f^{\mathcal{U}} \boldsymbol{d} := f\, \boldsymbol{d} \qquad P_\Gamma^{\mathcal{U}} \boldsymbol{d} := \Gamma \Rightarrow P\, \boldsymbol{d} \qquad \bot_\Gamma^{\mathcal{U}} := \Gamma \Rightarrow \dot\bot$$

The desired properties of $\mathcal{U}$ can be derived from the next lemma, which takes the shape of a normalisation-by-evaluation procedure [3,10].

Lemma 24. *In the universal Kripke model $\mathcal{U}$ the following hold.*

1. $\Gamma \Vdash_\sigma \varphi \rightarrow \Gamma \Rightarrow \varphi[\sigma]$
2. $(\forall \Gamma' \psi.\ \Gamma \subseteq \Gamma' \rightarrow \Gamma'; \varphi[\sigma] \Rightarrow \psi \rightarrow \Gamma' \Rightarrow \psi) \rightarrow \Gamma \Vdash_\sigma \varphi$

Proof. We prove (1) and (2) at once by induction on φ generalising Γ and σ. We only discuss the case of implications $\varphi \dot\to \psi$ in full detail.

1. Assuming $\forall \Gamma'.\ \Gamma \subseteq \Gamma' \to \Gamma' \Vdash_\sigma \varphi \to \Gamma' \Vdash_\sigma \psi$, one has to derive that $\Gamma \Rightarrow (\varphi \dot\to \psi)[\sigma]$. Per (IR) and inductive hypothesis (2) for ψ it suffices to show $\Gamma, \varphi[\sigma] \Vdash_\sigma \psi$. Applying the inductive hypothesis (2) for φ and the assumption, it suffices to show that $\Gamma'; \varphi[\sigma] \Rightarrow \theta[\sigma]$ implies $\Gamma' \Rightarrow \theta[\sigma]$ for any $\Gamma, \varphi[\sigma] \subseteq \Gamma'$ and θ, which holds per (C).
2. Assuming $\forall \Gamma' \theta.\ \Gamma \subseteq \Gamma' \to \Gamma'; (\varphi \dot\to \psi)[\sigma] \Rightarrow \theta \to \Gamma' \Rightarrow \theta$ one has to deduce $\Gamma' \Vdash_\sigma \varphi$ entailing $\Gamma' \Vdash_\sigma \psi$ for any $\Gamma \subseteq \Gamma'$. Because of the inductive hypothesis (2) for ψ it suffices to show $\Delta; \psi[\sigma] \Rightarrow \theta$ implying $\Delta \Rightarrow \theta$ for any $\Gamma' \subseteq \Delta$. By using the assumption, $\Delta \Rightarrow \theta$ reduces to $\Delta; (\varphi \dot\to \psi)[\sigma] \Rightarrow \theta$. This follows by (IL), as the assumption $\Gamma' \Vdash_\sigma \varphi$ implies $\Delta \Rightarrow \varphi[\sigma]$ per inductive hypothesis (2). $\square$

Corollary 25. $\mathcal{U}$ *is exploding and satisfies* $\Gamma \Vdash_\sigma \varphi$ *iff* $\Gamma \Rightarrow \varphi[\sigma]$.

Proof. Suppose that $\Gamma \Rightarrow \bot$, then (2) of Lemma 24 yields that $\Gamma \Vdash_\sigma \varphi$ for arbitrary φ. Thus $\mathcal{U}$ is exploding. The claimed equivalence then follows by (1) of Lemma 24 and soundness of LJT. $\square$

Being universal, $\mathcal{U}$ witnesses completeness for exploding Kripke models:

Fact 26. *1.* $\Gamma \Vdash_e \varphi$ *implies* $\Gamma \Rightarrow \varphi$.
2. In the $\to, \forall$*-fragment,* $\Gamma \Vdash \varphi$ *implies* $\Gamma \Rightarrow \varphi$.

Proof. 1. Since $\Gamma \Vdash_{id}^{\mathcal{U}} \Gamma$ we have that $\Gamma \Vdash_e \varphi$ implies $\Gamma \Vdash_{id}^{\mathcal{U}} \varphi$ and hence $\Gamma \Rightarrow \varphi$.
2. In the minimal fragment, $\bot$ remains uninterpreted and hence imposes no condition on the models. Hence $\mathcal{U}$ yields the completeness in this case.

Before we move on to completeness for standard models, we illustrate how the previous fact already establishes the cut rule for LJT.

Lemma 27. *If* $\Gamma \Rightarrow \varphi$ *and* $\Gamma; \varphi \Rightarrow \psi$, *then* $\Gamma \Rightarrow \psi$.

Proof. By the translation given in Fact 22, we obtain a derivation $\Gamma \vdash \psi$ from the two assumptions. This can be turned into $\Gamma \Rightarrow \psi$ using soundness (Fact 21) and completeness (Fact 26).

We now construct the universal standard model $\mathcal{C}$ as a refinement of $\mathcal{U}$. As standard models require that $\bot_v^{\mathcal{K}}$ implies $\bot$ for any v, the model $\mathcal{U}$ has to be restricted to the consistent contexts, those which do not prove $\bot$.

Definition 28. The model $\mathcal{C}$ over the domain $\mathbb{T}$ of terms is defined on the consistent contexts $\Gamma \not\Rightarrow \bot$ preordered by inclusion $\subseteq$. Further, we set:

$$f^{\mathcal{C}}\,\boldsymbol{d} := f\,\boldsymbol{d} \qquad\qquad P_\Gamma^{\mathcal{C}}\,\boldsymbol{d} := \neg\neg(\Gamma \Rightarrow P\,\boldsymbol{d}) \qquad\qquad \bot_\Gamma^{\mathcal{C}} := \bot$$

Note that $\mathcal{C}$ is obviously standard and that we weakened the interpretation of atoms to doubly negated provability. This admits the following normalisation-by-evaluation procedure for doubly negated sequents:

Lemma 29. *In the universal Kripke model $\mathcal{C}$ the following hold.*

1. $\Gamma \Vdash_\sigma \varphi \to \neg\neg(\Gamma \Rightarrow \varphi[\sigma])$
2. $(\forall\, \Gamma'\psi.\ \Gamma \subseteq \Gamma' \to \Gamma' ; \varphi[\sigma] \Rightarrow \psi \to \neg\neg(\Gamma' \Rightarrow \psi)) \to \Gamma \Vdash_\sigma \varphi$

Proof. We prove (1) and (2) at once by induction on φ generalising Γ and σ. Most cases are completely analogous to those in Lemma 24. Therefore we only discuss the crucial case (1) for implications $\varphi \dot{\to} \psi$.

1. Assuming $\Gamma \Vdash_\sigma \varphi \dot{\to} \psi$ we need to derive $\neg\neg(\Gamma \Rightarrow \varphi[\sigma] \dot{\to} \psi[\sigma])$. So we assume $\neg(\Gamma \Rightarrow \varphi[\sigma] \dot{\to} \psi[\sigma])$ and derive a contradiction. Because of the negative goal, we may assume that either $\Gamma, \varphi[\sigma]$ is consistent or not. In the positive case, we proceed as in Lemma 24 since the extended context is a node in $\mathcal{C}$. On the other hand, if $\Gamma, \varphi[\sigma] \Rightarrow \dot{\perp}$, then $\Gamma, \varphi[\sigma] \Rightarrow \psi[\sigma]$ by (E) and hence $\Gamma \Rightarrow \varphi[\sigma] \dot{\to} \psi[\sigma]$ by (IR), contradicting the assumption. $\quad\square$

Corollary 30. $\mathcal{C}$ *satisfies* $\Gamma \Vdash_\sigma \varphi$ *iff* $\neg\neg(\Gamma \Rightarrow \varphi[\sigma])$.

Proof. The first direction is (1) of Lemma 29 and the converse follows with (2) since $\neg\neg(\Gamma \Rightarrow \varphi[\sigma])$ and $\Gamma' ; \varphi[\sigma] \Rightarrow \psi$ for $\Gamma' \supseteq \Gamma$ together imply $\neg\neg(\Gamma' \Rightarrow \psi)$ via the cut rule established in Lemma 27. $\quad\square$

The advantage of the additional double negations is that, in contrast to the proof in [23], we only need a single application of stability to derive completeness. Thus we can prove the completeness of $\Gamma \vdash \varphi$ admissible in Sect. 4.

Fact 31. *1.* $\Gamma \Vdash \varphi$ *implies* $\Gamma \Rightarrow \varphi$, *provided that* $\Gamma \Rightarrow \varphi$ *is stable.*
2. $\Gamma \Vdash \varphi$ *implies* $\Gamma \vdash \varphi$, *provided that* $\Gamma \vdash \varphi$ *is stable.*

Proof. 1. Since $\Gamma \Vdash \varphi$ implies $\neg\neg(\Gamma \Rightarrow \varphi)$, we can conclude $\Gamma \Rightarrow \varphi$ per stability.
2. Since $\Gamma \Rightarrow \varphi$ iff $\Gamma \vdash \varphi$ per soundness and completeness (Facts 21 and 26). $\quad\square$

Conversely, unrestricted completeness requires the stability of classical ND.

Fact 32. *Completeness of* $\Gamma \Rightarrow \varphi$ *implies stability of* $\Gamma \vdash_c \varphi$.

Proof. Assume completeness of $\Gamma \Rightarrow \varphi$ and suppose $\neg\neg(\Gamma \vdash_c \varphi)$. We prove $\Gamma \vdash_c \varphi$, so it suffices to show $\Gamma, \dot{\neg}\varphi \vdash_c \dot{\perp}$. Employing a standard double negation translation φ^N on formulas φ, it is equivalent to establish $(\Gamma, \dot{\neg}\varphi)^N \Rightarrow \dot{\perp}$. Applying completeness, however, we may assume a standard model $\mathcal{K}$ with $\mathcal{K} \Vdash_\rho (\Gamma, \dot{\neg}\varphi)^N$ and derive a contradiction. Hence we conclude $\Gamma \vdash_c \varphi$ and so $\Gamma^N \Vdash \varphi^N$ from $\neg\neg(\Gamma \vdash_c \varphi)$ and soundness, in conflict to $\mathcal{K} \Vdash_\rho (\Gamma, \dot{\neg}\varphi)^N$. $\quad\square$

Thus, the completeness of intuitionistic ND is similar to the classical case.

Theorem 33. *1. Completeness of* $\Gamma \vdash \varphi$ *is equivalent to* $\mathsf{MP_L}$.
2. Completeness of $\mathcal{T} \vdash \varphi$ *for enumerable* $\mathcal{T}$ *implies* MP.
3. Completeness of $\mathcal{T} \vdash \varphi$ *for arbitrary* $\mathcal{T}$ *implies* EM.

4   On Markov's Principle

We show that the stability of $\Gamma \vdash_c \varphi$ and $\Gamma \vdash \varphi$ is equivalent to an object-level version of Markov's principle referencing procedures in a concrete model of computation. For formalisation purposes, we will use the call-by-value λ-calculus L [42,17] as model of computation. Since on paper the same proofs can be carried out for any model of computation we will not go into details of L. We only need two notions: first, L-enumerability [15; Definition 6], which is defined like synthetic enumerability, but where the enumerator is an L-computable function. Secondly, the halting problem for L, defined as $\mathcal{E}s :=$ *"the term s terminates"*.

We define the object-level Markov's principle $\mathsf{MP_L}$ as stability of $\mathcal{E}$:

$$\mathsf{MP_L} := \forall s. \; \neg\neg\mathcal{E}s \rightarrow \mathcal{E}s$$

$\mathsf{MP_L}$ can also be phrased similarly to MP with a condition on the sequence:

Lemma 34 ([17; Theorem 45]). $\mathsf{MP_L}$ *is equivalent to*

$$\forall f : \mathbb{N} \rightarrow \mathbb{B}. \, \mathsf{L}\text{-}computable \, f \rightarrow \neg\neg(\exists n. \, f \, n = \mathsf{tt}) \rightarrow \exists n. \, f \, n = \mathsf{tt}.$$

Corollary 35. MP *implies* $\mathsf{MP_L}$.

We show Lemma 8, i.e. that $\mathsf{MP_L}$ is equivalent to both the stability of $\vdash_c$ and $\vdash$ for finite contexts, thereby establishing that completeness of provability for standard Tarski and Kripke semantics for finite theories is equivalent to $\mathsf{MP_L}$.

Lemma 36 ([14; Fact 2.16]). *Let p and q be predicates. If p many-one reduces to q (i.e. $\exists f. \forall x. \, px \leftrightarrow q(fx)$, written $p \preceq q$) and q is stable, then p is stable.*

Thus, in order to prove the equivalence of the stability of $\mathcal{E}$, $\Gamma \vdash \varphi$, and $\Gamma \vdash_c \varphi$, it suffices to give many-one reductions between them. We start with the two simpler reductions:

Lemma 37. $\vdash_c \preceq \; \vdash$, *and thus stability of $\Gamma \vdash \varphi$ implies the stability of $\Gamma \vdash_c \varphi$.*

Proof. Using a standard double-negation translation proof. $\square$

Lemma 38. $\mathcal{E} \preceq \; \vdash_c$, *and thus stability of $\Gamma \vdash_c \varphi$ implies $\mathsf{MP_L}$.*

Proof. $\mathcal{E}$ reduces to the halting problem of Turing machines [55], which reduces to the Post correspondence problem [13], which in turn reduces to $\vdash_c$ by adapting [14; Corollary 3.49]. $\square$

Since $p \preceq \mathcal{E}$ for all L-enumerable predicates p [15; Theorem 7], it suffices to give an L-computable enumeration of type $\mathbb{N} \rightarrow \mathcal{L}(\mathbb{F})$ of provable formulas $\vdash \varphi$. Note that we continue to assume signatures to be (synthetically) enumerable and do *not* have to restrict to L-enumerability, which is enabled by the following signature extension lemma:

Lemma 39. *Let ι be an invertible embedding from Σ to Σ'. Then $\vdash \varphi$ over Σ if and only if $\vdash \iota\varphi$ over Σ', where $\iota\varphi$ is the recursive application of ι to formulas.*

Proof. $\Gamma \vdash \varphi \rightarrow \iota\Gamma \vdash \iota\varphi$ follows trivially by induction. For the inverse direction, we show that Kripke models M over Σ can be extended to Kripke models ιM over Σ s.t. $\rho, u \Vdash_M \varphi \leftrightarrow \rho, u \Vdash_{\iota M} \iota\varphi$. Then $\iota\Gamma \vdash \iota\varphi \rightarrow \Gamma \vdash \varphi$ follows from soundness and completness w.r.t. exploding models. $\square$

Lemma 40. $\Gamma \vdash \varphi$ *is* L-*enumerable for any enumerable signature Σ.*

Proof. Since Σ is enumerable, it can be injectively embedded via ι into the maximal signature $\Sigma_{\max} := (\mathbb{N}^2, \mathbb{N}^2)$ where the arity functions are just the second projections. Since $\mathbb{N}^2$ is L-enumerable, terms and formulas over $\Sigma_{\max}$ are also L-enumerable, and thus provability over $\Sigma_{\max}$ is L-enumerable. By Lemma 39 we obtain that provability over Σ is L-enumerable. $\square$

Corollary 41. $\vdash \preceq \mathcal{E}$, *and thus* MP$_L$ *implies the stability of $\Gamma \vdash \varphi$.*

We conclude the section with observations on independence and admissible of several statements in Coq's type theory pCuIC. By *independence* of a statement P, we mean that neither P nor $\neg P$ is provable in pCuIC without assumptions. By *admissibility* of a statement $\forall x.\ P(x) \rightarrow Q(x)$ we mean that whenever $P(t)$ is provable in pCuIC for a concrete term t without assumptions, $Q(t)$ is as well. Pédrot and Tabareau [41] show MP independent (Corollary 41) and admissible (Theorem 33). This transports to MP$_L$ as well as stability of deduction systems and completeness with respect to model-theoretic semantics.

Theorem 42. *The following are all independent and admissible in pCuIC:*

1. MP$_L$
2. *Stability of both $\Gamma \vdash_c \varphi$ and $\Gamma \vdash \varphi$.*
3. *Completeness of $T \vdash_c \varphi$ for enumerable T w.r.t. standard Tarski semantics.*
4. *Completeness of $\Gamma \vdash_c \varphi$ w.r.t. standard Tarski semantics.*
5. *Completeness of $\Gamma \vdash_c \varphi$ w.r.t. standard Tarski semantics.*

Proof. We exemplarily show (1) and (4), the other proofs are similar.

For (1), MP$_L$ is consistent since it is a consequence of EM. Lemma 40 in [41] shows that no theory conservative over the calculus of inductive constructions (CIC) can prove both the independence of premise rule IP and MP, by turning these assumptions into a decider for the halting problem of the untyped term language of CIC. One can adapt the proof to show that pCuIC cannot prove both IP and MP$_L$, by constructing a decider for the L-halting problem instead, which yields a contradiction as well. The admissibility of MP$_L$ follows from the admissibility of MP since a single application of MP suffices to derive MP$_L$.

For (4), independence follows directly from (1) and Theorem 15. For admissibility, assume that $\Gamma \vDash \varphi$ is provable in pCuIC. By Fact 13, $\neg\neg(\Gamma \vdash_c \varphi)$ is provable in pCuIC. Thus by (2), $\Gamma \vdash_c \varphi$ is provable in pCuIC. $\square$

5   Algebraic Semantics

In contrast to the model-theoretic semantics discussed in Sect. 3, algebraic semantics are not based on models interpreting the non-logical symbols but on algebras suitable for interpreting the logical connectives of the syntax. A formula is valid if it is satisfied by all algebras and completeness follows from the observation that deduction systems have the corresponding algebraic structure. Following [46], we discuss complete Heyting and Boolean algebras coinciding with intuitionistic and classical ND, respectively. We consider all formulas $\varphi : \mathbb{F}$.

Definition 43. A *Heyting algebra* consists of a preorder $(\mathcal{H}, \leq)$ and operations

$$0 : \mathcal{H}, \qquad \sqcap : \mathcal{H} \to \mathcal{H} \to \mathcal{H}, \qquad \sqcup : \mathcal{H} \to \mathcal{H} \to \mathcal{H}, \qquad \Rightarrow : \mathcal{H} \to \mathcal{H} \to \mathcal{H}$$

for bottom, meet, join, and implication satisfying the following properties:

1. $0 \leq x$
2. $z \sqcap x \leq y \leftrightarrow z \leq x \Rightarrow y$
3. $z \leq x \wedge z \leq y \leftrightarrow z \leq x \sqcap y$
4. $x \leq z \wedge y \leq z \leftrightarrow x \sqcup y \leq z$

Moreover, $\mathcal{H}$ is *complete* if there is a constant $\bigsqcap : (\mathcal{H} \to \mathbb{P}) \to \mathcal{H}$ for arbitrary meets satisfying $(\forall y \in P.\, x \leq y) \leftrightarrow x \leq \bigsqcap P$. Then $\mathcal{H}$ also has arbitrary joins $\bigsqcup P := \bigsqcap(\lambda x.\, \forall y \in P.\, y \leq x)$ satisfying $(\forall y \in P.\, y \leq x) \leftrightarrow \bigsqcup P \leq x$.

Arbitrary meets and joins indexed by a function $F : I \to \mathcal{H}$ on a type I are defined by $\bigsqcap_i F\, i := \bigsqcap(\lambda x.\, \exists i.\, x = F\, i)$ and $\bigsqcup_i F\, i := \bigsqcup(\lambda x.\, \exists i.\, x = F\, i)$, respectively. As we do not require $\leq$ to be antisymmetric in order to avoid quotient constructions, we establish equational facts about Heyting algebras only up to equivalence $x \equiv y := x \leq y \wedge y \leq x$ rather than actual equality.

Note that every Heyting algebra embeds into its down set algebra consisting of the sets $x \Downarrow := \lambda y.\, y \leq x$. The *MacNeille completion* [36] adding arbitrary meets and joins is a refinement of this embedding.

Fact 44. *Every Heyting algebra $\mathcal{H}$ embeds into a complete Heyting algebra $\mathcal{H}_c$, i.e. there is a function $f : \mathcal{H} \to \mathcal{H}_c$ with $x \leq y \leftrightarrow f\, x \leq_c f\, y$ and:*

1. $f\, 0 \equiv 0_c$
2. $f\, (x \Rightarrow y) \equiv f\, x \Rightarrow_c f\, y$
3. $f\, (x \sqcap y) \equiv f\, x \sqcap_c f\, y$
4. $f\, (x \sqcup y) \equiv f\, x \sqcup_c f\, y$

Proof. Given a set $X : \mathcal{H} \to \mathbb{P}$, we define the sets $\mathfrak{L} X := \lambda x.\, \forall y \in X.\, x \leq y$ of lower bounds and $\mathfrak{U} X := \lambda x.\, \forall y \in X.\, y \leq x$ of upper bounds of X. We say that a set X is down-complete if $\mathfrak{L}(\mathfrak{U} X) \subseteq X$. Note that in particular down sets $x \Downarrow$ are down-complete and that down-complete sets are downwards closed, i.e. satisfy $x \in X$ whenever $x \leq y$ for some $y \in X$.

Now consider the type $\mathcal{H}_c := \Sigma X.\, \mathfrak{L}(\mathfrak{U} X) \subseteq X$ of down-complete sets preordered by set inclusion $X \subseteq Y$. It is immediate by construction that the operation $\bigsqcap_c P := \bigcap P$ defines arbitrary meets in $\mathcal{H}_c$. Moreover, it is easily verified that further setting

$$0_c := 0\Downarrow \quad X \sqcap_c Y := X \cap Y \quad X \sqcup_c Y := \mathfrak{L}(\mathfrak{U}(X \cup Y)) \quad X \Rightarrow_c Y := \lambda x.\, \forall y \in X.\, x \sqcap y \in Y$$

turns $\mathcal{H}_c$ into a (hence complete) Heyting algebra. The only non-trivial case is implication, where $X \Rightarrow_c Y \equiv \bigsqcap_c (\lambda Z. \exists x \in X. Z \equiv (\lambda y. y \sqcap x \in Y))$ is a helpful characterisation to show that $X \Rightarrow_c Y$ is down-complete whenever Y is.

Finally, $x\!\!\Downarrow$ clearly is a structure preserving embedding as specified.   □

We now define how formulas can be evaluated in a complete Heyting algebra.

Definition 45. Given a complete Heyting algebra $\mathcal{H}$ we extend interpretations $[\![\_]\!] : \forall P : \mathcal{P}_\Sigma. \mathbb{T}^{|P|} \to \mathcal{H}$ of atoms to formulas using size recursion by

$$[\![\dot{\bot}]\!] := 0 \qquad\qquad [\![\varphi\dot{\wedge}\psi]\!] := [\![\varphi]\!] \sqcap [\![\psi]\!] \qquad [\![\dot{\forall}\varphi]\!] := \bigsqcap_t [\![\varphi[t]]\!]$$

$$[\![\varphi\dot{\to}\psi]\!] := [\![\varphi]\!] \Rightarrow [\![\psi]\!] \qquad [\![\varphi\dot{\vee}\psi]\!] := [\![\varphi]\!] \sqcup [\![\psi]\!] \qquad [\![\dot{\exists}\varphi]\!] := \bigsqcup_t [\![\varphi[t]]\!]$$

and to contexts by $[\![\Gamma]\!] := \bigsqcap \lambda x. \exists \varphi \in \Gamma. x = [\![\varphi]\!]$. A formula φ is *valid in* $\mathcal{H}$ whenever $x \le [\![\varphi]\!]$ for all $x : \mathcal{H}$.

We first show that intuitionistic ND is sound for this semantics.

Fact 46. $\Gamma \vdash \varphi$ *implies* $\forall \sigma. [\![\Gamma[\sigma]]\!] \le [\![\varphi[\sigma]]\!]$ *in every complete Heyting algebra.*

Proof. By induction on $\Gamma \vdash \varphi$, all cases but (DE) and (EE) are trivial.   □

Corollary 47. $\Gamma \vdash \varphi$ *implies* $[\![\Gamma]\!] \le [\![\varphi]\!]$ *in every complete Heyting algebra.*

Next turning to completeness, a strategy reminiscent to the case of Kripke semantics can be employed by exhibiting a universal structure, the so-called *Lindenbaum algebra*, that exactly coincides with provability.

Fact 48. *The type* $\mathbb{F}$ *of formulas together with the preorder* $\varphi \vdash \psi$ *and the logical connectives as corresponding algebraic operations forms a Heyting algebra.*

We write $\mathcal{L}$ for the Lindenbaum algebra (Fact 48) and $\mathcal{L}_c$ for its MacNeille completion (Fact 44). Formulas are evaluated in $\mathcal{L}_c$ according to Definition 45 using the syntactic atom interpretation $[\![P\,t]\!] := (P\,t)\!\!\Downarrow$.

Lemma 49. *Evaluating* φ *in* $\mathcal{L}_c$ *yields the set of all* ψ *with* $\psi \vdash \varphi$, *i.e.* $[\![\varphi]\!] \equiv \varphi\!\!\Downarrow$.

Proof. By size induction on φ. The case for atoms is by construction and the cases for all connectives but the quantifiers are immediate since $\Downarrow$ preserves the structure of $\mathcal{L}$ as specified in Fact 44.   □

Theorem 50. *If* φ *is valid in every complete Heyting algebra, then* $\vdash \varphi$.

Proof. If φ is valid, then Lemma 49 implies that $\psi \vdash \varphi$ forall ψ. By for instance choosing $\psi := \dot{\bot}\dot{\to}\dot{\bot}$ we can derive $\vdash \varphi$ since $\vdash \dot{\bot}\dot{\to}\dot{\bot}$.   □

A Heyting algebra is *Boolean* if it satisfies $(x \Rightarrow y) \Rightarrow x \le x$ for all x and y.

Theorem 51. *If* φ *is valid in every complete Boolean algebra, then* $\vdash_c \varphi$.

Proof. Analogous to the intuitionistic case, using that the Lindenbaum algebra over $\varphi \vdash_c \psi$ and hence its MacNeille completion are Boolean.   □

6   Game Semantics

Dialogues are games modeling a proponent defending the validity of a formula against an opponent. In the terminology of Felscher [11], the dialogues we consider in this section are the intuitionistic E-dialogues, generalised over their local rules $(\mathbb{F}, \mathbb{F}^a, \mathcal{A}, \triangleright, \mathcal{D}_-)$. Given abstract types for formulas $\mathbb{F}$ and attacks $\mathcal{A}$, the relation $a \mid \psi \triangleright \varphi$ states that a player may attack $\varphi : \mathbb{F}$ with $a : \mathcal{A}$ by possibly admitting a unique $\psi : \mathcal{O}(\mathbb{F})$. If $\psi = \emptyset$, no admission is made. Each $a : \mathcal{A}$ has an associated set $\mathcal{D}_a$ of formulas that may be admitted to fend off a. Special rules restrict when the proponent may admit atomic formulas, members of the set $\mathbb{F}^a$. We write $a \triangleright \varphi$ for $a \mid \emptyset \triangleright \varphi$. The local rules of first-order logic are given below with atomic formulas $\mathbb{F}^a := \{P\,t \mid P : \mathcal{P}_\Sigma\}$.

$$a_\vee \triangleright \varphi \dot\vee \psi \quad \mathcal{D}_{a_\vee} = \{\varphi, \psi\} \qquad a_\to \mid \ulcorner \varphi \urcorner \triangleright \varphi \dot\to \psi \quad \mathcal{D}_{a_\to} = \{\psi\} \qquad a_L \triangleright \varphi \dot\wedge \psi \quad \mathcal{D}_{a_L} = \{\varphi\}$$

$$a_t \triangleright \dot\forall \varphi \quad \mathcal{D}_{a_t} = \{\varphi[t]\} \qquad\qquad\qquad a_\perp \triangleright \dot\perp \quad \mathcal{D}_{a_\perp} = \{\} \qquad a_R \triangleright \varphi \dot\wedge \psi \quad \mathcal{D}_{a_R} = \{\psi\}$$

$$a_\exists \triangleright \dot\exists \varphi \quad \mathcal{D}_{a_\exists} = \{\varphi[t] \mid t : \mathbb{T}\}$$

In contrast to their usual presentation as sequences of alternating moves, we define dialogues as state transition systems over elements (A_o, c) of the type $\mathcal{L}(\mathbb{F}) \times \mathcal{A}$ containing the opponent's admissions (A_o) and last attack (c). The proponent opens each round by picking a move. She can defend against the opponent's attack c by admitting a justified defense formula $\varphi \in \mathcal{D}_c$, meaning $\varphi \in \mathbb{F}^a$ implies $\varphi \in A_o$. Alternatively, she can launch an attack a against any of the opponent's admissions if the admission resulting from a is justified.

$$\frac{\varphi \in \mathcal{D}_c \quad \text{justified } A_o\,\varphi}{(A_o, c) \rightsquigarrow_p \varphi} \; \text{PD} \qquad\qquad \frac{\varphi \in A_o \quad a \mid \psi \triangleright \varphi \quad \text{justified } A_o\,\psi}{(A_o, c) \rightsquigarrow_p (a, \varphi)} \; \text{PA}$$

Given such a move m, the opponent reacts to it by transforming the state s into s' (written as $s\,;\,m \rightsquigarrow_o s'$). The opponent may attack the proponent's defense formula (OA), defend against her attack (OD) or counter her attack by attacking her admission (OC). We define $\ulcorner \varphi \urcorner :: A := \varphi :: A$ and $\emptyset :: A := A$.

$$\frac{c' \mid \psi \triangleright \varphi}{(A_o, c)\,;\,\varphi \rightsquigarrow_o (\psi :: A_o, c')} \; \text{OA} \qquad\qquad \frac{\psi \in \mathcal{D}_a}{(A_o, c)\,;\,(a, \varphi) \rightsquigarrow_o (\psi :: A_o, c)} \; \text{OD}$$

$$\frac{a \mid \ulcorner \psi \urcorner \triangleright \varphi \quad c' \mid \theta \triangleright \psi}{(A_o, c)\,;\,(a, \varphi) \rightsquigarrow_o (\theta :: A_o, c')} \; \text{OC}$$

A formula φ is then considered E-valid if it is non-atomic and for all $c \mid \psi \triangleright \varphi$, there is a winning strategy Win $([\psi], c)$ as defined below.

$$\frac{s \rightsquigarrow_p m \quad \forall s'.\; s\,;\,m \rightsquigarrow_o s' \to \text{Win } s'}{\text{Win } s}$$

Following the strategy of [47], we first prove the soundness and completeness of the sequent calculus LJD which is defined in terms of the same notions as the dialogues. Indeed, as witnessed in the proofs of soundness and completeness, derivations of LJD are isomorphic to winning strategies, the R- and L-rule corresponding to a proponent defense and attack, their premises matching the possible opponent responses to each move. The statement $\Gamma \Rightarrow_D \mathcal{S}$ means that the context Γ entails the disjunction of the formulas contained in the set $\mathcal{S}$.

$$\frac{\varphi \in \mathcal{S} \quad \text{justified } \Gamma \varphi \quad \forall a \,|\, \psi \triangleright \varphi. \ \Gamma, \psi \Rightarrow_D \mathcal{D}_a}{\Gamma \Rightarrow_D \mathcal{S}} \ \text{R}$$

$$\frac{\varphi \in \Gamma \quad \text{justified } \Gamma \psi \quad a \,|\, \psi \triangleright \varphi \quad \forall \theta \in \mathcal{D}_a. \ \Gamma, \theta \Rightarrow_D \mathcal{S} \quad \forall a' \,|\, \tau \triangleright \psi. \ \Gamma, \tau \Rightarrow_D \mathcal{D}_{a'}}{\Gamma \Rightarrow_D \mathcal{S}} \ \text{L}$$

Theorem 52. *Any formula φ is E-valid if and only if one can derive $[] \Rightarrow_D \{\varphi\}$.*

Proof. Win $(A_o, c) \to A_o \Rightarrow_D \mathcal{D}_c$ holds per induction on Win (A_o, c). From this, completeness follows with an application of the R-rule, transforming Win $([\psi], c)$ for any $c \,|\, \psi \triangleright \varphi$ into $[\psi] \Rightarrow_D \mathcal{D}_c$. Soundness can be proven symmetrically.   □

To arrive at a more traditional soundness and completeness result, we show that one can translate between derivations in LJD and the intuitionistic sequent calculus LJ deriving sequents $\Gamma \Rightarrow_J \varphi$ as defined in Definition 58 of Appendix A.

Lemma 53. *One can derive $\Gamma \Rightarrow_D \{\varphi\}$ if and only if one can derive $\Gamma \Rightarrow_J \varphi$.*

Proof. Completeness is generalised as below and shown per induction on $\Gamma \Rightarrow_D \mathcal{S}$:

$$\Gamma \Rightarrow_D \mathcal{S} \ \to \ \forall \varphi. \ (\forall \psi, \Gamma \subseteq \Gamma'. \ \Gamma' \Rightarrow_J \psi \to \Gamma' \Rightarrow_J \varphi) \to \Gamma \Rightarrow_J \varphi$$

Soundness follows analogously from $\Gamma \Rightarrow_J \varphi \to \forall \sigma. \ \Gamma[\sigma] \Rightarrow_D \{\varphi[\sigma]\}$.   □

Corollary 54. *Any formula φ is E-valid if and only if one can derive $[] \Rightarrow_J \varphi$.*

7   Discussion

We have analysed the completeness of common deduction systems for first-order logic with regards to various explanations of logical validity. Model-theoretic semantics are the most direct implementation of the idea that terms represent objects of a domain of discourse. Particularly in a formal meta-theory such as constructive type theory, model-theoretic completeness justifies the common practice to verify consequences of a first-order axiomatisation by studying models satisfying corresponding meta-level axioms. However, model-theoretic semantics typically do not admit constructive completeness and, if not generalised to exploding models, require Markov's principle as soon as falsity is involved. Contrarily, evidence for the validity of a first-order formula in algebraic semantics and game semantics can be algorithmically transformed into syntactic derivations.

Of course, there are more semantics than the selection studied in this paper. For instance, there are hybrid variants such as interpreting both terms in a model and logical operations in an algebra, or dialogues with atomic formulas represented as underlying games. More generally, there are entirely different approaches like realisability semantics or proof-theoretic semantics, all coming with interesting completeness problems worth analysing in constructive type theory. Ideas for future work are outlined after a brief summary of related work.

7.1 Related Work

Our analysis of completeness in constructive type theory was motivated by previous work [14], carried out in Wehr's bachelor's thesis [54], and is directly influenced by multiple prior works. In their analysis of Henkin's proof, Herbelin and Ilik [22] give a constructive model existence proof and the constructivisation of completeness via exploding models. Herbelin and Lee [23] demonstrate the constructive Kripke completeness proof for minimal models and mention how to extend the approach to standard and exploding models. Scott [46] establishes completeness of free logic interpreted in a hybrid semantics comprising model-theoretic and algebraic components. Urzyczyn and Sørensen [47] give a proof of dialogue completeness via generalised dialogues for classical propositional logic.

The first proof that the completeness of intuitionistic first-order logic entails Markov's principle was given by Kreisel [28], although he attributes the proof idea to Gödel. The proof has since inspired a range of works deriving related non-constructivity results for different kinds of completeness [29,33,38,40,39]. By almost exclusively focusing our analysis on the $\forall, \rightarrow, \bot$-fragment, we did not concern ourselves with the contributions of $\exists$ and $\vee$ to the non-constructivity of completeness. Krivtsov's [31,32] work has the exact opposite focus: His analysis reveals that completeness with regards to exploding Tarski and Beth models, for full classical and intuitionistic first-order logic, respectively, are equivalent to the weak fan theorem. Another noteworthy work is that of Berardi [2], who analyses which abstract notions of models admit constructive completeness.

The completeness of first-order logic has been formalised in many interactive theorem provers such as Isabelle/HOL [4,43,44], NuPRL [6,52], Mizar [5], Lean [19], and Coq [23,24]. Among them, [6] and [24] share our focus on the constructivity of completeness. Constable and Bickford [6] give a constructive proof of completeness for the BHK-realisers of full intuitionistic first-order logic in NuPRL. Their proof is fully constructive when realisers are restricted to be normal terms, requiring Brouwer's fan theorem when lifting that restriction. In his PhD thesis [24], Ilik formalises multiple constructive proofs of first-order completeness in Coq. Especially noteworthy are the highly non-standard, constructivised Kripke models for full classical and intuitionistic first-order logic he presents in Chapters 2 and 3.

7.2 Future Work

We plan to further extend our constructive analysis and Coq library to all logical connectives and to uncountable signatures, both relying on additional logical assumptions. Subsequently, it would be interesting to study other aspects of model theory in the setting of constructive type theory, for instance the Löwenheim-Skolem theorems or first-order axiomatisations of arithmetic and set theory. Another idea is to analyse the completeness of second-order logic interpreted in Henkin semantics, as this formalism suffices to express the higher-order axiomatisation of set theory studied in [27]. Lastly, we conjecture that MP_L is strictly weaker than MP, but are not aware of a proof.

Acknowledgments. We thank Kathrin Stark for adapting Autosubst according to our needs, Fabian Kunze for helping with technicalities during the formalisation of Corollary 41, and Hugo Herbelin for fruitful discussion and pointers to relevant work.

A   Overview of Deduction Systems

Definition 55. Intuitionistic natural deduction is defined as follows:

$$\frac{\varphi \in \Gamma}{\Gamma \vdash \varphi}\,C \qquad \frac{\Gamma \vdash \dot\bot}{\Gamma \vdash \varphi}\,E \qquad \frac{\Gamma, \varphi \vdash \psi}{\Gamma \vdash \varphi \dot\to \psi}\,II \qquad \frac{\Gamma \vdash \varphi \dot\to \psi \quad \Gamma \vdash \varphi}{\Gamma \vdash \varphi}\,IE$$

$$\frac{\Gamma \vdash \varphi \quad \Gamma \vdash \psi}{\Gamma \vdash \varphi \dot\wedge \psi}\,CI \qquad \frac{\Gamma \vdash \varphi \dot\wedge \psi}{\Gamma \vdash \varphi}\,CE_1 \qquad \frac{\Gamma \vdash \varphi \dot\wedge \psi}{\Gamma \vdash \psi}\,CE_2$$

$$\frac{\Gamma \vdash \varphi}{\Gamma \vdash \varphi \dot\vee \psi}\,DI_1 \qquad \frac{\Gamma \vdash \psi}{\Gamma \vdash \varphi \dot\vee \psi}\,DI_2 \qquad \frac{\Gamma \vdash \varphi \dot\vee \psi \quad \Gamma, \varphi \vdash \theta \quad \Gamma, \psi \vdash \theta}{\Gamma \vdash \theta}\,DE$$

$$\frac{\uparrow\Gamma \vdash \varphi}{\Gamma \vdash \dot\forall \varphi}\,AI \qquad \frac{\Gamma \vdash \dot\forall \varphi}{\Gamma \vdash \varphi[t]}\,AE \qquad \frac{\Gamma \vdash \varphi[t]}{\Gamma \vdash \dot\exists \varphi}\,EI \qquad \frac{\Gamma \vdash \dot\exists \varphi \quad \uparrow\Gamma, \varphi \vdash \uparrow\psi}{\Gamma \vdash \psi}\,EE$$

We write $\vdash \varphi$ whenever φ is intuitionistically provable from the empty context.

Definition 56. Classical natural deduction is defined as follows:

$$\frac{\varphi \in \Gamma}{\Gamma \vdash_c \varphi}\,C \qquad \frac{\Gamma \vdash_c \dot\bot}{\Gamma \vdash_c \varphi}\,E \qquad \frac{\Gamma, \varphi \vdash_c \psi}{\Gamma \vdash_c \varphi \dot\to \psi}\,II \qquad \frac{\Gamma \vdash_c \varphi \dot\to \psi \quad \Gamma \vdash_c \varphi}{\Gamma \vdash_c \varphi}\,IE$$

$$\frac{\Gamma \vdash_c \varphi \quad \Gamma \vdash_c \psi}{\Gamma \vdash_c \varphi \dot\wedge \psi}\,CI \qquad \frac{\Gamma \vdash_c \varphi \dot\wedge \psi}{\Gamma \vdash_c \varphi}\,CE_1 \qquad \frac{\Gamma \vdash_c \varphi \dot\wedge \psi}{\Gamma \vdash_c \psi}\,CE_2$$

$$\frac{\Gamma \vdash_c \varphi}{\Gamma \vdash_c \varphi \dot\vee \psi}\,DI_1 \qquad \frac{\Gamma \vdash_c \psi}{\Gamma \vdash_c \varphi \dot\vee \psi}\,DI_2 \qquad \frac{\Gamma \vdash_c \varphi \dot\vee \psi \quad \Gamma, \varphi \vdash_c \theta \quad \Gamma, \psi \vdash_c \theta}{\Gamma \vdash_c \theta}\,DE$$

$$\frac{\uparrow\Gamma \vdash_c \varphi}{\Gamma \vdash_c \dot\forall \varphi}\,AI \qquad \frac{\Gamma \vdash_c \dot\forall \varphi}{\Gamma \vdash_c \varphi[t]}\,AE \qquad \frac{\Gamma \vdash_c \varphi[t]}{\Gamma \vdash_c \dot\exists \varphi}\,EI \qquad \frac{\Gamma \vdash_c \dot\exists \varphi \quad \uparrow\Gamma, \varphi \vdash_c \uparrow\psi}{\Gamma \vdash_c \psi}\,EE$$

$$\frac{}{\Gamma \vdash_c ((\varphi \dot\to \psi) \dot\to \varphi) \dot\to \varphi}\,P$$

We write $\vdash_c \varphi$ whenever φ is classically provable from the empty context.

Definition 57. The intuitionistic sequent calculus LJT is defined as follows:

$$\frac{}{\Gamma ; \varphi \Rightarrow \varphi}\,A \qquad \frac{\Gamma ; \varphi \Rightarrow \psi \quad \varphi \in \Gamma}{\Gamma \Rightarrow \psi}\,C \qquad \frac{\Gamma \Rightarrow \varphi \quad \Gamma ; \psi \Rightarrow \theta}{\Gamma ; \varphi \dot\to \psi \Rightarrow \theta}\,IL$$

$$\frac{\Gamma, \varphi \Rightarrow \psi}{\Gamma \Rightarrow \varphi \dot\to \psi}\,IR \qquad \frac{\Gamma ; \varphi[t] \Rightarrow \psi}{\Gamma ; \dot\forall \varphi \Rightarrow \psi}\,AL \qquad \frac{\uparrow\Gamma \Rightarrow \varphi}{\Gamma \Rightarrow \dot\forall \varphi}\,AR \qquad \frac{\Gamma \Rightarrow \dot\bot}{\Gamma \Rightarrow \varphi}\,E$$

Definition 58. The intuitionistic sequent calculus LJ is defined as follows:

$$\frac{}{\Gamma, \varphi \Rightarrow_J \varphi} \; A \qquad \frac{\Gamma, \varphi, \varphi \Rightarrow_J \psi}{\Gamma, \varphi \Rightarrow_J \psi} \; C \qquad \frac{\Gamma \Rightarrow_J \psi}{\Gamma, \varphi \Rightarrow_J \psi} \; W$$

$$\frac{\Gamma, \psi, \varphi, \Gamma' \Rightarrow_J \theta}{\Gamma, \varphi, \psi, \Gamma' \Rightarrow_J \theta} \; P \qquad \frac{\Gamma \Rightarrow_J \bot}{\Gamma \Rightarrow_J \varphi} \; E \qquad \frac{\Gamma \Rightarrow_J \varphi \quad \Gamma, \psi \Rightarrow_J \theta}{\Gamma, \varphi \dot\to \psi \Rightarrow_J \theta} \; IL$$

$$\frac{\Gamma, \varphi \Rightarrow_J \psi}{\Gamma \Rightarrow_J \varphi \dot\to \psi} \; IR \qquad \frac{\Gamma, \varphi, \psi \Rightarrow_J \theta}{\Gamma, \varphi \dot\wedge \psi \Rightarrow_J \theta} \; CL \qquad \frac{\Gamma \Rightarrow_J \varphi \quad \Gamma \Rightarrow_J \psi}{\Gamma \Rightarrow_J \varphi \dot\wedge \psi} \; CR$$

$$\frac{\Gamma, \varphi \Rightarrow_J \theta \quad \Gamma, \psi \Rightarrow_J \theta}{\Gamma, \varphi \dot\vee \psi \Rightarrow_J \theta} \; DL \qquad \frac{\Gamma \Rightarrow_J \varphi}{\Gamma \Rightarrow_J \varphi \dot\vee \psi} \; DR_1 \qquad \frac{\Gamma \Rightarrow_J \psi}{\Gamma \Rightarrow_J \varphi \dot\vee \psi} \; DR_2$$

$$\frac{\Gamma, \varphi[t] \Rightarrow_J \psi}{\Gamma, \dot\forall \varphi \Rightarrow_J \psi} \; AL \qquad \frac{\uparrow\Gamma \Rightarrow_J \varphi}{\Gamma \Rightarrow_J \dot\forall \varphi} \; AR \qquad \frac{\uparrow\Gamma, \varphi \Rightarrow_J \uparrow\psi}{\Gamma, \dot\exists \varphi \Rightarrow_J \psi} \; EL \qquad \frac{\Gamma \Rightarrow_J \varphi[t]}{\Gamma \Rightarrow_J \dot\exists \varphi} \; ER$$

B   Notes on the Coq Formalisation

Our formalisation consists of about 7500 lines of code, with an even split between specification and proofs. The code is structured as follows.

Section	Specification	Proofs
Preliminaries Autosubst	169	53
Preliminaries for $\mathbb{F}^*$	680	599
Tarski Semantics	655	682
Kripke Semantics	342	255
On Markov's Principle	593	978
Preliminaries for $\mathbb{F}$	523	430
Heyting Semantics	297	456
Dialogue Semantics	312	488
Total	3571	3941

In general, we find that Coq provides the ideal grounds for formalising projects like ours. It has external libraries supporting the formalisation of syntax, enough automation to support the limited amounts we need and allows constructive reverse mathematics due to its axiomatic minimality.

In the remainder of the section, we elaborate on noteworthy design choices of the formalisation.

Formalisation of Binders. There are various competing techniques to formalise binders in proof assistants. In first-order logic, binders occur in quantification. The chosen technique especially affects the definition of deduction systems and can considerably ease or impede proofs of standard properties like weakening.

We opted for a de Bruijn representation of variables and binders with parallel substitutions. The Autosubst 2 tool [49] provides convenient automation for the definition of and proofs about this representation of syntax.

Notably, our representation then results in very straightforward proofs for weakening with only 5 lines. In contrast, using other representations for binders results in considerably more complicated weakening proofs, e.g. 150 lines in an approach using names [14] and 95 lines in an approach using traced syntax [23].

Also note that first-order logic has the simplest structure of binders possible: Since quantifiers range over terms, but terms do not contain binders, we do not need a prior notion of renaming, as usually standard in de Bruijn presentations of syntax. This observation results in more compact code (because usually, every statement on substitutions has to be proved for renamings first, with oftentimes the same proof) and was incorporated into Autosubst 2, which now does not generate renamings if they are not needed. Furthermore, we remark that the HOAS encoding of such simple binding structures results in a strictly positive inductive type and would thus be in principle definable in Coq.

Formalisation of Signatures. Our whole development is parametrised against a signature, defined as a typeclass in Coq:

```
Class Signature := B_S { Funcs : Type; fun_ar  : Funcs -> nat ;
Preds : Type; pred_ar : Preds -> nat }.
```

We implement term and predicate application using the dependent vector type. While the vector type is known to cause issues in dependent programming, in this instance it was the best choice. Recursion on terms is accepted by Coq's guardness checker, and while the generated induction principle (as is always the case for nested inductives) is too weak, a sufficient version can easily be implemented by hand:

```
Inductive vec_in (A : Type) (a : A) : forall n, vector A n -> Type :=
| vec_inB n (v : vector A n) : vec_in a (cons a v)
| vec_inS a' n (v :vector A n) : vec_in a v -> vec_in a (cons a' v).

Lemma strong_term_ind (p : term -> Type) :
(forall x, p (var_term x)) ->
(forall F v, (forall t, vec_in t v -> p t) -> p (Func F v)) ->
forall (t : term), p t.
```

Syntactic Fragments. There are essentially four ways to formalise the syntactic fragment $\mathbb{F}^*$. First, we could parametrise the type of formulas with tags, as done in [14] and second, we could use well-explored techniques for modular syntax [26,9]. However, both of these approaches would not be compatible with the

Autosubst tool. Additionally, modular syntax would force users of our developed library for first-order logic to work on the peculiar representation of syntax using containers or functors instead of regular inductive types.

The third option is to only define the type $\mathbb{F}$, and then define a predicate on this formulas characterising the fragment $\mathbb{F}^*$. This approach introduces many additional assumptions in almost all statements, decreasing their readability and yielding many simple but repetitive proof obligations. Furthermore, we would have to parameterise natural deduction over predicates as well, in order for the (IE) rule to not introduce terms e.g. containing $\exists$ when only deductions over $\mathbb{F}^*$ should be considered.

To make the formalisation as clear and reusable as possible, we chose the fourth and most simple possible approach: We essentially duplicate the contents of Sect. 2 for both $\mathbb{F}^*$ and $\mathbb{F}$, resulting in two independent developments on top of the two preliminary parts.

Parametrised Deduction Systems. When defining the minimal, intuitionistic, and classical versions of natural deduction, a similar issue arises. Here, we chose to use one single predicate definition, where the rules for explosion and Peirce can be enabled or disabled using tags, which are parameters of the predicate.

```
Inductive peirce := class | intu.
Inductive bottom := expl | lconst.
Inductive prv : forall (p : peirce) (b : bottom),
list (form) -> form -> Prop := (* ... *).
```

We can then define all considered variants of ND by fixing those parameters:

```
Notation "A ⊢CE phi" := (@prv class expl A phi) (at level 30).
Notation "A ⊢CL phi" := (@prv class lconst A phi) (at level 30).
Notation "A ⊢IE phi" := (@prv intu expl A phi) (at level 30).
```

This definition allows us to give for instance a general weakening proof, which can then be instantiated to the different versions. Similarly, we can give a parametrised soundness proof, and depending on the parameters fix required properties on the models used in the definition of validity.

Object Tactics. At several parts of our developments we have to build concrete ND derivations. This can always be done by explicitly applying the constructors of the ND predicate, which however becomes tedious quickly. We thus developed object tactics reminiscent of the tactics available in Coq. The tactic `ointros` for instance applies the (II) rule, whereas the tactic `oapply` can apply hypotheses, i.e. combine the rules (IE) and (C). All object tactics are in the file `FullND.v`.

Extraction to λ-Calculus. The proof that completeness of provability w.r.t. standard Tarski and Kripke semantics is equivalent to MP_L crucially relies on an L-enumeration of provable formulas. While giving a Coq enumeration is easy using techniques described in [14], the translation of any function to a model of computation is considered notoriously hard. We use the framework by Forster and Kunze [16] which allows the automated translation of Coq functions to L.

Using the framework was mostly easy and spared us considerable formalisation effort. However, the framework covers only simple types, whereas our representation of both terms and formulas contains the dependent vector type. We circumvent this problem by defining a non-dependent term type term' and a predicate wf characterising exactly the terms in correspondence with our original type of terms.

```
Inductive term' := var_term' : nat -> term' | Func' (name : nat)
| App' : term' -> term' -> term'.
```

```
Inductive varornot := isvar | novar.
Inductive wf : varornot -> term' -> Prop :=
| wf_var n : wf isvar (var_term' n)
| wf_fun f : wf novar (Func' f)
| wf_app v s t : wf v s -> wf novar t -> wf novar (App' s t).
```

We then define a formula type form' based on term' and a suitable deduction system. One can give a bijection between well-formed non-dependent terms term' and dependent terms term and prove the equivalence of the corresponding deduction systems under this bijection.

Functions working on term' and form' were easily extracted to L using the framework, yielding an L-enumerability proof for ND essentially with no manual formalisation effort.

Library of Formalised Undecidable Problems in Coq. We take the formalisation of synthetic undecidability from [14], which is part of the Coq library of formalised undecidable problems [12]. The reduction from L-halting to provability is factored via Turing machines, Minsky machines, binary stack machines and the Post correspondence problem (PCP), all part of the library as well.

Equations Package. Defining non-structurally recursive functions is sometimes considered hard in Coq and other proof assistants based on dependent type theory. One such example is the function [[_]] used to embed formulas into Heyting algebras (Definition 45). We use the Equations package [48] to define this function by recursion on the size of the formula, ignoring terms. The definition then becomes entirely straightforward and the provided simp tactic, while sometimes a bit premature, enables compact proofs.

References

1. Bauer, A.: First steps in synthetic computability theory. Electron. Notes Theor. Comput. Sci. **155**, 5–31 (2006). Proceedings of the 21st Annual Conference on Mathematical Foundations of Programming Semantics (MFPS XXI)
2. Berardi, S.: Intuitionistic completeness for first order classical logic. J. Symbolic Logic **64**(1), 304–312 (1999)

3. Berger, U., Schwichtenberg, H.: An inverse of the evaluation functional for typed lambda-calculus. In: 1991 Proceedings Sixth Annual IEEE Symposium on Logic in Computer Science, pp. 203–211. IEEE (1991)
4. Blanchette, J.C., Popescu, A., Traytel, D.: Unified classical logic completeness. In: Demri, S., Kapur, D., Weidenbach, C. (eds.) IJCAR 2014. LNCS (LNAI), vol. 8562, pp. 46–60. Springer, Cham (2014). https://doi.org/10.1007/978-3-319-08587-6_4
5. Braselmann, P., Koepke, P.: Gödel's completeness theorem. Formalized Math. **13**(1), 49–53 (2005)
6. Constable, R., Bickford, M.: Intuitionistic completeness of first-order logic. Ann. Pure Appl. Logic **165**(1), 164–198 (2014)
7. Coquand, T., Mannaa, B.: The independence of Markov's principle in type theory. Logical Methods Comput. Sci. **13**(3), 18605974 (2017). arXiv: 1602.04530
8. de Bruijn, N.G.: Lambda calculus notation with nameless dummies, a tool for automatic formula manipulation, with application to the Church-Rosser theorem. Indagationes Mathematicae (Proceedings) **75**(5), 381–392 (1972)
9. Delaware, B., d S Oliveira, B.C., Schrijvers, T.: Meta-theory à la carte. In: ACM SIGPLAN Notices, vol. 48, pp. 207–218. ACM (2013)
10. Dybjer, P., Filinski, A.: Normalization and partial evaluation. In: Barthe, G., Dybjer, P., Pinto, L., Saraiva, J. (eds.) APPSEM 2000. LNCS, vol. 2395, pp. 137–192. Springer, Heidelberg (2002). https://doi.org/10.1007/3-540-45699-6_4
11. Felscher, W.: Dialogues, strategies, and intuitionistic provability. Ann. Pure Appl. Logic **28**(3), 217–254 (1985)
12. Forster, Y., Larchey-Wendling, D., Dudenhefner, A., Heiter, E., Kirst, D., Kunze, F., Gert, S., Spies, S., Wehr, D., Wuttke, M.: A Coq Library of Undecidable Problems (2019). https://github.com/uds-psl/coq-library-undecidability
13. Forster, Y., Heiter, E., Smolka, G.: Verification of PCP-related computational reductions in Coq. In: Avigad, J., Mahboubi, A. (eds.) ITP 2018. LNCS, vol. 10895, pp. 253–269. Springer, Cham (2018). https://doi.org/10.1007/978-3-319-94821-8_15
14. Forster, Y., Kirst, D., Smolka, G.: On synthetic undecidability in Coq, with an application to the Entscheidungsproblem. In: International Conference on Certified Programs and Proofs, pp. 38–51. ACM (2019)
15. Forster, Y., Kunze, F.: Verified extraction from Coq to a Lambda-Calculus. In: Coq Workshop, vol. 2016 (2016)
16. Forster, Y., Kunze, F.: A certifying extraction with time bounds from Coq to call-by-value Lambda Calculus. In: Harrison, J., O'Leary, J., Tolmach, A. (eds.) 10th International Conference on Interactive Theorem Proving, volume 141 of Leibniz International Proceedings in Informatics (LIPIcs), Dagstuhl, Germany, pp. 17:1–17:19. Schloss Dagstuhl-Leibniz-Zentrum fuer Informatik (2019)
17. Forster, Y., Smolka, G.: Weak call-by-value lambda calculus as a model of computation in Coq. In: Ayala-Rincón, M., Muñoz, C.A. (eds.) ITP 2017. LNCS, vol. 10499, pp. 189–206. Springer, Cham (2017). https://doi.org/10.1007/978-3-319-66107-0_13
18. Gödel, K.: Die Vollständigkeit der Axiome des logischen Funktionenkalküls. Monatshefte für Mathematik und Physik **37**, 349–360 (1930)
19. Han, J., van Doorn, F.: A formalization of forcing and the consistency of the failure of the continuum hypothesis. In: International Conference on Interactive Theorem Proving. Springer, Heidelberg (2019). https://doi.org/10.4230/LIPIcs.ITP.2019.19
20. Hasenjaeger, G.: Eine Bemerkung zu Henkin's Beweis für die Vollständigkeit des Prädikatenkalküls der Ersten Stufe. J. Symbolic Logic **18**(1), 42–48 (1953)

21. Henkin, L.: The completeness of the first-order functional calculus. J. Symbolic Logic **14**(3), 159–166 (1949)
22. Herbelin, H., Ilik, D.: An analysis of the constructive content of Henkin's proof of Gödel's completeness theorem. Draft (2016)
23. Herbelin, H., Lee, G.: Forcing-based cut-elimination for gentzen-style intuitionistic sequent calculus. In: Ono, H., Kanazawa, M., de Queiroz, R. (eds.) WoLLIC 2009. LNCS (LNAI), vol. 5514, pp. 209–217. Springer, Heidelberg (2009). https://doi.org/10.1007/978-3-642-02261-6_17
24. Ilik, D.: Constructive completeness proofs and delimited control. Ph.D. thesis, Ecole Polytechnique X (2010)
25. Ishihara, H.: Reverse mathematics in Bishop's constructive mathematics. Philosophia Scientae **6**, 43–59 (2006)
26. Keuchel, S., Schrijvers, T.: Generic datatypes à la carte. In: ACM SIGPLAN Workshop on Generic Programming, pp. 13–24. ACM (2013)
27. Kirst, D., Smolka, G.: Categoricity results and large model constructions for second-order ZF in dependent type theory. J. Autom. Reasoning **63**, 415–438 (2018)
28. Kreisel, G.: On weak completeness of intuitionistic predicate logic. J. Symbolic Logic **27**(2), 139–158 (1962)
29. Kreisel, G., Troelstra, A.S.: Formal systems for some branches of intuitionistic analysis. Ann. Math. Logic **1**(3), 229–387 (1970)
30. Krivine, J.-L.: Une preuve formelle et intuitionniste du théorème de complétude de la logique classique. Bull. Symbolic Logic **2**(4), 405–421 (1996)
31. Krivtsov, V.N.: An intuitionistic completeness theorem for classical predicate logic. Studia Logica **96**(1), 109–115 (2010)
32. Krivtsov, V.N.: Semantical completeness of first-order predicate logic and the weak fan theorem. Studia Logica **103**(3), 623–638 (2015)
33. Leivant, D.: Failure of completeness properties of intuitionistic predicate logic for constructive models. Annales scientifiques de l'Université de Clermont. Mathématiques **60**(13), 93–107 (1976)
34. Lorenzen, P.: Logik und Agon. Atti del XII Congresso Internazionale di Filosofia **4**, 187–194 (1960)
35. Lorenzen, P.: Ein dialogisches Konstruktivitätskriterium. In: Proceedings of the Symposium on Foundations of Mathematics (Warsaw, 2–9 September 1959), pp. 193–200 (1961)
36. MacNeille, H.M.: Partially ordered sets. Trans. Am. Math. Soc. **42**(3), 416–460 (1937)
37. Mannaa, B., Coquand, T.: The independence of Markov's principle in type theory. Logical Methods Comput. Sci. **13**(3:10), 1–28 (2017)
38. McCarty, C.: Constructive validity is nonarithmetic. J. Symbolic Logic **53**, 1036–1041 (1988)
39. McCarty, C.: Completeness and incompleteness for intuitionistic logic. J. Symbolic Logic **73**(4), 1315–1327 (2008)
40. McCarty, D.C., et al.: Incompleteness in intuitionistic metamathematics. Notre Dame J. Formal Logic **32**(3), 323–358 (1991)
41. Pédrot, P.-M., Tabareau, N.: Failure is not an option. In: Ahmed, A. (ed.) ESOP 2018. LNCS, vol. 10801, pp. 245–271. Springer, Cham (2018). https://doi.org/10.1007/978-3-319-89884-1_9
42. Plotkin, G.D.: Call-by-name, call-by-value and the lambda-calculus. Theor. Comput. Sci. **1**(2), 125–159 (1975)

43. Ridge, T., Margetson, J.: A mechanically verified, sound and complete theorem prover for first order logic. In: Hurd, J., Melham, T. (eds.) TPHOLs 2005. LNCS, vol. 3603, pp. 294–309. Springer, Heidelberg (2005). https://doi.org/10.1007/11541868_19
44. Schlichtkrull, A.: Formalization of the resolution calculus for first-order logic. J. Autom. Reason. **61**(1–4), 455–484 (2018)
45. Schumm, G.F.: A Henkin-style completeness proof for the pure implicational calculus. Notre Dame J. Formal Logic **16**(3), 402–404 (1975)
46. Scott, D.: The algebraic interpretation of quantifiers: intuitionistic and classical. In: Ehrenfeucht, V.M.A., Srebrny, M. (eds.) Andrzej Mostowski and Foundational Studies. IOS Press (2008)
47. Sørensen, M.H., Urzyczyn, P.: Sequent calculus, dialogues, and cut elimination. Reflections Type Theor. λ-Calculus Mind 253–261 (2007). http://www.cs.ru.nl/barendregt60/essays/
48. Sozeau, M., Mangin, C.: Equations reloaded: high-level dependently-typed functional programming and proving in Coq. Proc. ACM Program. Lang. **3**(ICFP), 86 (2019)
49. Stark, K., Schäfer, S., Kaiser, J.: Autosubst 2: reasoning with multi-sorted de Bruijn terms and vector substitutions. In: International Conference on Certified Programs and Proofs, pp. 166–180. ACM (2019)
50. The Coq Proof Assistant (2019). http://coq.inria.fr
51. Timany, A., Sozeau, M.: Cumulative inductive types in Coq. In: Kirchner, H. (ed.) International Conference on Formal Structures for Computation and Deduction, volume 108 of Leibniz International Proceedings in Informatics (LIPIcs), Dagstuhl, Germany,, pp. 29:1–29:16. Schloss Dagstuhl-Leibniz-Zentrum fuer Informatik (2018)
52. Underwood, J.: Aspects of the computational content of proofs. Technical report, Cornell University (1994)
53. Veldman, W.: An intuitiomstic completeness theorem for intuitionistic predicate logic 1. J. Symbolic Logic **41**(1), 159–166 (1976)
54. Wehr, D.: A constructive analysis of first-order completeness theorems in Coq. Bachelor's thesis, Saarland University (2019)
55. Wuttke, M.: Verified programming of turing machines in Coq. Bachelor's thesis, Saarland University (2018)

On the Constructive Truth and Falsity in Peano Arithmetic

Hirohiko Kushida$^{(\boxtimes)}$

Computer Science, Graduate Center, City University of New York, New York, USA
hkushida@gc.cuny.edu

Abstract. Recently, Artemov [4] offered the notion of constructive truth and falsity in the spirit of Brouwer-Heyting-Kolmogorov semantics and its formalization, the Logic of Proofs. In this paper, we provide a complete description of constructive truth and falsity for Friedman's constant fragment of Peano Arithmetic. For this purpose, we generalize the constructive falsity to n-constructive falsity where n is any positive natural number. We also establish similar classification results for constructive truth and n-constructive falsity of Friedman's formulas. Then, we discuss 'extremely' independent sentences in the sense that they are classically true but neither constructively true nor n-constructive false for any n.

Keywords: Peano arithmetic · Modal logic · Foundations of mathematics

1 Introduction

In the second incompleteness theorem, Gödel proved the impossibility to prove an arithmetical sentence, $Con(PA) = \forall x \neg Proof(x, 0 = 1)$, which is interpreted to mean a formalization of consistency of Peano Arithmetic, PA: *For all x, x is not a code of a proof of $0 = 1$.* The formalization is concerned with arithmetization of the universal quantifier in the statement and the arithmetization cannot rule out the interpretability of the quantifier to range over both standard and nonstandard numbers. In a recent paper [4], Artemov pointed out that it is too strong to capture fairly Hilbert's program on finitary consistency proof for arithmetic; it asked for a finitary proof that in a formal arithmetic *no finite sequence of formulas is a derivation of a contradiction.* Then, Artemov considered the notion of constructive consistency, $CCon(PA)$, and demonstrated that it is actually provable in PA

Moreover, the generalization of constructive consistency was offered in [4] in the spirit of Brouwer-Heyting-Kolmogorov (BHK) semantics and its formalization, the Logic of Proofs (LP): constructive falsity with its counterpart, the constructive truth. (On the family of systems called Justification Logics including the Logic of Proofs, we can refer to [2,3,6,7,14,15]). Constructive truth is taken to be provability in a finitary system, PA, along with the motto of constructivism: truth is provability.

© Springer Nature Switzerland AG 2020
S. Artemov and A. Nerode (Eds.): LFCS 2020, LNCS 11972, pp. 75–84, 2020.
https://doi.org/10.1007/978-3-030-36755-8_5

Definition 1. *An arithmetical sentence A is* constructively false *if* PA *proves: for any x, there is a proof that 'x is not a proof of A'.*

This is also viewed as the result of a refinement of the interpretation of negation and implication in the BHK semantics by the framework of the Logic of Proofs, which is compliant with the Kreisel 'second clause' criticism. (Cf. [7])

On the other hand, the letterless fragment of the logic of provability GL has been an object of modal logical study of Peano Arithmetic, PA, since Friedman's 35th problem in [13]. A letterless sentence is one built up from a constant for falsity $\bot$, boolean connectives, and the modality $\Box$. Boolos [9] showed that there is a specific normal form for these sentences and the fragment is decidable, which was an answer to the Friedman's question.

Following Boolos [11], we call the counterpart of letterless sentences in PA *constant* sentences. Formally, they are built from the sentence $0 = 1$, a suitable provability predicate $Prov_{PA}(*)$ and boolean connectives. Any arithmetical interpretations convert a letterless sentence to the same constant sentence in PA. Here, for the sake of simplicity, we write $\bot$ to mean $0 = 1$ and $\Box(*)$ to mean a fixed provability predicate of PA.

In this paper, we are primarily concerned with the constant sentences in PA; in Sect. 2, we provide a complete delineation of the constant sentences in terms of the notions of constructive truth and falsity. Then, it turns out natural to generalize constructive falsity to n-constructive falsity, where n is any positive natural number. Also, for each n, we provide classification results for constructive truth and n-constructive falsity for constant sentences.

The 'constructive' liar sentence was introduced and discussed in [4] along with the Rosser sentence. In Sect. 3, we generalize both of these two kinds of arithmetical sentences, and specify the logical status of them on the basis of generalized constructive falsity. Also, we clarify which constant sentences can be the generalized Rosser sentences.

In Sect. 4, we offer the notion of 'extreme' independence from PA for arithmetical sentences A: both they and their negation are neither provable in PA nor belong to n-constructive falsity for any n. We show that there is an extremely independent arithmetical sentence but no constant sentence is extremely independent.

2   The Constant Sentences of Peano Arithmetic

In [4], Artemov clarified the status of some constant sentences on classical and constructive truth and falsity: $Con(PA)$ is classically true and constructively false. $0 = 1$ is classically false and constructively false. $\neg Con(PA)$ is classically false and neither constructively true nor constructively false. Then, it is natural to ask a general question: under which condition a constant sentence is said to be constructively true or constructively false.

First of all, we generalize the notion of constructive falsity to n-constructive falsity $(n \geq 1)$. Put $cf^n(F) = \forall x \Box^n \neg (x : F)$ for each $n \geq 1$, where $\Box^n = \overbrace{\Box \cdots \Box}^{n}$ and $(* : *)$ is a fixed proof predicate.

Definition 2. *An arithmetical sentence A is* n-constructive false *if and only if* PA *proves the sentence* $cf^n(A)$.

The original constructive falsehood is the special case with $n = 1$.

Theorem 1. *(Normal Form Theorem)* $\vdash_{PA} cf^n(F) \leftrightarrow .\Box F \to \Box^n \bot$.

Proof. Work in PA. Suppose $\Box F$, that is, $\exists x(x : F)$ holds. Then, for some y, we have $y : F$. By applying Σ_1-completeness n times, we obtain $\Box^n(y : F)$.[1] On the other hand, suppose $\forall x \Box^n \neg (x : F)$. Then, $\Box^n \neg (y : F)$ holds. Hence, we obtain $\Box^n \bot$. Thus, $\forall x \Box^n \neg (x : F) \to .\Box F \to \Box^n \bot$. For the other direction, obviously, $\Box^n \bot \to \Box^n \neg (x : F)$. By generalization, $\Box^n \bot \to \forall x \Box^n \neg (x : F)$. On the other hand, by applying Σ_1-completeness n times, for any x, $\neg(x : F) \to \Box^n \neg (x : F)$. By predicate calculus, $\neg \exists x(x : F) \to \forall x \Box^n \neg (x : F)$, that is, $\neg \Box F \to \forall x \Box^n \neg (x : F)$. Therefore, $\neg \Box F \vee \Box^n \bot. \to \forall x \Box^n \neg (x : F)$. ∎

Here we observe some simple facts.

(F1) If A is n-constructively false and PA proves $B \to A$, B is also n-constructively false.

(F2) If A is n-constructively false and $n \leq m$, A is m-constructively false.

(F3) If PA is n-consistent, that is, PA does not prove $\Box^n \bot$, then no n-constructively false sentence is constructively true.

We say that a sentence is n-constructively false *at the smallest* if and only if it is n-constructively false but not m-constructively false sentence for any $m < n$.

We introduce the following three types of arithmetical sentences.

(α)-sentences: of the form $\Box^n \bot \to \Box^m \bot$ $(0 \leq n \leq m)$
(β, n)-sentences: of the form $\Box^m \bot \to \Box^{n-1} \bot$ $(1 \leq n \leq m)$
(γ, n)-sentences: of the form: $\Box^{n-1} \bot$ $(1 \leq n)$

Lemma 1. *(1)* (β, n)- *and* (γ, n)-sentences are n-constructively false at the smallest.
(2) (α)-sentences are constructively true.

Proof. (2) is immediate. For (β, n)-sentences, consider the formula $\Box(\Box^m \bot \to \Box^{n-1} \bot) \to \Box^k \bot$ with $0 \leq n \leq m$. This is provably equivalent in PA to $\Box^n \bot \to \Box^k \bot$. Therefore, PA proves it if and only if $k \geq n$, in terms of Gödelean incompleteness theorems. The proof is similar for (γ, n)-sentences. ∎

By $(\beta\gamma, n)$-sentence we mean a conjunction of (β, a)- and (γ, b)-sentences such that n is the minimum of all such a's and b's. In particular, when it consists only of (β, a)-sentences, it is called a (β^+, n) sentence.

---

[1] A detailed proof of the provable Σ_1-completeness is found on pp. 46–49 of [11].

Lemma 2. *$(\beta\gamma, n)$-sentences are n-constructively false at the smallest.*

Proof. Temporarily, let (β, n_i) and (γ, m_i) denote a (β, n_i)- and a (γ, m_i)-sentence, respectively. Consider the following sentence.

$$(*) \qquad \Box(\bigwedge_i(\beta, n_i) \wedge \bigwedge_j(\gamma, m_j)) \to \Box^k \bot$$

where $n = min_{i,j}(n_i, m_j)$. By using derivability conditions on the provability predicate $\Box$, this is provably equivalent in PA to the following.

$$\bigwedge_i \Box(\beta, n_i) \wedge \bigwedge_j \Box(\gamma, m_j) \to \Box^k \bot.$$

Furthermore, we can execute the following transformations, keeping equivalence in PA.

$$\bigwedge_i \Box^{n_i} \bot \wedge \bigwedge_j \Box^{m_j} \bot \to \Box^k \bot;$$
$$\Box^n \bot \to \Box^k \bot.$$

Thus, in terms of Gödelean incompleteness theorems, $(*)$ is provable in PA if and only if $k \geq n$. ∎

Lemma 3. *Any constant sentence is provably in PA equivalent to an (α)-sentence or a $(\beta\gamma, n)$-sentence for some $n \geq 1$.*

Proof. Boolos' normal form theorem for constant sentences in [11] states that any constant sentence is equivalent in PA to a boolean combination of $\Box^n \bot$. By propositional transformation, it is further equivalent to a conjunction of sentences of the form of $(\alpha), (\beta, n)$ and (γ, m). If it contains only conjuncts which are (α)-sentences, it is equivalent to an (α)-sentence. Suppose that it is of the form $X \wedge Y$ where X contains no (α)-sentence and Y contains only (α)-sentences. As $X \wedge Y$ is equivalent in PA to X, it is a $(\beta\gamma, n)$ sentence with some n. ∎

Theorem 2. *Any constant sentence is provably in PA equivalent to a constructively true sentence or an n-consistently false sentence for some n.*

Proof. Derived by Lemmas 2, 3. ∎

Theorem 3. *Let A be any constant sentence and n be any positive natural number. Suppose that PA is n-consistent. Then, we have the following.*

(1) A is n-constructively false and classically true, if and only if, A is provably in PA equivalent to a (β^+, m)-sentence for some $m \leq n$.
(2) A is n-constructively false and classically false, if and only if, A is provably in PA equivalent to a (γ, m)-sentence for some $m \leq n$.
(3) A is constructively true, if and only if, A is provably in PA equivalent to an (α)-sentence.

Proof. The 'if' directions in (1–3) are immediate by Lemma 2. For the 'only if' direction. (3) is obvious. We prove (1, 2). Suppose that A is m-constructively false at the smallest for some $m \leq n$. By (F3), A is not constructively true and so, is not an (α)-sentence. Since A is constant, by Lemma 3, A is equivalent to a $(\beta\gamma, a)$-sentence for some $a \geq 1$. By Lemma 2, $a = m$.

Now, if it is classically true, A is equivalent to (β^+, m)-sentence; if it is classically false, A is equivalent to a conjunction of (γ, m_i)-sentences where $min_i(m_i) = m$, which is equivalent to a (γ, m)-sentence, that is, $\Box^{m-1}\bot$.   ∎

3   Generalized 'Constructive' Liar Sentences and Rosser Sentences

In [4], Artemov offered a constructive version, L, of 'Liar Sentence' by applying the diagonal lemma:

$$\vdash_{\mathsf{PA}} L \leftrightarrow \forall x \Box \neg (x : L)$$
$$\leftrightarrow (\Box L \to \Box \bot)$$

And he pointed out that L is classically true but neither constructively true nor constructively false. We show that L is 2-constructively false and $\neg L$ is (1-)constructively false.

We shall introduce a general version of 'Constructive Liar Sentence'. For each $n \geq 1$, L_n is provided by the following.

$$\vdash_{\mathsf{PA}} L_n \leftrightarrow \forall x \Box^n \neg (x : L_n)$$
$$\leftrightarrow (\Box L_n \to \Box^n \bot)$$

The existence of L_n, we call *n-constructive liar*, is guaranteed by the diagonal lemma.

Theorem 4. *(1) L_n is classically true and $(n+1)$-constructively false at the smallest.*
(2) $\neg L_k$ is classically false and 1-constructively false. ($k \geq 1$)

Proof. For (1). Suppose that L_n is not true. Then, $\Box L_n$ is not true and $\Box L_n \to \Box^n \bot$ is true. This means L_n is true by definition of L_n. Hence, a contradiction.

Next, again by definition, PA proves $\Box[L_n \to (\Box L_n \to \Box^n \bot)]$ and so $\Box L_n \to \Box\Box^n \bot$. This means L_n is $(n+1)$-constructively false. To show that $(n+1)$ is the smallest, suppose that PA proves $\Box L_n \to \Box^n \bot$, that is, $\Box(\Box L_n \to \Box^n \bot) \to \Box^n \bot$. Then, PA also proves $\Box\Box^n \bot \to \Box^n \bot$, which is impossible in terms of Gödelean incompleteness theorems.

The proof for (2) is similar.   ∎

How about Gödelean Liar Sentence? It is considered to be $Con(PA)$, that is, $\neg\Box\bot$. We can generalize this as follows: n-*Gödelean Liar Sentence*, or n-*Liar Sentence* is defined to be $Con(PA^n)$, which is well known to be equivalent to $\neg\Box^n\bot$.[2] About this, we already know its status from the result of the previous section. $Con(PA^n)$ is a $(\beta,1)$-sentence and, by Lemma 1, it is 1-constructively false at the smallest. As to $\neg Con(PA^n)$, it is equivalent to $\Box^n\bot$, which is a $(\gamma, n+1)$-sentence and, by Lemma 1, $(n+1)$-constructively false at the smallest.

In [4], Artemov pointed out that the Rosser sentence, R, is classically true and constructively false; $\neg R$ is classically false and constructively false. Therefore, the result of Rosser's incompleteness theorem is said to have been the discovery of such a sentence which is 1-constructively false and the negation of which is also 1-constructively false.

Here again, we can make a generalization: an arithmetical sentence R_n is an n-*Rosser sentence* if both R_n and $\neg R_n$ are n-constructively false at the smallest $(n \geq 1)$. This condition is equivalent to the following: PA proves

$$\neg\Box^k\bot \to (\neg\Box R_n \wedge \neg\Box\neg R_n)$$

for any $k \geq n$ and does not for any $k < n$. The original Rosser sentence R is an instance of 1-Rosser sentence R_1. It is well-konwn that such an R_n can be constructed in PA.

Now, we can naturally ask: is it possible to construct constant n-Rosser sentences?

Lemma 4. *Let A be any constant sentence containing the provability predicate $\Box$. If A is n-constructively false, $\neg A$ is 1-constructively false.*

Proof. If A is classically true, by Theorem 3, $\neg A$ is equivalent to the form: $\bigvee_i(\Box^{k_i}\bot \wedge \neg\Box^{a_i}\bot)$ where for each i, $a_i < n$ and $a_i < k_i$. Note that in PA, $\bigvee_i(\Box^{k_i}\bot \wedge \neg\Box^{a_i}\bot)$ implies $\neg\Box^{min_i(a_i)}\bot$. We have a derivation in PA:

$$\begin{aligned}
\Box(\bigvee_i(\Box^{k_i}\bot \wedge \neg\Box^{a_i}\bot)) &\to \Box(\neg\Box^{min_i(a_i)}\bot) \\
&\to \Box(\Box^{min_i(a_i)}\bot \to \bot) \\
&\to \Box(\Box\bot \to \bot) \\
&\to \Box\bot
\end{aligned}$$

If A is classically false, by Theorem 3, $\neg A$ is equivalent to the form: $\neg\Box^a\bot$ with $a < n$. By the hypothesis, $a \neq 0$. We have a derivation in PA:

$$\begin{aligned}
\Box\neg\Box^a\bot &\to \Box(\Box^a\bot \to \bot) \\
&\to \Box(\Box\bot \to \bot) \\
&\to \Box\bot
\end{aligned}$$

Thus, in any case, $\neg A$ is 1-constructively false. ∎

Theorem 5. *Let A be any constant sentence containing the provability predicate $\Box$. Then, the following are equivalent.*

---

[2] As usual, PA^n is defined: $PA^1 = PA$; $PA^{n+1} = PA + Con(PA^n)$.

(1) A is an n-Rosser sentence for some n;
(2) A is a 1-Rosser sentence;
(3) A is 1-constructively false.

Proof. Proofs from (2) to (1) and from (2) to (3) are immediate.

From (1) to (2): If (1) holds, both A and $\neg A$ are both n-constructively false and, by Lemma 4, $n = 1$.

From (3) to (2): If (3) holds, by Lemma 4, $\neg A$ is 1-constructively false. Then, (2) holds. ∎

By Theorem 5, constant sentences can be n-Rosser sentences only when $n = 1$. Of course, we can weaken the definition of n-Rosser sentences: R_n is a *weak n-Rosser sentence* if and only if both R_n and $\neg R_n$ are n-constructively false (not necessarily at the smallest).

Corollary 1. *Any constant sentence containing the provability predicate $\square$ is a weak n-Rosser sentence for some n, unless it is constructively true.*

Proof. For any constant sentence A containing $\square$, if A is not constructively true, by Theorem 2, A is n-constructively false for some $n \geq 1$. By Lemma 4, $\neg A$ is 1-constructively, therefore, n-constructively false. ∎

Also, we obtain a relationship between n-constructive liar sentences and n-Rosser sentences.

Corollary 2. *(1) No one of constructive liar sentences and the negation of them is an n-Rosser sentence for any n.*
(2) For any $n \geq 1$, any n-constructive liar sentence L_n is a weak $(n+1)$-Rosser sentence.

Proof. Derived by Theorem 4. ∎

Here is a table to sum up some of the results from Sects. 2, 3.

	Classically True	Classically False
⋮	⋮	⋮
n-const. false	$L_{n-1}, \quad R_n$ $\square^m \bot \to \square^{n-1} \bot \ (m \geq n)$	$\square^{n-1} \bot$
⋮	⋮	⋮
3-const. false	$L_2, \quad R_3$ $\square^m \bot \to \square^2 \bot \ (m \geq 3)$	$\dot\square^2 \bot$
2-const. false	$L_1, \quad R_2$ $\square^m \bot \to \square^1 \bot \ (m \geq 2)$	$\square \bot$
1-const. false	R_1 $\neg \square^m \bot \ (m \geq 1)$	$\neg L_i, \ \neg R_i \ (1 \leq i)$ $\square^m \bot \wedge \neg \square^{n-1} \bot \ (m \geq n \geq 2)$ $\bot$

4 'Extremely' Independent Sentences

We showed that any constant sentence is n-constructively false for some n, unless it is constructively true (Theorem 2). This implies that well-known constant Gödelean sentences such as $Con(PA^n)$ and $\neg Con(PA^n)$ are m-constructively false for some m.

How about the 'Reflection Principles'? For any sentence A, let $Ref(A)$ denote $\Box A \to A$ (what we call the local Reflection Principle for A). We claim the following.

Theorem 6. *For any sentence A, $\neg Ref(A)$ is 2-constructively false.*

Proof. In PA, we have the following derivation.

$$\Box(\Box A \wedge \neg A) \to \Box\Box A \wedge \Box\neg A$$
$$\to \Box\Box A \wedge \Box\Box\neg A$$
$$\to \Box\Box\bot$$

This finishes the proof.   ∎

We note that the above argument does not generally hold for what we call the uniform Reflection Principle.

In addition, it is known that $Ref(A^{\Pi_1})$ for Π_1-sentences A^{Π_1} is provably equivalent to $Con(PA)$ in PA (cf. [17]). Hence, $Ref(A^{\Pi_1})$ is 1-constructively false and $\neg Ref(A^{\Pi_1})$ is 2-constructive false.

These results raise the question of the status of independence of a kind of Gödelean sentences (such as $Con(PA)$, $Ref(A)$, and other constant ones) from PA. So, we can naturally ask if there is a 'truly' independent arithmetical sentence from PA or not. We consider stronger notions of independence.

Definition 3. *1. An arithmetical sentence A is* strongly independent *from PA if and only if A is neither constructively true nor n-constructively false for any n.*

2. An arithmetical sentence A is extremely independent *from PA if and only if both A and $\neg A$ are strongly independent from PA.*

Note that if a sentence A is extremely independent, so is $\neg A$.

Theorem 7. *No instance of the local Reflection Principle is extremely independent from PA.*

Proof. Derived by Theorem 6.   ∎

Theorem 8. *No arithmetical constant sentence is strongly nor extremely independent from* PA.

Proof. Derived by Theorem 2. ∎

In [4], Artemov showed that there is an arithmetical sentence A such that both A and $\neg A$ are not 1-constructively false by using the uniform arithmetical completeness for the modal logic GL. We extend this result to our general setting.

Proposition 1. *(Uniform Arithmetical Completeness for* GL*) There is an arithmetical interpretation $*$ such that for any formula A of modal logic, $\vdash_{GL} A$ iff $\vdash_{PA} A^*$.*

This was established independently in [1, 8, 10, 16, 19].

Theorem 9. *There is an extremely independent sentence.*

Proof. Fix a propositional variable p. It is easily seen that for any positive natural number n, $\nvdash_{GL} \Box p \to \Box^n \bot$ and $\nvdash_{GL} \Box \neg p \to \Box^n \bot$. (This can be proved by an argument of Kripke completemess or the arithmetical completeness for GL.) Therefore, by the above proposition, there is an arithmetical sentence F such that for any positive natural number n, $\nvdash_{PA} \Box F \to \Box^n \bot$ and $\nvdash_{PA} \Box \neg F \to \Box^n \bot$. This sentence F is extremely independent from PA. ∎

Corollary 3. *There is an instance of the local Reflection Principle which is strongly independent from* PA.

Proof. In the proof of Theorem 9, we obtain the sentence F such that for any positive natural number n, $\nvdash_{PA} \Box F \to \Box^n \bot$. This sentence is equivalent to $\Box(\Box F \to F) \to \Box^n \bot$. Therefore, $Ref(F) = \Box F \to F$ is the desired instance. ∎

Theorem 8 could signify a limitation of the expressibility of arithmetical constant sentences, as contrasted with Theorem 9.

5 Concluding Remark

In this paper, we reported some results on the notion of constructive truth and falsity in PA, which was recently introduced in Artemov [4]. In particular, we showed some theorems on the relationship of those notions and the 'constant' fragment of PA, which has been actively studied since Friedman [13].

As is easily observed, an arithmetical sentence is n-constructively false if and only if its unprovability in PA is provable in PA extended with $Con(\mathsf{PA}^n)$. As an extension of the work of this paper, a natural research problem would be to examine whether or not things change in an essential way, if we are permitted to talk about extensions of PA in the well-known 'transfinite progression' since Turing [18].

References

1. Artemov, S.: Extensions of arithmetic and modal logics (in Russian). Ph.D. Thesis, Moscow State University, Steklov Mathematical Institute (1979)
2. Artemov, S.: Operational modal logic. Technical report MSI 95–29, Cornell University (1995)
3. Artemov, S.: Explicit provability and constructive semantics. Bull. Symb. Log. **7**(1), 1–36 (2001)
4. Artemov, S.: The Provability of Consistency. arXiv preprint arXiv:1902.07404 (2019)
5. Artemov, S., Beklemishev, L.: Provability logic. In: Gabbay, D., Guenthner, F. (eds.) Handbook of Philosophical Logic. Handbook of Philosophical Logic, vol. 13, 2nd edn, pp. 189–360. Springer, Dordrecht (2005). https://doi.org/10.1007/1-4020-3521-7_3
6. Artemov, S., Fitting, M.: Justification Logic. The Stanford Encyclopedia of Philosophy (2012)
7. Artemov, S., Fitting, M.: Justification Logic: Reasoning with Reasons. Cambridge University Press, Cambridge (2019)
8. Avron, A.: On modal systems having arithmetical interpretations. J. Symb. Log. **49**, 935–942 (1984)
9. Boolos, G.: On deciding the truth of certain statements involving the notion of consistency. J. Symb. Log. **41**, 779–781 (1976)
10. Boolos, G.: Extremely undecidable sentences. J. Symb. Log. **47**(1), 191–196 (1982)
11. Boolos, G.: The Logic of Provability. Cambridge University Press, Cambridge (1993)
12. Feferman, S.: Arithmetization of metamathematics in a general setting. Fundamenta mathematicae **49**(1), 35–92 (1960)
13. Friedman, H.: One hundred and two problems in mathematical logic. J. Symb. Log. **40**, 113–129 (1975)
14. Kuznets, R., Studer, T.: Weak arithmetical interpretations for the logic of proofs. Log. J. IGPL **24**(3), 424–440 (2016)
15. Kuznets, R., Studer, T.: Logics of Proofs and Justifications. College Publications (2019)
16. Montagna, F.: On the diagonalizable algebra of Peano arithmetic. Bollettino della Unione Math. Italiana **16**(5), 795–812 (1979)
17. Smorýnski, C.: Self-Reference and Modal Logic. Springer, Heidelberg (1985). https://doi.org/10.1007/978-1-4613-8601-8
18. Turing, A.M.: Systems of logic based on ordinals. Proc. London Math. Soc. **2**(1), 161–228 (1939)
19. Visser, A.: Aspects of diagonalization and provability. Ph.D. dissertation, Drukkerij Elinkwijk (1981)

Belief Expansion in Subset Models

Eveline Lehmann$^{(\boxtimes)}$ and Thomas Studer

Institute of Computer Science, University of Bern, Bern, Switzerland
{eveline.lehmann,thomas.studer}@inf.unibe.ch

Abstract. Subset models provide a new semantics for justifcation logic. The main idea of subset models is that evidence terms are interpreted as sets of possible worlds. A term then justifies a formula if that formula is true in each world of the interpretation of the term.

In this paper, we introduce a belief expansion operator for subset models. We study the main properties of the resulting logic as well as the differences to a previous (symbolic) approach to belief expansion in justification logic.

Keywords: Justification logic · Subset models · Belief expansion

1 Introduction

Justification logic is a variant of modal logic where the $\Box$-modality is replaced with a family of so-called evidence terms, i.e. instead of formulas $\Box F$, justification logic features formulas of the form $t : F$ meaning F *is known for reason t* [7,8,19].

The first justification logic, the Logic of Proofs, has been developed by Artemov [1,2] in order to provide intuitionistic logic with a classical provability semantics. Thus evidence terms represent proofs in a formal system like Peano arithmetic. By *proof* we mean a Hilbert-style proof, that is a sequence of formulas

$$F_1, \ldots, F_n \tag{1}$$

where each formula is either an axiom or follows by a rule application from formulas that occur earlier in the sequence. A justification formula $t : F$ holds in this arithmetical semantics if F occurs in the proof represented by t. Observe that F need not be the last formula in the sequence (1), but can be any formula F_i in it, i.e. we think of proofs as multi-conclusion proofs [2,18].

After the Logic of Proofs has been introduced, it was observed that terms can not only represent mathematical proofs but evidence in general. Using this interpretation, justification logic provides a versatile framework for epistemic logic [3,4,9,11,13,15]. In order to obtain a semantics of evidence terms that fits this general reading, one has to ignore the order of the sequence (1). That is evidence terms are interpreted simply as sets of formulas.

This is anticipated in both Mkrtychev models [22] as well as Fitting models [14]. The former are used to obtain a decision procedure for justification logic

© Springer Nature Switzerland AG 2020
S. Artemov and A. Nerode (Eds.): LFCS 2020, LNCS 11972, pp. 85–97, 2020.
https://doi.org/10.1007/978-3-030-36755-8_6

where one of the main steps is to keep track of which (set of) formulas a term justifies, see, e.g., [19,24]. The latter provide first epistemic models for justification logic where each possible world is equipped with an evidence function that specifies which terms serve as possible evidence for which (set of) formulas in that world.

Artemov [5] conceptually addresses the problem of the logical type of justifications. He claims that in the logical setting, justifications are naturally interpreted as sets of formulas. He introduces so-called modular models, which are based on the basic interpretation of justifications as sets of propositions and the convenience assumption of

$$\text{justification yields belief.} \tag{JYB}$$

That means if a term justifies a formula (i.e., the formula belongs to the interpretation of the term), then that formula is believed (i.e., true in all accessible possible worlds) [16]. Note that (JYB) has been dropped again in more recent versions of modular models [8].

So let us consider models for justification logic that interpret terms as sets of formulas. A belief change operator on such a model will operate by changing those sets of formulas (or introducing new sets, etc.). Dynamic epistemic justification logics have been studied, e.g., in [12,13,17,23]. Kuznets and Studer [17], in particular, introduce a justification logic with an operation for belief expansion. Their system satisfies a Ramsey principle as well as minimal change. In fact, their system meets all AGM postulates for belief expansion.

In their model, the belief expansion operation is monotone: belief sets can only get larger, i.e.,

$$\text{belief expansion always only adds new beliefs.} \tag{2}$$

This is fine for first-order beliefs. Indeed, one of the AGM postulates for expansion requires that beliefs are persistent. However, as we will argue later, this behavior is problematic for higher-order beliefs.

In this paper, we present an alternative approach that behaves better with respect to higher-order beliefs. It uses subset models for justification logics. This is a recently introduced semantics [20,21] that interprets terms not as sets of formulas but as sets of possible worlds. There, a formula $t{:}A$ is true if the interpretation of t is a subset of the truth-set of A, i.e., A is true in each world of the interpretation of t. Intuitively, we can read $t{:}A$ as A is believed and t justifies this belief. Subset models lead to new operations on terms (like intersection). Moreover, they provide a natural framework for probabilistic evidence (since the interpretation of a term is a set of possible worlds, we can easily measure it). Hence they support aggregation of probabilistic evidence [6,20]. They also naturally contain non-normal worlds and support paraconsistent reasoning.

It is the aim of this paper to equip subset models with an operation for belief expansion similar to [17]. The main idea is to introduce justification terms $\mathsf{up}(A)$ such that after a belief expansion with A, we have that A is believed and $\mathsf{up}(A)$ (representing the expansion operation on the level of terms) justifies this belief.

Semantically, the expansion A is dealt with by intersecting the interpretation of $\mathsf{up}(A)$ with the truth-set of A. This provides a better approach to belief expansion than [17] as (2) will hold for first-order beliefs but it will fail in general.

The paper is organized as follows. In the next section we introduce the language and a deductive system for JUS, a justification logic with belief expansion and subset models. Then we present its semantics and establish soundness of JUS. Section 5 is concerned with persistence properties of first-order and higher-order beliefs. Further we prove a Ramsey property for JUS. Finally, we conclude the paper and mention some further work.

2   Syntax

Given a set of countably many constants c_i, countably many variables x_i, and countably many atomic propositions P_i, terms and formulas of the language of JUS are defined as follows:

- Evidence terms
  - Every constant c_i and every variable x_i is an atomic term. If A is a formula, then $\mathsf{up}(A)$ is an atomic term. Every atomic term is a term.
  - If s and t are terms and A is a formula, then $s \cdot_A t$ is a term.
- Formulas
  - Every atomic proposition P_i is a formula.
  - If A, B, C are formulas, and t is a term, then $\neg A$, $A \to B$, $t : A$ and $[C]A$ are formulas.

The annotation of the application operator may seem a bit odd at first. However, it is often used in dynamic epistemic justification logics, see, e.g. [17, 23].

The set of atomic terms is denoted by ATm, the set of all terms is denoted by Tm. The set of atomic propositions is denoted by Prop and the set of all formulas is denoted by $\mathsf{L_{JUS}}$. We define the remaining classical connectives, $\bot$, $\wedge$, $\vee$, and $\leftrightarrow$, as usual making use of the law of double negation and de Morgan's laws.

The intended meaning of the justification term $\mathsf{up}(A)$ is that after an update with A, this act of updating serves as justification to believe A. Consequently, the justification term $\mathsf{up}(A)$ has no specific meaning before the update with A happens.

Definition 1 (Set of Atomic Subterms). *The set of atomic subterms of a term or formula is inductively defined as follows:*

- $\mathsf{atm}(t) := \{t\}$ *if t is a constant or a variable*
- $\mathsf{atm}(\mathsf{up}(C)) := \{\mathsf{up}(C)\} \cup \mathsf{atm}(C)$
- $\mathsf{atm}(s \cdot_A t) := \mathsf{atm}(s) \cup \mathsf{atm}(t) \cup \mathsf{atm}(A)$
- $\mathsf{atm}(P) := \emptyset$ *for $P \in \mathsf{Prop}$*
- $\mathsf{atm}(\neg A) := \mathsf{atm}(A)$
- $\mathsf{atm}(A \to B) := \mathsf{atm}(A) \cup \mathsf{atm}(B)$
- $\mathsf{atm}(t : A) := \mathsf{atm}(t) \cup \mathsf{atm}(A)$
- $\mathsf{atm}([C]A) := \mathsf{atm}(A) \cup \mathsf{atm}(C)$.

Definition 2. *We call a formula A* up-independent *if for each subformula $[C]B$ of A we have that $\mathsf{up}(C) \notin \mathsf{atm}(B)$.*

Using Definition 1, we can control that updates and justifications are independent. This is of importance to distinguish cases where updates change the meaning of justifications and corresponding formulas from cases where the update does not affect the meaning of a formula.

We will use the following notation: τ denotes a finite sequence of formulas and ϵ denotes the empty sequence. Given a sequence $\tau = C_1, \ldots C_n$ and a formula A, the formula $[\tau]A$ is defined by

$$[\tau]A = [C_1]\ldots[C_n]A \text{ if } n > 0 \quad \text{and} \quad [\epsilon]A := A.$$

The logic JUS has the following axioms and rules where τ is a finite (possibly empty) sequence of formulas:

1. $[\tau]A$ for all propositional tautologies A (Taut)
2. $[\tau](t : (A \to B) \wedge s : A \leftrightarrow t \cdot_A s : B)$ (App)
3. $[\tau]([C]A \leftrightarrow A)$ if $[C]A$ is up-independent (Indep)
4. $[\tau]([C]\neg A \leftrightarrow \neg[C]A)$ (Funct)
5. $[\tau]([C](A \to B) \leftrightarrow ([C]A \to [C]B))$ (Norm)
6. $[\tau][A]\mathsf{up}(A) : A$ (Up)
7. $[\tau](\mathsf{up}(A) : B \to [A]\mathsf{up}(A) : B)$ (Pers)

A constant specification CS for JUS is any subset

$$\mathsf{CS} \subseteq \{(c, \ [\tau_1]c_1 : [\tau_2]c_2 : \ldots : [\tau_n]c_n : A) \ |$$
$$n \geq 0, c, c_1, \ldots, c_n \text{ are constants,}$$
$$\tau_1, \ldots, \tau_n \text{ are sequences of formulas,}$$
$$A \text{ is an axiom of JUS}\}$$

$\mathsf{JUS_{CS}}$ denotes the logic JUS with the constant specification CS. The rules of $\mathsf{JUS_{CS}}$ are Modus Ponens and Axiom Necessitation:

$$\frac{A \qquad A \to B}{B} \text{ (MP)} \qquad\qquad \frac{}{[\tau]c : A} \text{ (AN)} \quad \text{if } (c, A) \in \mathsf{CS}$$

Before establishing some basic properties of $\mathsf{JUS_{CS}}$, let us briefly discuss its axioms. The direction from left to right in axiom (App) provides an internalization of modus ponens. Because of the annotated application operator, we also have the other direction, which is a minimality condition. It states that a justification represented by a complex term can only come from an application of modus ponens.

Axiom (Indep) roughly states that an update with a formula C can only affect the truth of formulas that contain certain update terms.

Axiom (Funct) formalizes that updates are functional, i.e. the result of an update is uniquely determined.

Axiom (Norm), together with Lemma 1, states that $[C]$ is a normal modal operator for each formula C.

Axiom (Up) states that after a belief expansion with A, the formula A is indeed believed and $\mathrm{up}(A)$ justifies that belief.

Axiom (Pers) is a simple persistency property of update terms.

Definition 3. *A constant specification* CS *is called* axiomatically appropriate *if*

1. *for each axiom A, there is a constant c with $(c, A) \in$ CS and*
2. *for any formula A and any constant c, if $(c, A) \in$ CS, then for each sequence of formulas τ there exists a constant d with*

$$(d, [\tau]c : A) \in \mathsf{CS}.$$

The first clause in the previous definition is the usual condition for an axiomatically appropriate constant specification (when the language includes the !-operation). Here we also need the second clause in order to have the following two lemmas, which establish that necessitation is admissible in JUS$_{\mathsf{CS}}$. Both are proved by induction on the length of derivations.

Lemma 1. *Let* CS *be an arbitrary constant specification. For all formulas A and C we have that if A is provable in* JUS$_{\mathsf{CS}}$, *then $[C]A$ is provable in* JUS$_{\mathsf{CS}}$.

Lemma 2 (Constructive Necessitation). *Let* CS *be an axiomatically appripriate constant specification. For all formulas A we have that if A is provable in* JUS$_{\mathsf{CS}}$, *then there exists a term t such that $t : A$ is provable in* JUS$_{\mathsf{CS}}$.

We will also need the following auxiliary lemma.

Lemma 3. *Let* CS *be an arbitrary constant specification. For all terms s, t and all formulas A, B, C,* JUS$_{\mathsf{CS}}$ *proves:*

$$[C]t : (A \to B) \land [C]s : A \;\leftrightarrow\; [C]t \cdot_A s : B$$

3   Semantics

Now we are going to introduce subset models for the logic JUS$_{\mathsf{CS}}$. In order to define a valuation function on these models, we will need the following measure for the length of formulas.

Definition 4 (Length). *The* length *of a term or formula is inductively defined by:*

$$\ell(t) := 1 \; \textit{if } t \in \mathsf{ATm} \qquad\qquad \ell(s \cdot_A t) := \ell(s) + \ell(t) + \ell(A) + 1$$
$$\ell(P) := 1 \; \textit{if } P \in \mathsf{Prop} \qquad\qquad \ell(A \to B) := \ell(A) + \ell(B) + 1$$
$$\ell(\neg A) := \ell(A) + 1 \qquad\qquad\quad \ell(t : A) := \ell(t) + \ell(A) + 1$$
$$\ell([B]A) := \ell(B) + \ell(A) + 1$$

Definition 5 (Subset Model). *We define a* subset model

$$\mathcal{M} = (W, W_0, V_1, V_0, E)$$

for JUS *by:*

- W *is a set of objects called* worlds.
- $W_0 \subseteq W$, $W_0 \neq \emptyset$.
- $V_1 : (W \setminus W_0) \times \mathsf{L_{JUS}} \to \{0,1\}$.
- $V_0 : W_0 \times \mathsf{Prop} \to \{0,1\}$.
- $E : W \times \mathsf{Tm} \to \mathcal{P}(W)$ *such that for $\omega \in W_0$ and all $A \in \mathsf{L_{JUS}}$:*

$$E(\omega, s \cdot_A t) \subseteq E(\omega, s) \cap E(\omega, t) \cap W_{MP},$$

where W_{MP} is the set of all deductively closed worlds, formally given by

$$W_{MP} := W_0 \cup W_{MP}^1 \quad where$$
$$W_{MP}^1 := \{\omega \in W \setminus W_0 \mid$$
$$\forall A, B \in \mathsf{L_{JUS}} \ ((V_1(\omega, A) = 1 \ and \ V_1(\omega, A \to B) = 1)$$
$$implies \ V_1(\omega, B) = 1)\}.$$

We call W_0 the set of *normal* worlds. The worlds in $W \setminus W_0$ are called *non-normal* worlds. W_{MP} denotes the set of worlds where the valuation function (see the following definition) is closed under modus ponens.

In normal worlds, the laws of classical logic hold, whereas non-normal worlds may behave arbitrarily. In a non-normal world we may have that both P and $\neg P$ hold or we may have that neither P nor $\neg P$ holds. We need non-normal worlds to take care of the hyperintensional aspects of justification logic. In particular, we must be able to model that constants do not justify all axioms. In normal worlds, all axioms hold. Thus we need non-normal worlds to make axioms false.

Let $\mathcal{M} = (W, W_0, V_1, V_0, E)$ be a subset model. We define the *valuation function* $V_{\mathcal{M}}$ for $\mathcal{M}$ and the *updated model* $\mathcal{M}^C$ for any formula C simultaneously. For $V_{\mathcal{M}}$, we often drop the subscript $\mathcal{M}$ if it is clear from the context.

We define $V : W \times \mathsf{L_{JUS}} \to \{0,1\}$ as follows by induction on the length of formulas:

1. Case $\omega \in W \setminus W_0$. We set $V(\omega, F) := V_1(\omega, F)$;
2. Case $\omega \in W_0$. We define V inductively by:
   (a) $V(\omega, P) := V_0(\omega, P)$ for $P \in \mathsf{Prop}$;
   (b) $V(\omega, t : F) := 1$ iff $E(\omega, t) \subseteq \{v \in W \mid V(v, F) = 1\}$ for $t \in \mathsf{ATm}$;
   (c) $V(\omega, s \cdot_F r : G) = 1$ iff $V(\omega, s : (F \to G)) = 1$ and $V(\omega, r : F) = 1$;
   (d) $V(\omega, \neg F) = 1$ iff $V(\omega, F) = 0$;
   (e) $V(\omega, F \to G) = 1$ iff $V(\omega, F) = 0$ or $V(\omega, G) = 1$;
   (f) $V(\omega, [C]F) = 1$ iff $V_{\mathcal{M}^C}(\omega, F) = 1$ where $V_{\mathcal{M}^C}$ is the valuation function for the updated model $\mathcal{M}^C$.

The following notation for the truth set of F will be convenient:

$$[[F]]_{\mathcal{M}} := \{v \in W \mid V_{\mathcal{M}}(v, F) = 1\}.$$

The updated model $\mathcal{M}^C = (W^{\mathcal{M}^C}, W_0^{\mathcal{M}^C}, V_1^{\mathcal{M}^C}, V_0^{\mathcal{M}^C}, E^{\mathcal{M}^C})$ is given by:

$$W^{\mathcal{M}^C} := W \qquad W_0^{\mathcal{M}^C} := W_0 \qquad V_1^{\mathcal{M}^C} := V_1 \qquad V_0^{\mathcal{M}^C} := V_0$$

and

$$E^{\mathcal{M}^C}(\omega, t) := \begin{cases} E^{\mathcal{M}}(\omega, t) \cap [[C]]_{\mathcal{M}^C} & \text{if } \omega \in W_0 \text{ and } t = \mathsf{up}(C) \\ E^{\mathcal{M}}(\omega, t) & \text{otherwise} \end{cases}$$

The valuation function for complex terms is well-defined.

Lemma 4. *For a subset model $\mathcal{M}$ with a world $\omega \in W_0$, $s, t \in \mathsf{Tm}$, $A, B \in \mathsf{L_{JUS}}$, we find that*

$$V(\omega, s \cdot_A t : B) = 1 \quad \text{implies} \quad E(\omega, s \cdot_A t) \subseteq [[B]]_{\mathcal{M}}.$$

Proof. The proof is by induction on the structure of s and t:

- base case $s, t \in \mathsf{ATm}$:
  Suppose $V(\omega, s \cdot_A t : B) = 1$. Case 2c of the definition of V in Definition 5 for normal worlds yields that

$$V(\omega, s : (A \to B)) = 1 \text{ and } V(\omega, t : A) = 1.$$

  With case 2b from the same definition we obtain

$$E(\omega, s) \subseteq [[A \to B]]_{\mathcal{M}} \text{ and } E(\omega, t) \subseteq [[A]]_{\mathcal{M}}.$$

  Furthermore the definition of E for normal worlds guarantees that

$$E(\omega, s \cdot_A t) \subseteq E(\omega, s) \cap E(\omega, t) \cap W_{MP}.$$

  So for each $v \in E(\omega, s \cdot_A t)$ there is $V(v, A \to B) = 1$ and $V(v, A) = 1$ and $v \in W_{MP}$ and hence either by the definition of W_{MP}^1 or by case 2e of the definition of V in normal worlds there is $V(v, B) = 1$. Therefore $E(\omega, s \cdot_A t) \subseteq [[B]]_{\mathcal{M}}$.
- $s, t \in \mathsf{Tm}$ but at least one of them is not atomic: w.l.o.g. suppose $s = r \cdot_C q$.
  Suppose $V(\omega, s \cdot_A t : B) = 1$ then $V(\omega, s : (A \to B)) = 1$ and $V(\omega, t : A) = 1$. Since $s = r \cdot_C q$ and $\omega \in W_0$ we obtain

$$V(\omega, r : (C \to (A \to B))) = 1 \text{ and } V(\omega, q : C) = 1$$

  and by I.H. that

$$E(\omega, r) \subseteq [[C \to (A \to B)]]_{\mathcal{M}} \text{ and } E(\omega, q) \subseteq [[C]]_{\mathcal{M}}.$$

With the same reasoning as in the base case we obtain

$$E(\omega, s) = E(\omega, r \cdot_C q) \subseteq [[A \to B]]_{\mathcal{M}}.$$

If t is neither atomic, the argumentation works analoguously and since we have then shown both $E(\omega, s) \subseteq [[A \to B]]_{\mathcal{M}}$ and $E(\omega, t) \subseteq [[A]]_{\mathcal{M}}$, the conclusion is the same as in the base case.   □

Remark 1. The opposite direction to Lemma 4 need not hold. Consider a model $\mathcal{M}$ and a formula $s \cdot_A t : B$ with atomic terms s and t such that

$$V_{\mathcal{M}}(\omega, s \cdot_A t : B) = 1$$

and thus also $E(\omega, s \cdot_A t) \subseteq [[B]]_{\mathcal{M}}$. Now consider a model $\mathcal{M}'$ which is defined like $\mathcal{M}$ except that

$$E'(\omega, s) := E(\omega, t) \quad \text{and} \quad E'(\omega, t) := E(\omega, s).$$

We observe the following:

1. We have $E'(\omega, s \cdot_A t) = E(\omega, s \cdot_A t)$ as the condition

$$E'(\omega, s \cdot_A t) \subseteq E'(\omega, s) \cap E'(\omega, t) \cap W_{MP}$$

   still holds since intersection of sets is commutative. Therefore

$$E'(\omega, s \cdot_A t) \subseteq [[B]]_{\mathcal{M}'}$$

   holds.
2. However, it need not be the case that

$$E'(\omega, s) \subseteq [[A \to B]]_{\mathcal{M}'} \text{ and } E'(\omega, t) \subseteq [[A]]_{\mathcal{M}'}.$$

   Therefore $V_{\mathcal{M}'}(\omega, s : (A \to B)) = 1$ and $V_{\mathcal{M}'}(\omega, t : A) = 1$ need not hold and thus also $V_{\mathcal{M}'}(\omega, s \cdot_A t : B) = 1$ need not be the case anymore.

Definition 6 (CS-Model). *Let* CS *be a constant specification. A subset model* $\mathcal{M} = (W, W_0, V_1, V_0, E)$ *is called a* CS*-subset model or a subset model for* JUS$_{CS}$ *if for all* $\omega \in W_0$ *and for all* $(c, A) \in$ CS *we have*

$$E(\omega, c) \subseteq [[A]]_{\mathcal{M}}.$$

We observe that updates respect CS-subset models.

Lemma 5. *Let* CS *be an arbitrary constant specification and let* $\mathcal{M}$ *be a* CS-*subset model. We find that* $\mathcal{M}^C$ *is a* CS-*subset model for any formula* C.

4   Soundness

Definition 7 (Truth in Subset Models). *Let*

$$\mathcal{M} = (W, W_0, V_1, V_0, E)$$

be a subset model, $\omega \in W$, and $F \in \mathsf{L_{JUS}}$. We define the relation $\Vdash$ as follows:

$$\mathcal{M}, \omega \Vdash F \quad \textit{iff} \quad V_{\mathcal{M}}(\omega, F) = 1.$$

Theorem 1 (Soundness). *Let CS be an arbitrary constant specification. Let $\mathcal{M} = (W, W_0, V_1, V_0, E)$ be a CS-subset model and $\omega \in W_0$. For each formula $F \in \mathsf{L_{JUS}}$ we have that*

$$\mathsf{JUS_{CS}} \vdash F \quad \textit{implies} \quad \mathcal{M}, \omega \Vdash F.$$

Proof. As usual by induction on the length of the derivation of F. We only show the case where F is an instance of axiom (Indep).

By induction on $[C]A$ we show that for all ω

$$\mathcal{M}^C, \omega \Vdash A \qquad \text{iff} \qquad \mathcal{M}, \omega \Vdash A.$$

We distinguish the following cases.

1. A is an atomic proposition. Trivial.
2. A is $\neg B$. By I.H.
3. A is $B \to D$. By I.H.
4. A is $t : B$. Subinduction on t:
    (a) t is a variable or a constant. Easy using I.H. for B.
    (b) t is a term $\mathsf{up}(D)$. By assumption, we have that $C \neq D$. Hence this case is similar to the previous case.
    (c) t is a term $r \cdot_D s$. We know that $t : B$ is equivalent to

$$r : (D \to B) \land s : D.$$

   Using I.H. twice, we find that

$$\mathcal{M}^C, \omega \Vdash r : (D \to B) \quad \text{and} \quad \mathcal{M}^C, \omega \Vdash s : D$$

   if and only if

$$\mathcal{M}, \omega \Vdash r : (D \to B) \quad \text{and} \quad \mathcal{M}, \omega \Vdash s : D.$$

   Now the claim follows immediately.
5. A is $[D]B$. Making use of the fact that A is up-independent, this case also follows using I.H. $\qquad \square$

5 Basic Properties

We first show that first-order beliefs are persistent in JUS. Let F be a formula that does not contain any justification operator. We have that if t is a justification for F, then, after any update, this will still be the case. Formally, we have the following lemma.

Lemma 6. *For any term t and any formulas A and C we have that if A does not contain a subformula of the form $s : B$, then*

$$t : A \ \rightarrow \ [C]t : A$$

is provable.

Proof. We proceed by induction on the complexity of t and distinguish the following cases:

1. Case t is atomic and $t \neq \mathsf{up}(C)$. Since A does not contain any evidence terms, the claim follows immediately from axiom (Indep).
2. Case $t = \mathsf{up}(C)$. This case is an instance of axiom (Pers).
3. Case $t = r \cdot_B s$. From $r \cdot_B s : A$ we get by (App)

$$s : B \quad \text{and} \quad r : (B \to A).$$

By I.H. we find
$$[C]s : B \quad \text{and} \quad [C]r : (B \to A).$$

Using Lemma 3 we conclude $[C]r \cdot_B s : A$. □

Let us now investigate higher-order beliefs. We argue that persistence should not hold in this context. Consider the following scenario. Suppose that you are in a room together with other people. Further suppose that no announcement has been made in that room. Therefore, it is not the case that P is believed because of an announcement. Formally, this is expressed by

$$\neg\mathsf{up}(P) : P. \tag{3}$$

We find that
$$\text{the fact that you are in that room} \tag{4}$$
justifies your belief in (3). Let the term r represent (4). Then we have

$$r : \neg\mathsf{up}(P) : P. \tag{5}$$

Now suppose that P is publicly announced in that room. Thus we have in the updated situation
$$\mathsf{up}(P) : P. \tag{6}$$

Moreover, the fact that you are in that room justifies now your belief in (6). Thus we have $r : \mathsf{up}(P) : P$ and hence in the original situation we have

$$[P]r : \mathsf{up}(P) : P \tag{7}$$

and (5) does no longer hold after the announcement of P.

The following lemma formally states that persistence fails for higher-oder beliefs.

Lemma 7. *There exist formulas $r : B$ and A such that*

$$r : B \rightarrow [A]r : B$$

is not provable.

Proof. Let B be the formula $\neg\mathsf{up}(P) : P$ and consider the subset model

$$\mathcal{M} = (W, W_0, V_1, V_0, E)$$

with

$$W := \{\omega, \upsilon\} \quad W_0 := \{\omega\} \quad V_1(\upsilon, P) = 0 \quad V_0(\omega, P) = 1$$

and

$$E(\omega, r) = \{\omega\} \quad E(\omega, \mathsf{up}(P)) = \{\omega, \upsilon\}.$$

Hence $[[P]]_{\mathcal{M}} = \{\omega\}$ and thus $E(\omega, \mathsf{up}(P)) \not\subseteq [[P]]_{\mathcal{M}}$. Since $\omega \in W_0$, this yields $V(\omega, \mathsf{up}(P) : P) = 0$. Again by $\omega \in W_0$, this implies

$$V(\omega, \neg\mathsf{up}(P) : P) = 1.$$

Therefore $E(\omega, r) \subseteq [[\neg\mathsf{up}(P) : P]]_{\mathcal{M}}$ and using $\omega \in W_0$, we get

$$\mathcal{M}, \omega \Vdash r : \neg\mathsf{up}(P) : P.$$

Now consider the updated model $\mathcal{M}^P$. We find that

$$E^{\mathcal{M}^P}(\omega, \mathsf{up}(P)) = \{\omega\}$$

and thus $E^{\mathcal{M}^P}((\omega, \mathsf{up}(P))) \subseteq [[P]]_{\mathcal{M}^P}$. Further, using $\omega \in W_0^{\mathcal{M}^P}$ we get

$$V_{\mathcal{M}^P}(\mathsf{up}(P) : P) = 1$$

and thus $V_{\mathcal{M}^P}(\neg\mathsf{up}(P) : P) = 0$. That is $\omega \notin [[(\neg\mathsf{up}(P) : P]]_{\mathcal{M}^P}$. We have $E^{\mathcal{M}^P}(\omega, r) = \{\omega\}$ and, therefore, $E^{\mathcal{M}^P}(\omega, r) \not\subseteq [[(\neg\mathsf{up}(P) : P]]_{\mathcal{M}^P}$.

With $\omega \in W_0^{\mathcal{M}^P}$ we get $\mathcal{M}^P, \omega \nVdash r : \neg\mathsf{up}(P) : P$. We conclude

$$\mathcal{M}, \omega \nVdash [P]r : \neg\mathsf{up}(P) : P.$$

$\square$

Next, we show that $\mathsf{JUS}_{\mathsf{CS}}$ proves an explicit form of the Ramsey axiom

$$\Box(C \rightarrow A) \leftrightarrow [C]\Box A$$

from Dynamic Doxastic Logic.

Lemma 8. *Let the formula* $[C]s : (C \to A)$ *be up-independent. Then* $\mathsf{JUS_{CS}}$
proves

$$s : (C \to A) \; \leftrightarrow \; [C]s \cdot_C \mathsf{up}(C) : A. \tag{8}$$

Proof. First observe that by (Up), we have $[C]\mathsf{up}(C) : C$.

Further, since $[C]s : (C \to A)$ is up-independent, we find by (Indep) that

$$s : (C \to A) \leftrightarrow [C]s : (C \to A).$$

Finally we obtain (8) using Lemma 3.   □

Often, completeness of public announcement logics is established by showing
that each formula with announcements is equivalent to an announcement-free
formula. Unfortunately, this approach cannot be employed for $\mathsf{JUS_{CS}}$ although
(8) provides a reduction property for certain formulas of the form $[C]t : A$. The
reason is the hyperintensionality of justification logic [8,21], i.e. justification logic
is not closed under substitution of equivalent formulas. Because of this, the proof
by reduction cannot be carried through in $\mathsf{JUS_{CS}}$, see the discussion in [10].

6   Conclusion

We have introduced the justification logic JUS for subset models with belief
expansion. We have established basic properties of the deductive system and
shown its soundness. We have also investigated persitence properties for first-
order and higher-order beliefs.

The next step is, of course, to obtain a completeness result for subset models
with updates. We suspect, however, that the current axiomatization of JUS is
not strong enough. The proof of Lemma 7 shows that persistence of higher-
order beliefs fails in the presence of a negative occurence of an up-term. Thus we
believe that we need a more subtle version of axiom (Indep) that distinguishes
between positive and negative occurences of terms. Introducing polarities for
term occurences, like in Fitting's realization procedure [14], may help to obtain
a complete axiomatization.

Acknowledgements. This work was supported by the Swiss National Science Foundation grant 200020_184625.

References

1. Artemov, S.N.: Operational modal logic. Technical report MSI 95–29, Cornell University (1995)
2. Artemov, S.N.: Explicit provability and constructive semantics. Bull. Symb. Log.
   7(1), 1–36 (2001)
3. Artemov, S.N.: Justified common knowledge. TCS **357**(1–3), 4–22 (2006). https://
   doi.org/10.1016/j.tcs.2006.03.009
4. Artemov, S.N.: The logic of justification. RSL **1**(4), 477–513 (2008). https://doi.
   org/10.1017/S1755020308090060

5. Artemov, S.N.: The ontology of justifications in the logical setting. Stud. Log. **100**(1–2), 17–30 (2012). https://doi.org/10.1007/s11225-012-9387-x

6. Artemov, S.N.: On aggregating probabilistic evidence. In: Artemov, S., Nerode, A. (eds.) LFCS 2016, vol. 9537, pp. 27–42. Springer, Cham (2016). https://doi.org/10.1007/978-3-319-27683-0-3

7. Artemov, S.N., Fitting, M.: Justification logic. In: Zalta, E.N. (ed.) The Stanford Encyclopedia of Philosophy, fall 2012 edn. (2012). http://plato.stanford.edu/archives/fall2012/entries/logic-justification/

8. Artemov, S.N., Fitting, M.: Justification Logic: Reasoning with Reasons. Cambridge University Press, Cambridge (2019)

9. Baltag, A., Renne, B., Smets, S.: The logic of justified belief, explicit knowledge, and conclusive evidence. APAL **165**(1), 49–81 (2014). https://doi.org/10.1016/j.apal.2013.07.005

10. Bucheli, S., Kuznets, R., Renne, B., Sack, J., Studer, T.: Justified belief change. In: Arrazola, X., Ponte, M. (eds.) LogKCA-10, pp. 135–155. University of the Basque Country Press (2010)

11. Bucheli, S., Kuznets, R., Studer, T.: Justifications for common knowledge. Appl. Non-Class. Log. **21**(1), 35–60 (2011). https://doi.org/10.3166/JANCL.21.35-60

12. Bucheli, S., Kuznets, R., Studer, T.: Partial realization in dynamic justification logic. In: Beklemishev, L.D., de Queiroz, R. (eds.) WoLLIC 2011. LNCS (LNAI), vol. 6642, pp. 35–51. Springer, Heidelberg (2011). https://doi.org/10.1007/978-3-642-20920-8-9

13. Bucheli, S., Kuznets, R., Studer, T.: Realizing public announcements by justifications. J. Comput. Syst. Sci. **80**(6), 1046–1066 (2014). https://doi.org/10.1016/j.jcss.2014.04.001

14. Fitting, M.: The logic of proofs, semantically. APAL **132**(1), 1–25 (2005). https://doi.org/10.1016/j.apal.2004.04.009

15. Kokkinis, I., Maksimović, P., Ognjanović, Z., Studer, T.: First steps towards probabilistic justification logic. Log. J. IGPL **23**(4), 662–687 (2015). https://doi.org/10.1093/jigpal/jzv025

16. Kuznets, R., Studer, T.: Justifications, ontology, and conservativity. In: Bolander, T., Braüner, T., Ghilardi, S., Moss, L. (eds.) Advances in Modal Logic, vol. 9, pp. 437–458. College Publications (2012)

17. Kuznets, R., Studer, T.: Update as evidence: belief expansion. In: Artemov, S., Nerode, A. (eds.) LFCS 2013. LNCS, vol. 7734, pp. 266–279. Springer, Heidelberg (2013). https://doi.org/10.1007/978-3-642-35722-0-19

18. Kuznets, R., Studer, T.: Weak arithmetical interpretations for the logic of proofs. Log. J. IGPL **24**(3), 424–440 (2016)

19. Kuznets, R., Studer, T.: Logics of Proofs and Justifications. College Publications (2019)

20. Lehmann, E., Studer, T.: Subset models for justification logic. In: Iemhoff, R., Moortgat, M., de Queiroz, R. (eds.) WoLLIC 2019. LNCS, vol. 11541, pp. 433–449. Springer, Heidelberg (2019). https://doi.org/10.1007/978-3-662-59533-6-26

21. Lehmann, E., Studer, T.: Exploring subset models for justification logic (submitted)

22. Mkrtychev, A.: Models for the logic of proofs. In: Adian, S., Nerode, A. (eds.) LFCS 1997. LNCS, vol. 1234, pp. 266–275. Springer, Heidelberg (1997). https://doi.org/10.1007/3-540-63045-7-27

23. Renne, B.: Multi-agent justification logic: communication and evidence elimination. Synthese **185**(S1), 43–82 (2012). https://doi.org/10.1007/s11229-011-9968-7

24. Studer, T.: Decidability for some justification logics with negative introspection. JSL **78**(2), 388–402 (2013). https://doi.org/10.2178/jsl.7802030

Finitism, Imperative Programs and Primitive Recursion

Daniel Leivant[1,2(✉)] (iD)

[1] SICE, Indiana University, Bloomington, USA
leivant@indiana.edu
[2] IRIF, Université Paris-Diderot, Paris, France

Abstract. The finitistic philosophy of mathematics, critical of referencing infinite totalities, has been associated from its inception with primitive recursion. That kinship was not initially substantiated, but is widely assumed, and is supported by Parson's Theorem, which may be construed as equating finitistic reasoning with finitistic computing.

In support of identifying **PR** with finitism we build on the generic framework of [7] and articulate a finitistic theory of finite partial-structures, and a generic imperative programming language for modifying them, equally rooted in finitism. The theory is an abstract generalization of Primitive Recursive Arithmetic, and the programming language is a generic generalization of first-order recurrence (primitive recursion). We then prove an abstract form of Parson's Theorem that links the two.

Keywords: Finitism · Primitive recursive arithmetic · Finite partial functions · Primitive recursive functions · Imperative programming · Loop variant · Parson's theorem

1 Introduction

Primitive recursive arithmetic was invented in 1923 by Skolem, as a necessary first proposal for a general notion of numeric algorithms rooted in finitism [14]. Hilbert's Programme, striving to show that finitistic theorems are provable using finitistic means [5], was dashed by Gödel's First Incompleteness Theorem, but the issue of delineating finitistic mathematics and relating it to finitistic forms of programming has remained relevant, with the influential essay *Finitism* by William Tait [15] proposing to identify finitistic mathematics with primitive recursive arithmetic.

Our aim here is to offer a context for discussing finitism which is broader than natural numbers, recurrence equations, and numeric induction. We use the generic framework presented in [7] to propose finitistic restrictions to both reasoning about data, and computing over it.

We take finite partial (fp) structures, built exclusively from fp-functions, as a generic notion of basic data. This simply formalizes the intuition that computational data in general, and in inductive data (e.g. $\mathbb{N}$) in particular, is internally structured. In [7] we presented a second order theory for fp-structures,

S. Artemov and A. Nerode (Eds.): LFCS 2020, LNCS 11972, pp. 98–110, 2020.
https://doi.org/10.1007/978-3-030-36755-8_7

and proved that it is mutually interpretable with Peano's Arithmetic. We modify that theory here by restricting the induction on fp-functions to finitistically meaningful formulas. In [8] we proposed an imperative programming language over fp-structures, providing a generic generalization of primitive recursion to fp-structures. Here we show that the finitistic version of the theory presented in [7] matches that programming language, by proving a generic form of Parson's Theorem: a mapping between fp-structure is finitistically provable iff it is defined by a program.

2 A Finitistic Theory of Finite Partial Structures

2.1 Finite Partial Functions

Basic data objects come in two forms: bare "points", such as nodes of a graph, versus elements of inductive data, such as natural numbers and strings over an alphabet. The former have no independent computational content, whereas the computational nature of the latter is conveyed by the recursive definition of the entire data type to which they belong, via the corresponding recurrence operators. This dichotomy is antithetical, though, to an ancient alternative approach that takes individual inductive data objects, such as natural numbers, to be finite structures on their own, whose computational behavior is governed by their internal makings [3,9]. Under this approach, computing over inductive data is reduced to operating over finite structures, and functions over inductive data are construed as mappings between finite structures.

Embracing this approach yields two benefits. First, we obtain a common "hardware-based" platform for programming not only within finite structures, but also for the transformation of inductive data. Moreover, inductive types are just one way of aggregating finite structures, so focusing on finite structures enhances genericity and generality.

Function finiteness is enforced by a principle of induction. Induction over finite *sets* has been articulated repeatedly (for example [16] and [9]). However, our interest in the kinship between reasoning and computing leads us to refer to finite partial-functions as the basic aggregate entity, rather than to finite sets or relations.

We posit a denumerable set A of *atoms,* i.e. unspecified and unstructured objects. To accommodate in due course non-denoting terms we extend A to a set $A_\perp =_{df} A \cup \{\perp\}$, where $\perp$ is a fresh object intended to denote "undefined". The elements of A are the *standard* elements of $A_\perp$.

A *(k-ary) fp-function* is a function $F : A_\perp^k \to A_\perp$ for which there is a finite $D_F \subset A^k$, the *support* of F, such that $F(\vec{a}) \in A$ for $\vec{a} \in D_F$, and $F(\vec{a}) = \perp$ for $\vec{a} \in A_\perp^k - D_A$. An *entry* of F is a tuple $\langle a_1 \ldots a_k, b \rangle$ where $b = F(a_1, \ldots, a_k) \neq \perp$. The *image* of F is the set $\{b \in A \mid b = F(\vec{a}) \text{ for some } \vec{a} \in A^k \}$.

Function partiality provides a natural representation of finite relations over A by partial functions, avoiding *ad hoc* constants. Namely, a finite k-ary relation R over A $(k > 0)$ is represented by the fp-function

$$\xi_R(a_1, \ldots, a_k) = \text{if } R(a_1, \ldots, a_k) \text{ then } a_1 \text{ else } \perp$$

Conversely, any partial k-ary function F over A determines the k-ary relation

$$R_F \;=\; \{\langle \vec{a} \rangle \in A^k \mid F(\vec{a}) \text{ is defined } \}$$

Thus we write $\vec{a} \in f$ for $f\vec{a} \neq \omega$.

2.2   Finite Partial Structures

A *vocabulary* is a finite set V of function-identifiers, referred to as V-ids, where each $\mathbf{f} \in V$ is assigned an *arity* $\mathfrak{r}(\mathbf{f}) \geqslant 0$. We optionally right-superscript an identifier by its arity, when convenient. We refer to nullary V-ids as *tokens* and to identifiers of positive arities as *pointers*. Our default is to use type-writer symbols for identifiers: $\mathsf{a}, \mathsf{b}, \mathsf{e}, ...$ for tokens and $\mathsf{s}, \mathsf{f}, \mathsf{g}, 0, 1...$ for pointers. The distinction between tokens and pointers is computationally significant, because (non-nullary) functions can serve as memory, whereas atoms cannot. For a vocabulary V, we write V^0 for the set of tokens, V^+ for the pointers, and V^k for the pointers of arity k.

An *fp-structure over* V, or more concisely a V-*structure,* is a mapping σ that to each $\mathbf{f} \in V^k$ assigns a k-ary fp-function $\sigma(\mathbf{f})$, said to be a *component* of σ. The intention is to identify isomorphic fp-structures, but this intention may be left implicit, sparing us perpetual references to equivalence classes. Note that a tuple $\vec{\sigma} = (\sigma_1, \ldots, \sigma_k)$ of fp-structures is representable as a single structure, defined as the union $\bigcup_{1 \leqslant i \leqslant k} \sigma_i$ over the disjoint union of the vocabularies V_i.

The *support* (respectively, *range*) of an fp-structure σ is the union of the supports (ranges) of its components, and its *scope* is the union of its support and its range.

2.3   Basic-Terms and Accessibility

Given a vocabulary V, the set $\mathbf{Tm}_V$ of *basic V-terms* is generated inductively as follows. Note that no variables are used. $\omega \in \mathbf{Tm}_V$; and if $\mathbf{f} \in V^k$ $(k \geqslant 0)$ and $\mathbf{t}_1, \ldots, \mathbf{t}_k \in \mathbf{Tm}_V$, then $\mathbf{ft}_1 \cdots \mathbf{t}_k \in \mathbf{Tm}_V$. A term $\mathbf{t}$ without ω is *standard*.

We write function application in formal terms without parentheses and commas, as in $\mathbf{f}xy$ or $\mathbf{f}\vec{x}$. Also, we implicitly posit that the arity of a function matches the number of arguments displayed; thus writing $f^k\vec{a}$ assumes that $\vec{a}$ is a vector of length k, and $f\vec{a}$ (with no superscript) that the vector $\vec{a}$ is as long as f's arity.

An atom $a \in A$ is V-*accessible in* σ if $a = \sigma(\mathbf{t})$ for some $\mathbf{t} \in \mathbf{Tm}_V$. A V-structure σ is *accessible* if every atom in the range of σ is V-accessible.

Given a V-structure σ the *value* of terms $\mathbf{t}$ in σ, denoted $\sigma(\mathbf{t})$, is defined by recurrence:

- $\sigma(\omega) = \bot$
- $\sigma(\mathbf{ft}_1 \cdots \mathbf{t}_k) = \sigma(\mathbf{f})(\sigma(\mathbf{t}_1), \ldots, \sigma(\mathbf{t}_k))$ $(\mathbf{f} \in V^k,\ k \geqslant 0)$

An atom $a \in A$ is V-*accessible in* σ if it is the value in σ of some V-term. A V-structure σ is *accessible* if every atom in the domain of σ is V-accessible (and therefore every atom in the range of σ is also accessible).

2.4   A Formal Language

Fp-structures for a vocabulary V are intended to be the object of formal reasoning as well as computing. For both ends, vocabularies broader than V are called for. We thus move on from a fixed vocabulary to a broader, generic, formal language. Our language of reference, $\mathcal{L}_\mathcal{A}$, is two-sorted with bound variables and parameters (i.e. free variables) for *atoms* and bound variables and parameters for *finite partial-functions* of arbitrary positive arity. We refer to the former as *A-variables*, using $u, v, \dots$ as syntactic-parameters for them, and to the latter as *F-variables*, with $f, g, h, \dots$ as syntactic-parameters. A *vocabulary* can be construed as finite set of reserved parameters.

The basic term-formation operator of $\mathcal{L}_\mathcal{A}$ is a form of value-assignment, geared to reasoning about, and programming for, transformations of fp-structures. We thus generate simultaneously *A-terms*, denoting atoms, and *F-terms*, denoting fp-functions. Contrary to the basic V-terms defined above for a fixed vocabulary V, we refer here to a function-extension operation.

- (Intial A-terms) $\boldsymbol{\omega}$ and A-variables are A-terms.
- (Initial F-terms) For each $k > 0$, $\boldsymbol{\emptyset}_k$ and k-ary function-variables are k-ary F-terms.
- (Application) If $\mathbf{F}$ is a k-ary F-term and $\mathbf{t}_1 \dots \mathbf{t}_k$ are A-terms, then $\mathbf{F}\,\mathbf{t}_1 \cdots \mathbf{t}_k$ is an A-term.
- (Extension) If $\mathbf{F}$ is a k-ary F-term, $\vec{\mathbf{t}}$ a vector of k A-terms, and $\mathbf{q}$ an A-term, then $\{\vec{\mathbf{t}} \mapsto \mathbf{q}\}\,\mathbf{F}$ is an F-term.

The *formulas* of $\mathcal{L}_\mathcal{A}$ are generated from formal equations $\mathbf{t} \simeq \mathbf{q}$ between A-terms $\mathbf{t}, \mathbf{q}$ using propositional connectives and quantifiers over A-variables and F-variables. We say that a formula without F-quantifiers is *elementary*.

An *fp-structure for* a formula (or a term) is an fp-structure for a vocabulary containing the parameters therein. If σ is a structure for the term $\mathbf{t}$, then the *value* of $\mathbf{t}$ in σ, denoted $\sigma(\mathbf{t})$, is defined by recurrence on $\mathbf{t}$, as above for V-terms, with the following clause added for the Extension operation:

$\star$ $\sigma(\{\vec{\mathbf{t}} \mapsto \mathbf{q}\}\,\mathbf{F})$ is identical to $\sigma(\mathbf{F})$, except that it maps $\sigma(\vec{\mathbf{t}})$ to $\sigma(\mathbf{q})$ *IF* $\sigma(\mathbf{F})(\sigma(\vec{\mathbf{t}}))$ is undefined and $\sigma(\vec{\mathbf{t}})$ is defined.

Note that extensions are interpreted differently from tradition assignments, in that they don't replace one entry of F by another one.

The *truth* of a formula φ in structures σ for φ is defined by recurrence on formulas:

$$\sigma \models \mathbf{t} \simeq \mathbf{q} \qquad \text{IFF} \qquad \sigma(\mathbf{t}) = \sigma(\mathbf{q})$$
$$\sigma \models \exists u \, \varphi \qquad \text{IFF} \qquad \sigma \cup \{u \mapsto a\}\rangle \models \varphi \text{ for some } a \in A$$
$$\sigma \models \exists f \, \varphi^k \qquad \text{IFF} \qquad \sigma \cup \{f \mapsto F\} \models \varphi \text{ for some fp-function } F : A^k \rightharpoonup A$$

The cases for propositional connectives and $\forall$ are analogous.

From [7] it can be shown that the truth of $\mathcal{L}_\mathcal{A}$ formulas, without restriction on quantifiers, is Turing-complete for HYP (the hyper-arithmetical subsets of $\mathbb{N}$). The situation is far simpler for elementary formulas:

PROPOSITION 1. *The truth of an elementary V-formula in accessible V-structures is decidable in time polynomial in the size of the structure.*

PROOF. For linearly ordered fp-structures one obtains for each fp-formula a log-space decision algorithm, as for first-order formulas in Finite Model Theory [6, Theorem 3.1]. The same proof-idea applies in the more general case, with functions of varying arities, provided one uses non-deterministic and co-non-deterministic scans for the A-quantifiers. For each formula such scan stays within the logspace hierarchy, so within alternating log-space, i.e. P-Time.

We can proceed instead by expanding the accessible input structure with a linear enumerator, an expansion that can be performed in PTime [8], and then apply the log-space decision with respect to the linear order obtained.   □

2.5   Definable Classes of Fp-Structures

A class $\mathfrak{C}$ of V-structures is *defined* by a V-formula ψ (with F-parameters $\vec{f}$) if for every structure σ for ψ we have

$$\sigma \in \mathfrak{C} \quad \text{IFF} \quad \sigma \models \psi$$

Recall that definability here is restricted to finite structures, so classical undefinability results for first-order logic do not apply.

We identify inductive data, such as elements of a free algebra, with finite partial-structures. For example, binary strings are finite partial-structure over the vocabulary with a constant e and unary function identifiers $\mathsf{0}$ and $\mathsf{1}$. Thus, the string $\mathsf{011}$ is taken to be the following partial-structure with four atoms, and where the partial-function denoted by $\mathsf{0}$ is defined for only one of the four:

$$\mathsf{e} \; \circ \xrightarrow{\;\;0\;\;} \circ \xrightarrow{\;\;1\;\;} \circ \xrightarrow{\;\;1\;\;} \circ$$

A function over $\mathbb{A}$ can thus be construed as a mapping between such finite structures.

The class of structures for an inductive data-type are, in fact, defined by elementary formulas. For example, the class of those structures for the vocabulary $\{\mathsf{z}^0, \mathsf{s}^1\}$ that represent natural numbers is defined by the elementary formula that states that s is injective, and every atom in the support of s is in its range, except z:

$$(\forall u, v \; \mathsf{s}u \simeq \mathsf{s}v \; \rightarrow \; u{\simeq}v) \quad \wedge \quad \forall u \, (u \simeq \boldsymbol{\omega} \; \vee \; \mathsf{s}u \simeq \mathsf{z} \; \vee \; \exists v \; \mathsf{s}v \simeq u)$$

The same idea can be extended, for any free algebra $\mathbb{A}$, into an elementary definition of the structures representing elements of $\mathbb{A}$.

From [7, Theorem 4] one obtains a definition of the structures that are accessible in a given vocabulary V, i.e. for which every element is denoted by a V-term. That definition uses function-quantifiers to express inductive definability, and is thus non-elementary.

3   A Finitistic Theory of Fp-Structures

3.1   On Axiomatizing Finiteness

The set $\mathfrak{F}^k$ of k-ary fp-functions is generated inductively by[1]

- *The empty k-ary function is in $\mathfrak{F}^k$;*
- *If $f \in \mathfrak{F}^k$, $\vec{u}, v \in A$, and $f(\vec{u}) = \bot$, then extending F with an entry $F\vec{u} = v$ yields an fp-function in $\mathfrak{F}_k$.*

This definition yields an Induction Schema for fp-functions:

$$(\forall f^k, \vec{u}, v \;\; \varphi[f] \;\wedge\; (f\vec{u} \simeq \omega) \;\rightarrow\; \varphi[\{\vec{u} \mapsto v\} f] \,)$$
$$\rightarrow \;\; \varphi[\emptyset_k] \;\rightarrow\; \forall f \, \varphi[f] \tag{1}$$

Since the components of fp-structures are fp-functions, without constraints that relate them, there is no need to articulate a separate induction principle for fp-structures. Indeed, every fp-structure σ for a vocabulary $V = \{f_1, \ldots, f_k\}$ is obtained by first generating the entries of $\sigma(f_1)$, then those for $\sigma(f_2)$, and so on. This would no longer be the case for an inductive definition of the class $\mathfrak{A}$ of *accessible* structures, whose components must be generated in tandem.

3.2   The Theory FST$_0$

In [7] we define a formal theory, **ST**, to reason about fp-structures and their transformation. **ST** is mutually interpretable with Peano Arithmetic, and is indeed not finitistic. We define now a finitistic sub-theory **FST$_0$** of **FST**.

We say that a formula of $\mathcal{L}_A$ is *F-existential* if it is of the form $\exists \vec{f} \, \psi$, where ψ is elementary. A formula is *concrete* if it has an F-existential prenex form. In particular, any formula with no F-$\forall$, and with no F-$\exists$ in the positive scope of an unbounded A-$\forall$, is concrete. Here an A-$\forall$ is *bounded* if it is of the form $\forall a \in \mathbf{F} \cdots$, where $\mathbf{F}$ is an F-term. Such formulas are provably concrete in the theory **FST$_0$** which we are about to describe, using the schema of Concrete-function-choice.

The axiom schemas of **FST$_0$** are the following templates, for all arities i, k, terms $\mathbf{t}$, and function-parameter (i.e. free variable) f. Recall that $\vec{u} \in f$ stands for $f\vec{u} \not\simeq \omega$.

[Strictness] $\qquad\qquad u_i \simeq \omega \;\;\rightarrow\;\; f u_1 \cdots u_k \simeq \omega \qquad (i = 1..k)$

[Empty-function] $\qquad\quad \emptyset_k u_1 \cdots u_k \simeq \omega$

[Function-extension] $\qquad \begin{array}{l} f\,\vec{u} \simeq \omega \;\;\rightarrow\;\; (\{\vec{u} \mapsto v\}f)\,\vec{u} \simeq v \\ \vec{w} \not\simeq \vec{u} \;\;\rightarrow\;\; (\{\vec{u} \mapsto v\}f)\,\vec{w} \simeq f\vec{w} \end{array}$

---

[1] Note that we generate here *all* finite structures, not a special subset which is somehow related to primitive recursion.

[Explicit-definition] $\exists g \forall \vec{u} \; (\; \vec{u} \in f \; \wedge \; g\vec{u} \simeq \mathbf{t} \;) \vee (\; \vec{u} \notin f \; \wedge \; g\vec{u} \simeq \boldsymbol{\omega} \;)$
This schema combines Zermelo's Separation Schema with an Explicit Definition principle: g is defined by the term $\mathbf{t}$, for arguments in the domain of f. Note that $\mathbf{t}$ refers (in all interesting cases) to $\vec{u}$; so $\mathbf{t}$ cannot be replaced by a variable w, since substitution of $\mathbf{t}$ for w would be illegal.

[Infinity] $\exists \vec{u} \; f^k \vec{u} \simeq \boldsymbol{\omega}$

Combined with the Function-extension schema Infinity implies that A is infinite, but it says nothing about functions being finite or not. For that we'll need Induction.

[Concrete-induction-rule] For every concrete formula φ,

$$\frac{\vdash \varphi[\emptyset] \quad \vdash \varphi[f] \; \rightarrow \; \varphi[\{\vec{u} \mapsto v\}f]}{\varphi[g]}$$

When φ is F-existential, say $\exists \vec{g}\, \varphi_0[f, \vec{g}]$, with φ_0 elementary, the second premise is concrete, since it can be rewritten, using fresh parameters $\vec{h}$, as

$$\exists \vec{g} \; (\varphi_0[f, \vec{h}] \rightarrow \varphi_0[f, \vec{g}])$$

[Concrete-function-choice] For every concrete formulas φ

$$\frac{\vdash \; \vec{x} \in f \; \rightarrow \; \exists g \; \varphi[\vec{x}, g]}{\exists h \; \forall \vec{x} \in f \; \varphi[\vec{x}, \; h_{\vec{x}}]}$$

where $\varphi[\vec{x}, \; h_{\vec{x}}]$ is $\varphi[\vec{x}, g]$ with every A-term $g\vec{\mathbf{t}}$ replaced by $h\vec{x}\vec{\mathbf{t}}$. As of this writing we don't know whether the schema of Concrete-function-choice is provable from the other axioms and rules. The treatment of finite choice in [12] suggests that it is not.

Note that all formulas in the templates above (not only φ) are concrete. We shall insist that all formulas used in $\mathbf{FST}_0$ are concrete. Alternatively, we could allow non-concrete formulas, and prove a cut-elimination theorem for a sequential calculus for $\mathbf{FST}_0$.

3.3 Some Derived Schemas

[Union] $\exists g \; \forall \vec{u} \; \vec{u} \in g \; \leftrightarrow \; (\vec{u} \in f_1 \vee \vec{u} \in f_2)$.
The proof is by Induction on f_2 and Extension.

[Composition]
$$\exists h^k \; \forall \vec{u} \; \; h\vec{u} \simeq f^i (g_1 \vec{u}) \cdots (g_i \vec{u}) \tag{2}$$

This follows from Explicit-definition and Union.

[Branching] $\exists h^k \; \forall \vec{u} \quad (\; f\vec{u} \not\simeq \boldsymbol{\omega} \wedge h\vec{u} \simeq f\vec{u} \;) \vee (\; f\vec{u} \simeq \boldsymbol{\omega} \wedge h\vec{u} \simeq g\vec{u} \;)$

The proof is by Induction using Update.

[Contraction]

$$\exists g \ g\vec{u} \simeq \omega \ \wedge \ \forall \vec{v} \ \vec{v} \not\simeq \vec{u} \ \rightarrow \ g\vec{v} \simeq f\vec{v} \tag{3}$$

This is the dual of Extension. The proof of (3) is by induction on f for the universal closure of (3) with respect to $\vec{u}$.

[Function-pairing]

$$\exists h^{k+1} \ \exists u, v \ \forall \vec{x} \ f\vec{x} \simeq hu\vec{x} \ \wedge \ g\vec{x} \simeq hv\vec{x} \tag{4}$$

Much of the expressive and proof theoretic power of arithmetic is due to the representation of finite sequences and finite sets of numbers by numbers. The Function-pairing schema provides a representation of two k-ary fp-functions by a single $(k+1)$-ary fp-function. Namely, f, g are "tagged" withing h by the tags a, b respectively. In other words, writing h_u for $\lambda \vec{x} \, hu\vec{x}$, we have $f = h_a$ and $g = h_b$. (We might require, in addition, that $hu\vec{x} = \bot$ for all atoms $u \neq a, b$, but this is inessential if we include a, b explicitly in the representation.)

[Concrete-atomic-choice] A more interesting form of tagging is provided by the following principles of *Choice*.[2]
For all concrete formulas φ,

$$(\forall \vec{u} \in f \ \exists y \ \varphi[\vec{x}, y]) \ \rightarrow \ \exists g \ \forall \vec{u} \in f \ \varphi[\vec{u}, g\vec{u}] \tag{5}$$

This is analogous to [12, Lemma 2(e)], and is straightforward by induction on f. Suppose the schema holds for f, yielding the function g. To show the schema true for $f' = \{\vec{u} \mapsto v\}f$ suppose it satisfies the premise

$$\forall \vec{x} \in f \ \exists y \ \varphi[\vec{x}, y] \tag{6}$$

Then f' satisfies (6) as well, yielding a g for the conclusion. Also, by (6) there is an atom y such that $\varphi[\vec{u}, y]$, and so the conclusion is satisfied by $\{\vec{u} \mapsto y\}g$ in place of g.
Note that the bounding condition in the choice schemas above is essential: even the simplest case

$$(\forall x^0 \exists y^0 \varphi[x, y]) \ \rightarrow \ \exists f^1 \forall x \ \varphi[x, fx] \tag{7}$$

is false already for $\varphi \equiv x \simeq y$, since the identity function over A is not finite, and so f cannot be finite.

---

[2] It is a special case of Concrete-function-choice above, but one for which we have a proof from the remaining axioms.

4   Imperative Programs for Structure Transformation

We introduced in [8] the programming language **STV** for transformation of fp-structures. **STV** is finitistic in that the iterative construct is bounded by the size of the fp-structures present, generalizing the loop programs of [10] for primitive recursive functions over $\mathbb{N}$. We summarize here the definition of **STV** and its main properties, and refer the reader to [8] for background, examples, and proofs.

4.1   Structure Revisions

We refer to the following three basic transformations of V-structures. In each case we indicate how an input structure σ is transformed by the operation into a structure σ' that differs from σ only as indicated.

1. An *extension* is a command $\mathbf{f}\,\mathbf{t}_1 \cdots \mathbf{t}_k \downarrow \mathbf{q}$ where the $\mathbf{t}_i$'s and $\mathbf{q}$ are all standard terms. The intent is that σ' is identical to σ, except that if $\sigma(\mathbf{f}\,\mathbf{t}_1 \cdots \mathbf{t}_k) = \bot$ then $\sigma'(\mathbf{f}\,\mathbf{t}_1 \cdots \mathbf{t}_k) = \sigma(\mathbf{q})$. Thus, σ' is identical to σ if $\sigma(\mathbf{f}\,\mathbf{t}_1 \cdots \mathbf{t}_k)$ *is* defined;
2. A *contraction,* the dual of an extension, is a command of the form $\mathbf{f}\mathbf{t}_1 \cdots \mathbf{t}_k \uparrow$. The intent is that $\sigma'(\mathbf{f})(\sigma(\mathbf{t}_1),\dots,\sigma(\mathbf{t}_k)) = \bot$, and if $\mathbf{f}\mathbf{t}_1 \cdots \mathbf{t}_k$ is defined then we say that the contraction is *active*. Note that this removes the entry $\langle \sigma(\mathbf{t}_1),\dots,\sigma(\mathbf{t}_k), \sigma(\mathbf{f}\mathbf{t}_1 \cdots \mathbf{t}_k)\rangle$ (if defined) from $\sigma(\mathbf{f})$ but not from $\sigma(\mathbf{g})$ for other identifiers $\mathbf{g}$.
3. An *inception* is a command of the form $\mathbf{c} \Downarrow$, where $\mathbf{c}$ is a token. A common alternative notation is $\mathbf{c} := \mathbf{new}$. The intent is that σ' is identical to σ, except that if $\sigma(\mathbf{c}) = \bot$, then $\sigma'(\mathbf{c})$ is an atom not in the scope of σ.

An extension is *active* if it increases the number of entries, and a contraction is *active* if it reduces it. We refer to extensions, contractions, and inceptions as *revisions*. The identifiers $\mathbf{c}$ and $\mathbf{f}$ in the templates above are the revision's *eigen-identifier*.

4.2   Imperative Programs for Generic PR Computing

Recall that the schema of *recurrence* over a free algebra $\mathbb{A} = \mathbb{A}(C)$, generated from a finite set C of constructors, consists of one equation per each $\mathbf{c} \in \mathsf{C}$:

$$f(\mathbf{c}(z_1, \cdots, z_k), \vec{x}) = g_{\mathbf{c}}(\vec{z}, \vec{x}, f_1, \dots, f_k) \tag{8}$$

where k is $\mathbf{c}$'s arity, and $f_i = f(z_i, \vec{x})$. The set $\mathbf{PR}(\mathbb{A})$ of *primitive recursive (PR)* functions over $\mathbb{A}$ is generated from the constructors of $\mathbb{A}$ by explicit definitions and recurrence over $\mathbb{A}$. The choice of $\mathbb{A}$ is immaterial, since for any non-trivial free algebras $\mathbb{A}$ and $\mathbb{B}$ we have that $\mathbf{PR}(\mathbb{A})$ is interpretable in $\mathbf{PR}(\mathbb{B})$.

A recurrence over a free algebra terminates because the recurrence argument is being consumed. A more generic rendition of the same idea is to adopt an iterative construct whose termination is controlled not only by guards (a qualitative condition), but also by data depletion (a quantitative condition). We define a

V-*variant* to be a finite set T of identifiers in V, to which we refer as T's *components*.fnThe phrase "variant" is borrowed from Dijkstra-style verification of structured imperative programs.

The programs of the programming language **STV** are generated as follows.

1. A revision is a program.
2. If P and Q are programs, then so is $P; Q$.
3. If G is a guard and P, Q are programs, then **if** $[G] \{P\} \{Q\}$ is a program.
4. If G is a guard, T a variant, and Q a program, then **do** $[G] [T] \{Q\}$ is a program.

The denotational semantics of **STV**-programs P is defined, as usual, via an inductively generated binary *yield relation* $\Rightarrow_P$ between an input V-structure and the corresponding output structure. Note that the *configurations* of an executions are simply fp-structures. The details are routine, except for the iteration construct. A program **do** $[G] [T] \{Q\}$ is executed by iterating Q, and entering a new pass only if two conditions are met: (a) G is true for the current configuration (i.e. fp-structure); and (b) the last pass through Q has decreased the total size of the variant.

Keeping track of the variant T along an execution of Q requires in general unbounded memory. This unpleasantness is alleviated by slightly modifying the syntax of the language and the semantic of **do** (see [8]).

The role of **STV** as a generic generalization of primitive recursion is elucidated by the following theorems.

THEOREM 2 [Soundness for **PR**] [8]. *Every program of* **STV** *runs in time and space that are primitive-recursive in the input's size.*

Turning to the completeness of **STV** for primitive recursion, we could prove that **STV** is complete for $\mathbf{PR}(\mathbb{N})$, and invoke the numeric coding of any free algebra. This, however, would fail to establish a direct representation of generic recurrence by **STV**-programs, which is one of the *raisons d'être* of **STV**. The following completeness theorems are proved in [8].

THEOREM 3. *For each free algebra* $\mathbb{A}$, *the collection of* **STV**-*programs is complete for* $\mathbf{PR}(\mathbb{A})$.

Theorem 3 establishes, for any free algebra $\mathbb{A}$ a simple and direct mapping from definitions in $\mathbf{PR}(\mathbb{A})$ to **STV**-programs, If we take **ST**-programs as the Turing-complete computation model of reference, the question remains as to whether every **ST**-program P running within primitive-recursive resources is directly mapped to an equivalent **STV**-program Q.

THEOREM 4. *Every* **ST** *program that runs in* **PR** *time can be effectively mapped into an* **STV** *program for the same function.*

5   An Abstract Form of Parson's Theorem

Parson's Theorem asserts that the Σ_1 Induction Axiom is conservative over Primitive Recursive Arithmetic for Π_2 formulas. In a computational form, the theorem states that the provablye recursive functions of Peano's Arithmetic with induction restricted to Σ_1 formulas are precisely the primitive-recursive functions. Since the proof that every **PR** function (over $\mathbb{N}$) is provable using Σ_1-Induction is fairly straightforward, the thrust of Parson's Theorem is in the converse, which in fact combines two results of different natures. The first states that the Σ_1-Induction Axiom is conservative for Π_2 formulas over the Σ_1-Induction *Rule*. This is a miniature form of Hilbert's programme, as it establishes the finitistic nature of Σ_1-Induction. The second result states that every function provable by the Σ_1-Induction Rule is **PR**, thereby equating proof-theoretic finitism with computational finitism. Indeed, **PR** is a quintessential form of finitistic computing [14,15], in contrast for example to recurrence in higher type, which refers to actual infinity (thereby yielding, for example, Ackermann's Function). We shall focus on generalizing this second part.

Parson's Theorem was proved independently by Charles Parsons [12,13] and Grigori Mints [11].[3] A number of proofs have been presented over the years, mostly by proof theoretic methods, such as the Dialectica interpretation [12], cut-free proofs [13], no-counterexample interpretations [11], and ordinal analysis [17]. Additional insight has been gleaned from model theoretic proofs, such as Herbrand saturated models [1], witness predicates [2], and Herbrand's theorem [4].

Defining function provability is unproblematic for computing over inductive data, but is less obvious when referring to fp-structures. Given a vocabulary $V = \{\mathbf{f}_1 \ldots \mathbf{f}_k\}$, write $\vec{g}^{\,V}$ for a vector of function-variables $g_1 \ldots g_k$ with $\mathfrak{r}(g_i) = \mathfrak{r}(\mathbf{f}_i)$. Let $\mathfrak{C}$ be a class of fp-structures over a vocabulary V, defined by a concrete formula $\psi[\vec{f}^{\,V}]$. We say that a mapping $\mathfrak{F}$ with domain $\mathfrak{C}$ is *defined in* **FST**$_0$ by a concrete formula $\varphi[\vec{f}; \vec{g}]$ if the formula

$$\psi \quad \rightarrow \quad \exists \vec{g}^{\,W} \; \varphi[\vec{f}, \vec{g}]$$

is provable in **FST**$_0$.

Let Φ be a mapping from a class $\mathfrak{C}$ to a class $\mathfrak{D}$, where $\mathfrak{C}$ is a class of accessible V-structures defined by a concrete formula ψ_C, and similarly for $\mathfrak{D}$. We say that Φ is *provable in* **FST**$_0$ if there is a concrete formula φ such that

$$\psi_C[\vec{f}] \quad \rightarrow \quad \exists \vec{g} \; \psi_D[\vec{g}] \wedge \varphi[\vec{f}, \vec{g}]$$

is provable in **FST**$_0$ (with $\vec{f}, \vec{g}$ as parameters for V-structures and W-structures, respectively).

In [7, Theorem 5] we proved that if P is an **STV**-program with V-structures as input and W-structures as output, where V is listed as $\vec{\mathbf{f}}$, then there is an

---

[3] Mints cites Parsons' paper, but mentions his own earlier unpublished presentations.

existential formula $\varphi[\vec{f}, \vec{g}]$ such that

$$\varphi[\vec{f}; \vec{g}] \qquad \text{IFF} \qquad \vec{f} \Rightarrow_P \vec{g} \tag{9}$$

The existentially quantified functions in φ_P define the computation trace of P on input $\vec{f}$. By formalizing the proof of [7, Theorem 5] in $\mathbf{FST}_0$, crucially using the rules of Concrete-induction and Concrete-function-choice, we obtain (details in the full paper).

PROPOSITION 5. *If a mapping Φ as above is computed by an $\mathbf{STV}$ program then it is provable in $\mathbf{FST}_0$.*

For the converse implication, we use structural induction on cut-free sequential proofs for $\mathbf{FST}_0$ to prove (details in the full paper).

LEMMA 6. *If $\exists \vec{g}\, \varphi[\vec{f}; \vec{g}]$ is provable in $\mathbf{FST}_0$ from concrete assumptions $\psi_i[\vec{f}; \vec{h}_i]$ $(i \in I)$, then there is an $\mathbf{STV}$-program P that maps input $(\vec{f}; \vec{h}_i)_{i\in I}$ into an output $\vec{g}_1; \ldots; \vec{g}_m$ such that*

$$\bigwedge_{i\in I} \psi_i[\vec{f}; \vec{h}_i] \;\rightarrow\; \bigvee_{j=1..m} \varphi[\vec{f}; \vec{g}_j] \tag{10}$$

□

Since φ is elementary, we can invoke Proposition 1 and extend P with a program that selects which of the disjuncts in (10) holds, yielding a generic form of the main implication of Parson's Theorem:

THEOREM 7. *If a mapping between fp-structures is provable in $\mathbf{FST}_0$, then it is computed by a program in $\mathbf{STV}$.*

References

1. Avigad, J.: Saturated models of universal theories. Ann. Pure Appl. Log. **118**(3), 219–234 (2002)
2. Buss, S.: First-order proof theory of arithmetic. In: Buss, S.R. (ed.) Handbook of Proof Theory, pp. 79–148. Elsevier, Amsterdam (1998)
3. Euclid: Elements. Dover, New York (1956)
4. Ferreira, F.: A simple proof of Parsons' theorem. Notre Dame J. Formal Log. **46**(1), 83–91 (2005)
5. Hilbert, D.: Über das unendliche. Math. Ann. **95**, 161–190 (1926)
6. Immerman, N.: Descriptive Complexity. Graduate Texts in Computer Science. Springer, New York (1998). https://doi.org/10.1007/978-1-4612-0539-5
7. Leivant, D.: A theory of finite structures. Logical methods in computer science (2019, to appear). Preliminary version: arXiv.org:1808.04949 (2018)
8. Leivant, D., Marion, J.: Primitive recursion in the abstract. Mathematical structures in computer science (2019, to appear). Preliminary version under the title Implicit complexity via structure transformation, in arXiv:1802.03115 (2018)

9. Mayberry, J.: The Foundations of Mathematics in the Theory of Sets. Encyclopedia of Mathematics, vol. 82. Cambridge University Press, Cambridge (2000)
10. Meyer, A., Ritchie, D.: The complexity of loop programs. In: Proceedings of the 1967 22nd National Conference, New York, pp. 465–469 (1967)
11. Mints, G.: Quantifier-free and one-quantifier systems. J. Soviet Math. **1**, 71–84 (1972)
12. Parsons, C.: On a number theoretic choice schema and its relation to induction. In: Intuitionism and Proof Theory, North-Holland, Amsterdam , pp. 459–473 (1970)
13. Parsons, C.: On n-quantier induction. J. Symb. Log. **37**(3), 466–482 (1972)
14. Skolem, T.: Einige bemerkungen zur axiomatischen begründung der mengenlehre. In: Matematikerkongressen in Helsingfors Den femte skandinaviske matematik-erkongressen (1922)
15. Tait, W.: Finitism. J. Philos. **78**, 524–546 (1981)
16. Takahashi, M.: An induction principle in set theory I. Yokohama Math. J. **17**, 53–59 (1969)
17. Takeuti, G.: Proof Theory. Amsterdam (1975)

Knowledge of Uncertain Worlds: Programming with Logical Constraints

Yanhong A. Liu$^{(\boxtimes)}$ and Scott D. Stoller

Computer Science Department, Stony Brook University,
Stony Brook, NY, USA
{liu,stoller}@cs.stonybrook.edu

Abstract. Programming with logic for sophisticated applications must deal with recursion and negation, which have created significant challenges in logic, leading to many different, conflicting semantics of rules. This paper describes a unified language, DA logic, for design and analysis logic, based on the unifying founded semantics and constraint semantics, that support the power and ease of programming with different intended semantics. The key idea is to provide meta-constraints, support the use of uncertain information in the form of either undefined values or possible combinations of values, and promote the use of knowledge units that can be instantiated by any new predicates, including predicates with additional arguments.

Keywords: Datalog · Unrestricted negation and quantification · Meta-constraints · Founded semantics · Constraint semantics · Knowledge unit

1 Introduction

Programming with logic has allowed many design and analysis problems to be expressed more easily and clearly at a high level. Examples include problems in program analysis, network management, security frameworks, and decision support. However, when sophisticated problems require reasoning with negation and recursion, possibly causing contradiction in cyclic reasoning, programming with logic has been a challenge. Many languages and semantics have been proposed, but they have different, conflicting underlying assumptions that are subtle and do not work for all problems.

This paper describes a unified language, DA logic, for design and analysis logic, for programming with logic using logical constraints. It supports logic rules with unrestricted negation in recursion, as well as unrestricted universal and existential quantification. It is based on the unifying founded semantics and constraint semantics, and it supports the power and ease of programming with different intended semantics without causing contradictions in cyclic reasoning.

- The language provides meta-constraints on predicates. These meta-constraints capture the different underlying assumptions of different logic language semantics.

© Springer Nature Switzerland AG 2020
S. Artemov and A. Nerode (Eds.): LFCS 2020, LNCS 11972, pp. 111–127, 2020.
https://doi.org/10.1007/978-3-030-36755-8_8

- The language supports the use of uncertain information in the results of different semantics, in the form of either undefined values or possible combinations of values.
- The language further supports the use of knowledge units that can be instantiated by any new predicates, including predicates with additional arguments.

Together, the language allows complex problems to be expressed clearly and easily, where different assumptions can be easily used, combined, and compared for expressing and solving a problem modularly, unit by unit.

We present examples from different games that show the power and ease of programming with DA logic.

2   Need of Easier Programming with Logic

We discuss the challenges of understanding and programming with negation and recursion. We use a small well-known example, the win-not-win game, for illustration.

Consider the following rule, called the win rule. It says that x is a winning position if there is a move from x to y and y is not a winning position.

```
win(x) ← move(x,y) ∧ ¬ win(y)
```

This seems to be a reasonable rule, because it captures the rule for winning for many games, including in chess for the King to not be captured, giving winning, losing, and draw positions. However, there could be potential problems. For example if there is a move(1,1) for some position 1, then the win rule would imply: win(1) if not win(1), and thus the truth value of win(1) becomes unclear.

Inductive Definitions. Instead of the single win rule, one could use the following three rules to determine the winning, losing, and draw positions.

```
win(x) ← ∃ y | move(x,y) ∧ lose(y)
lose(x) ← ∀ y | ¬ move(x,y) ∨ win(y)
draw(x) ← ¬ win(x) ∧ ¬ lose(x)
```

The first two rules form inductive definitions [6,14], avoiding the potential problems of the single win rule. The base case is the set of positions that have no moves to any other position and thus are losing positions. With winning and losing positions defined, the draw positions are those in cycles of moves that have no moves to losing positions.

However, clearly, these rules are much more cumbersome than the single win rule.

Well-Founded Semantics. Indeed, with well-founded semantics (WFS) [25], which computes a 3-valued model, the single win rule above gives win(x) being True, False, or Unknown for each x, corresponding to x being a winning, losing, or draw position, respectively. However, win(x) being 3-valued does not allow the three outcomes to be used as three predicates or sets for further computation; the three predicates defined by the three rules do allow this.

For example, there is no way to use the Unknown positions explicitly, say to find all reachable nodes following another kind of moves from draw positions. One might try to do it by adding two additional rules to the single win rule:

```
lose(x) ← ¬ win(x)
draw(x) ← ¬ win(x) ∧ ¬ lose(x)
```

However, the result is that `draw(x)` is False for all positions that `win(x)` is True or False, and is Unknown for all draw positions.

Stable Model Semantics. Stable model semantics (SMS) [13] computes a set of 2-valued models, instead of a single 3-valued model. It has been used for solving many constraint problems in answer set programming (ASP), because its set of 2-valued models can provide the set of satisfying solutions.

For example, for the single win rule, if besides winning and losing positions, there is a separate cycle of even length, say move(1,2) and move(2,1), then instead of `win` being Unknown for 1 and 2 as in WFS, SMS returns two models: one with `win` being True for 1 and other winning positions but not 2, and one with `win` being True for 2 and other winning positions but not 1. This is a very different interpretation of the win-not-win rule.

However, for the single rule above, when there are draw positions, SMS may also return just an empty set, that is, a set with no models at all. For example, if besides winning and losing positions, there is a separate cycle of moves of odd length, say simply move(1,1), then SMS returns simply the empty set. This is clearly undesired for the win-not-win game.

Founded Semantics and Constraint Semantics. Founded semantics and constraint semantics [18] unify different prior semantics. They allow different underlying assumptions to be specified for each predicate, and compute the desired semantics as a simple least fixed point to return a 3-valued model and, if there are undefined values, as constraint solving to return a set of 2-valued models.

For the win-not-win game, one can write the single win rule, with the default assumption that `win` is *complete*, that is, the win rule is the only rule that infers `win`, which is an implicit assumption underlying WFS and SMS.

- With founded semantics, the three rules that use inductive definitions can be automatically derived, and True, False, and Undefined positions for `win` are inferred, corresponding to the three predicates from inductive definitions and the 3-valued results from WFS.
- Then constraint semantics, if desired, computes all combinations of True and False values for the Undefined values for the draw positions, that satisfy all the rules as constraints. It equals SMS for the single win rule.

Both WFS and SMS also assume that if nothing is said about some p, then p is false. When this is not desired, some programming tricks are used to get around it. For example, with SMS, to allow p to be possibly true in some models, one can introduce some new q and two new rules as below, to make it possible that, in some models, p is true and q is false.

$$p \leftarrow \neg\, q$$
$$q \leftarrow \neg\, p$$

Founded semantics and constraint semantics allow p to be simply declared as *uncertain*.

Both WFS and SMS also assume that if all ways that can infer p require using p in the condition of some rule, then p is false. Founded semantics and constraint semantics allow this reasoning to be used where desired, by applying it if p is declared as *closed*.

Founded semantics and constraint semantics also allow unrestricted universal and existential quantifications and unrestricted nesting of Boolean conditions; these are not supported in WFS and SMS.

However, founded semantics and constraint semantics alone do not address how to use different semantics seamlessly in a single logic program.

Programming with Logical Constraints. Because different assumptions and semantics help solve different problems or different parts of a problem, easier programming with logic requires supporting all assumptions and semantics in a simple and integrated design.

This paper treats different assumptions as different meta-constraints for expressing a problem or parts of a problem, and support results from different semantics to be used easily and directly. For the win-not-win example:

- We name the positions for which win is true, false, and undefined in founded semantics using three predicates, win.T, win.F, and win.U, corresponding exactly to the inductively defined win, lose, and draw. These predicates can be used explicitly and directly for further reasoning, unlike with the truth values of WFS or founded semantics.
- We let CS be the constraint semantics of a set of rules and facts. For m ∈ CS, we use m.win(x) to denote the truth value of win(x) in model m. Predicate CS(m) means exactly m ∈ CS and can be used directly for further reasoning, unlike the set of models in SMS or constraint semantics.

Table 1 summarizes the meta-constraints that can be used to express different assumptions, corresponding declarations and resulting predicates in founded semantics and constraint semantics, and corresponding other prior semantics if all predicates use the same meta-constraint. Columns 2 and 4 are presented and proved in our prior work [18]. Columns 1 and 3 are introduced in DA logic.

More fundamentally, we must enable easy specification of problems with reusable parts and where different parts may use different assumptions and semantics. To that end, we support instantiation and re-use of existing parts, and allow predicates in any existing parts to be bound to other given predicates, including predicates with additional arguments.

Even with all this power, DA logic is decidable, because it does not include function symbols and is over finite domains.

Table 1. Meta-constraints and corresponding prior semantics.

Meta-constraint on predicate P	Founded/Constraint semantics		Other prior semantics
	Declarations on P	Resulting predicates	
certain(P)	certain	P.T, P.F	Stratified (Perfect, Inductive Definition)
open(P)	uncertain, not complete	P.T, P.F, P.U $m.P$ for $m \in K$.CS	First-Order Logic
complete(P)	uncertain, complete	As above	Fitting (Kripke-Kleene) Supported
closed(P)	uncertain, complete, closed	As above	WFS SMS

3   DA Logic

This section presents the syntax and informal meaning of DA logic, for design and analysis logic. The constructs described in the paragraphs on "Conjunctive rules with unrestricted negation", "Disjunction", and "Quantification" appear in our prior work on founded semantics and constraint semantics [18]. The other features are new.

Knowledge Unit. A *program* is a set of knowledge units. A *knowledge unit*, abbreviated as *kunit*, is a set of rules, facts, and meta-constraints, defined below. The definition of a kunit has the following form, where K is the name of the kunit, and *body* is a set of rules, facts, meta-constraints, and instantiations of other kunits:

```
kunit K:
    body
```

The scope of a predicate is the kunit in which it appears. Predicates with the same name, but appearing in different kunits, are distinct.

Example. A kunit for the single win rule is

```
kunit win_unit:
    win(x) ← move(x,y) ∧ ¬ win(y)
```

∎

Kunits provide structure and allow knowledge to be re-used in other contexts by instantiation, as described below.

Conjunctive Rules with Unrestricted Negation. We first present a simple core form of logic rules and then describe additional constructs that can appear in rules. The core form of a rule is the following, where any P_i may be preceded with¬:

$$Q(X_1, ..., X_a) \leftarrow P_1(X_{11}, ..., X_{1a_1}) \wedge ... \wedge P_h(X_{h1}, ..., X_{ha_h}) \qquad (1)$$

Q and the P_i are predicates, each argument X_k and X_{ij} is a constant or a variable, and each variable in the arguments of Q must also be in the arguments of some P_i. In arguments of predicates in example programs, we use numbers for constants and letters for variables.

If $h = 0$, there is no P_i or X_{ij}, and each X_k must be a constant, in which case $Q(X_1, ..., X_a)$ is called a *fact*. For the rest of the paper, "rule" refers only to the case where $h \geq 1$, in which case the left side of the backward implication is called the *conclusion*, the right side is called the *body*, and each conjunct in the body is called a *hypothesis*.

These rules have the same syntax as in Datalog with negation, but are used here in a more general setting, because variables can range over complex values, such as constraint models, as described below.

Predicates as Sets. We use a syntactic sugar in which a predicate P is also regarded as the set of x such that $P(x)$ holds. For example, we may write `move = {(1,2), (1,3)}` instead of the facts `move(1,2)` and `move(1,3)`; to ensure the equality holds, this shorthand is used only when there are no other facts or rules defining the predicate.

Disjunction. The hypotheses of a rule may be combined using disjunction as well as conjunction. Conjunction and disjunction may be nested arbitrarily.

Quantification. Existential and universal quantifications in the hypotheses of rules are written using the following notations:

$$\begin{array}{l} \exists X_1, ..., X_n \mid Y \text{ existential quantification} \\ \forall X_1, ..., X_n \mid Y \text{ universal quantification} \end{array} \qquad (2)$$

In quantifications of this form, the domain of each quantified variable is the set of all constants in the containing kunit. As syntactic sugar, a domain can be specified for a quantified variable, using a unary predicate regarded as a set. For example, $\exists$ `x` $\in$ `win | move(x,x)` is syntactic sugar for $\exists$ `x | win(x)` $\wedge$ `move(x,x)`, and $\forall$ `x in win | move(x,x)` is syntactic sugar for $\forall$ `x | ¬win(x)` $\vee$ `move(x,x)`.

Meta-constraints. Assumptions about predicates are indicated in programs using the meta-constraints in the first column of Table 1. Each meta-constraint specifies the declarations listed in the second column of Table 1. For example, if a kunit contains `open(P)`, we say that P is declared uncertain and incomplete in that kunit. In each kunit, at most one meta-constraint may be given for each predicate.

A predicate declared *certain* means that each assertion of the predicate has a unique true (T) or false (F) value. A predicate declared *uncertain* means that each assertion of the predicate has a unique true, false, or undefined (U) value. A predicate declared *complete* means that all rules with that predicate in the conclusion are given in the containing kunit. A predicate declared *closed* means that an assertion of the predicate is made false, called *self-false*, if inferring it to be true using the given rules and facts requires assuming itself to be true.

A predicate in the conclusion of a rule is said to be *defined* using the predicates or their negation in the hypotheses of the rule, and this defined-ness relation is transitive. A predicate must be declared uncertain (using one of the corresponding meta-constraints) if it is defined transitively using its own negation, or is defined using an uncertain predicate; otherwise, it may be declared certain or uncertain and is by default certain. A predicate may be declared complete or not only if it is uncertain, and it is by default complete. If a meta-constraint is not given for a predicate, these default declarations apply.

Using Kunits with Instantiation. The body of a kunit K_1 can use another kunit K using an instantiation of the form:

$$\text{use } K \ (P_1 = Q_1(Y_{1,1}, ..., Y_{1,b_1}), \ ..., \ P_n = Q_n(Y_{n,1}, ..., Y_{n,b_n})) \tag{3}$$

This has the same effect as applying the following substitution to the body of K and inlining the result in the body of K_1: for each i in $1..n$, replace each occurrence $P_i(X_1, ..., X_a)$ of predicate P_i with $Q_i(X_1, ..., X_a, Y_{i,1}, ..., Y_{i,b_i})$. Note that arguments of Q_i specified in the use construct are appended to the argument list of each occurrence of P_i in K, hence the number of such arguments must be $\text{arity}(Q_i) - \text{arity}(P_i)$. The check for having at most one meta-constraint per predicate, and the determination of default declarations, are performed after expansion of all use constructs. A kunit K_1 has a *use-dependency* on kunit K if K_1 uses K. The use-dependency relation must be acyclic.

Example. For the example kunit `win_unit` given earlier in this section, the following kunit is an instantiation of the win-not-win game with different predicates for moving and winning:

```
kunit win2_unit:
  use win_unit (move = move2, win = win2)
```

■

In some logic programming languages, including our prior work on founded semantics [18], a program is an unstructured set of rules and facts. The structure and re-use provided by kunits is vital for development of larger programs for practical applications.

Referencing Founded Semantics. The founded semantics of a predicate P can be referenced using special predicates $P.\text{T}$, $P.\text{F}$, and $P.\text{U}$. For each of the three truth values t, $P.t(c_1, ..., c_n)$ is true if $P(c_1, ..., c_n)$ has truth value t, and is false otherwise. To ensure that the semantics of P is fully determined before these predicates are used, these predicates cannot be used in rules defining P or any predicate on which P depends. Predicates that reference founded semantics are implicitly declared certain and can appear only in rule bodies.

When referencing the undefined part of a predicate, it is sometimes desirable to prune uninteresting values. For example, consider the rule `draw(x) ←win.U(x)`. If the kunit contains constants representing players as well as positions, and $\text{win}(X)$ is undefined when X is a player, and the user wants `draw` to hold only for positions, then the user could add to the rule an additional hypothesis `position(x)`, defined to hold only for positions.

Referencing Constraint Semantics. The constraint semantics of a kunit K can be referenced in another kunit K_1 using the special predicate K.CS, where K is the name of another kunit in the program. Using this special predicate in any rule in K_1 has the effect of adding all of the constraint models of K to the domain (that is, set of constants) of K_1. In other words, the possible values of variables in K_1 include the constraint models of K. The assertion K.CS(X) is true when X is a constraint model of K and is false for all other constants. The constraint models of a kunit K can be referenced using K.CS only if K does not reference its own founded semantics (using predicates such as P.U). When the value of a variable X is a constraint model of K, a predicate P of K can be accessed using the notation $X.P(...)$. If the value of X is not a constraint model, or P is not a predicate defined in that constraint model, then $X.P(...)$ is undefined, regardless of the arguments. Predicates that reference constraint semantics are implicitly declared certain and can appear only in rule bodies. A kunit K_1 has a *CS-dependency* on another kunit K if K_1 uses K.CS. The CS-dependency relation must be acyclic.

4   Formal Definition of Semantics of DA Logic

This section extends the definitions of founded semantics and constraint semantics in [18] to handle the new features of DA logic.

Handling kunits is relatively straightforward. Since each kunit defines a distinct set of predicates, the founded semantics of the program is simply a collection of the founded semantics of its kunits, and similarly for the constraint semantics. All **use** constructs in a kunit are expanded, as described in Sect. 3, before considering its semantics. Therefore, the constants, facts, rules, and meta-constraints of a kunit include the corresponding elements (appropriately instantiated) of the kunits it uses.

Handling references to founded semantics and constraint semantics requires changes in the definitions of domain, literal, interpretation, and dependency graph.

Handling disjunction, which is mentioned as an extension in [18] but not considered in the detailed definitions, requires changes in the definition of completion rules and the handling of closed predicates.

The paragraphs "Founded semantics of DA logic without closed declarations", "Least fixed point", and "Constraint semantics of DA logic" are essentially the same as in [18]; they are included for completeness.

Atoms, Literals, and Projection. Let π be a program. Let K be a kunit in π. A predicate is *intensional* in K if it appears in the conclusion of at least one rule in K; otherwise, it is *extensional* in K. The *domain* of K is the set of constants in K plus, for each kunit K_1 such that K_1.CS appears in K, the constraint models of K_1, computed as defined below. The requirement that the CS-dependency relation is acyclic ensures the constraint models of K_1 are determined before the semantics of K is considered.

An *atom* of K is a formula $P(c_1, ..., c_a)$ formed by applying a predicate P in K with arity a to a constants in the domain of K. A *literal* of K is a formula of the form $P(c_1, ..., c_a)$ or $P.\text{F}(c_1, ..., c_a)$, for any atom $P(c_1, ..., c_a)$ of K where P is a predicate that does not reference founded semantics or constraint semantics. These are called *positive literals* and *negative literals* for $P(c_1, ..., c_a)$, respectively. A set of literals is *consistent* if it does not contain positive and negative literals for the same atom. The *projection* of a kunit K onto a set S of predicates, denoted $Proj(K, S)$, contains all facts of K for predicates in S and all rules of K whose conclusions contain predicates in S.

Interpretations, Ground Instances, Models, and Derivability. An *interpretation I* of K is a consistent set of literals of K. Interpretations are generally 3-valued. For predicates that do not reference founded or constraint semantics, $P(c_1, ..., c_a)$ is *true* (T) in I if I contains $P(c_1, ..., c_a)$, is *false* (F) in I if I contains $P.\text{F}(c_1, ..., c_a)$, and is *undefined* (U) in I if I contains neither $P(c_1, ..., c_a)$ nor $P.\text{F}(c_1, ..., c_a)$. For the predicates that reference founded semantics, for each of the three truth values t, $P.t(c_1, ..., c_a)$ is true in I if $P(c_1, ..., c_a)$ has truth value t in I, and is false otherwise. For the predicates that reference constraint semantics, $K_1.\text{CS}(c)$ is true in I if c is a constraint model of K_1, as defined below, and is false otherwise; the requirement that the CS-dependency relation is acyclic ensures that the constraint models of K_1 are determined before the semantics of $K_1.\text{CS}(c)$ is considered. If c is a constraint model that provides a truth value for $P(c_1, ..., c_a)$, then $c.P(c_1, ..., c_a)$ has the same truth value in I that $P(c_1, ..., c_a)$ has in c, otherwise it is undefined. An interpretation I of K is *2-valued* if every atom of K is true or false in I, that is, no atom is undefined. Interpretations are ordered by set inclusion $\subseteq$.

A *ground instance* of a rule R is any rule that can be obtained from R by expanding universal quantifications into conjunctions over all constants in the domain, instantiating existential quantifications with constants, and instantiating the remaining variables with constants. An interpretation is a *model* of a kunit if it contains all facts in the kunit and satisfies all rules of the kunit, interpreted as formulas in 3-valued logic [10], that is, for each ground instance of each rule, if the body is true, then so is the conclusion. A collection of interpretations, one per kunit in a program π, is a *model* of π if each interpretation is a model of the corresponding kunit.

The *one-step derivability* operator T_K performs one step of inference using rules of K, starting from a given interpretation. Formally, $C \in T_K(I)$ iff C is a fact of K or there is a ground instance R of a rule in K with conclusion C such that the body of R is true in I.

Dependency Graph. The *dependency graph $DG(K)$* of kunit K is a directed graph with a node for each predicate of K that does not reference founded semantics and constraint semantics (including these predicates is unnecessary, because they cannot appear in conclusions), and an edge from Q to P labeled $+$ (respectively, $-$) if a rule whose conclusion contains Q has a positive (respectively, negative) hypothesis that contains P. If the node for predicate P is in a cycle containing only positive edges, then P has *circular positive dependency* in K; if it is in a cycle containing a negative edge, then P has *circular negative dependency* in K.

Founded Semantics of DA Logic Without Closed Declarations. We first define a version of founded semantics, denoted $Founded_0$, that does not take declarations of predicates as closed into account; below we extend the definition to handle those declarations. Intuitively, the *founded model* of a kunit K ignoring closed declarations, denoted $Founded_0(K)$, is the least set of literals that are given as facts or can be inferred by repeated use of the rules. We define $Founded_0(K) = LFPbySCC(NameNeg(Cmpl(K)))$, where functions $Cmpl$, $NameNeg$, and $LFPbySCC$, are defined as follows.

Completion. The completion function, $Cmpl(K)$, returns the *completed* version of K. Formally, $Cmpl(K) = AddInv(Combine(K))$, where $Combine$ and $AddInv$ are defined as follows.

The function $Combine(K)$ returns the kunit obtained from K by replacing the facts and rules defining each uncertain complete predicate Q with a single *combined rule* for Q that is logically equivalent to those facts and rules. The detailed definition of combined rule is the same as in [18], except generalized in a straightforward way to allow rule bodies to contain disjunction and quantifiers. Similar completion rules are used in [5,10].

The function $AddInv(K)$ returns the kunit obtained from K by adding, for each uncertain complete predicate Q, a *completion rule* that derives negative literals for Q. The completion rule for Q is obtained from the inverse of the combined rule defining Q (recall that the inverse of $C \leftarrow B$ is $\neg C \leftarrow \neg B$), by putting the body of the rule in negation normal form, that is, using equivalences of predicate logic to move negation inwards and eliminate double negations, so that negation is applied only to atoms.

Least Fixed Point. Explicit use of negation is eliminated before the least fixed point is computed, by applying the function $NameNeg$. The function $NameNeg(K)$ returns the kunit obtained from K by replacing each $\neg P(X_1, ..., X_a)$ with $P.\mathrm{F}(X_1, ..., X_a)$.

The function $LFPbySCC(K)$ uses a least fixed point to infer facts for each strongly connected component (SCC) in the dependency graph of K, as follows. Let $S_1, ..., S_n$ be a list of the SCCs in dependency order, so earlier SCCs do not depend on later ones; it is easy to show that any linearization of the dependency order leads to the same result for $LFPbySCC$. For convenience, we overload S_i to also denote the set of predicates in the SCC S_i. Define $LFPbySCC(K) = I_n$, where $I_0 = \emptyset$ and $I_i = AddNeg(LFP(T_{I_{i-1} \cup Proj(K,S_i)}), S_i)$ for $i \in 1..n$. $LFP(f)$ is the least fixed point of function f. The least fixed point is well-defined, because $T_{I_{i-1} \cup Proj(K,S_i)}$ is monotonic, because the kunit K was transformed by $NameNeg$ and hence does not contain negation. The function $AddNeg(I, S)$ returns the interpretation obtained from interpretation I by adding *completion facts* for certain predicates in S to I; specifically, for each such predicate P, for each combination of values $v_1, ..., v_a$ of arguments of P, if I does not contain $P(v_1, ..., v_a)$, then add $P.\text{F}(v_1, ..., v_a)$.

Founded Semantics of DA Logic with Closed Declarations. Informally, when an uncertain complete predicate of kunit K is declared *closed*, an atom A of the predicate is false in an interpretation I, called *self-false* in I, if every ground instance of rules that concludes A, or recursively concludes some hypothesis of that rule instance, has a hypothesis that is false or, recursively, is self-false in I. A formal definition of $SelfFalse_K(I)$, the set of self-false atoms of kunit K with respect to interpretation I, appears in [18]; it is the same as the definition of greatest unfounded set [25], except limited to closed predicates. The definition does not take disjunction into account, so each rule containing disjunction is put into disjunctive normal form (DNF) and then replaced with multiple rules (one per disjunct of the DNF) not containing disjunction, before determining the self-false atoms.

The founded semantics is defined by repeatedly computing the semantics given by $Founded_0$ (the founded semantics without closed declarations) and then setting self-false atoms to false, until a least fixed point is reached. For a set S of positive literals, let $\neg \cdot S = \{P.\text{F}(c_1, ..., c_a) \mid P(c_1, ..., c_a) \in S\}$. For a kunit K and an interpretation I, let $K \cup I$ denote K with the literals in I added to its body. Formally, the founded semantics is $Founded(K) = LFP(F_K)$, where $F_K(I) = Founded(K \cup I) \cup \neg \cdot SelfFalse_K(Founded(K \cup I))$.

Constraint Semantics of DA Logic. Constraint semantics is a set of 2-valued models based on founded semantics. A *constraint model* of K is a consistent 2-valued interpretation M of K such that M is a model of $Cmpl(K)$ and such that $Founded(K) \subseteq M$ and $\neg \cdot SelfFalse_K(M) \subseteq M$. Let $Constraint(K)$ denote the set of constraint models of K. Constraint models can be computed from $Founded(K)$ by iterating over all assignments of true and false to atoms that are undefined in $Founded(K)$, and checking which of the resulting interpretations satisfy all rules in $Cmpl(K)$ and satisfy $\neg \cdot SelfFalse_K(M) \subseteq M$.

Properties of DA Logic Semantics. The following theorems express the most important properties of the semantics.

Theorem 1. *The founded model and constraint models of a program π are consistent.*

Proof: First we consider founded semantics. Each kunit in the program defines a distinct set of predicates, so consistency can be established one kunit at a time, considering them in CS-dependency order. For each kunit K, the proof of consistency is a straightfoward extension of the proof of consistency of founded semantics [17, Theorem 1]. The extension is needed to show that consistency holds for the new predicates that reference founded semantics and constraint semantics.

For predicates that reference founded semantics, we prove this for each SCC S_i in the dependency graph for K; the proof is by induction on i. The predicates used in SCC S_i to reference founded semantics have the same truth values as the referenced predicates in earlier SCCs, and by the induction hypothesis, the interpretation computed for predicates in earlier SCCs is consistent.

For predicates that reference constraint semantics, the proof is by induction on the kunits in CS-dependency order. The predicates used in kunit K to reference constraint semantics have the same truth values as the referenced predicates in earlier kunits, and by the induction hypothesis, the interpretation computed for predicates in earlier kunits is consistent.

For constraint semantics, note that constraint models are consistent by definition. ∎

Theorem 2. *The founded model of a kunit K is a model of K and $Cmpl(K)$. The constraint models of K are 2-valued models of K and $Cmpl(K)$.*

Proof: The proof that $Founded(K)$ is a model of $Cmpl(K)$ is essentially the same as the proof that $Founded(\pi)$ is a model of $Cmpl(\pi)$ [17, Theorem 2], because the proof primarily depends on the behavior of $Cmpl$, $AddNeg$, and the one-step derivability operator, and they handle atoms of predicates that reference founded semantics and constraint semantics in exactly the same way as other atoms. Constraint models are 2-valued models of $Cmpl(K)$ by definition. Any model of $Cmpl(K)$ is also a model of K, because K is logically equivalent to the subset of $Cmpl(K)$ obtained by removing the completion rules added by $AddInv$. ∎

Theorem 3. *DA logic is decidable.*

Proof: DA logic has a finite number of constants from given facts, and has sets of finite nesting depths bounded by the depths of CS-dependencies. In particular, it has no function symbols to build infinite domains in recursive rules. Thus, DA logic is over finite domains and is decidable. ∎

5   Additional Examples

We present additional examples that show the power of our language. They are challenging or impossible to express and solve using prior languages and semantics. We use - - to prefix comments.

Same Different Games. The same win-not-win game can be over different kinds of moves, forming different games, as introduced with kunit instantiation. However, the fundamental winning, losing, or draw situations stay the same, parameterized by the moves. The moves could also be defined easily using another kunit instantiation.

Example. A new game can use winning, losing, draw positions defined by win_unit in Sect. 2, whose moves use paths defined by path_unit, whose edges use given links.

```
kunit path_unit:
  path(x,y) ← edge(x,y)
  path(x,y) ← edge(x,z) ∧ path(z,y)

kunit win_path_unit:
  link = {(1,2), (1,3), ...}    -- shorthand for link(1,2), link(1,3), ...
  use path_unit (edge = link)   -- instantiate path_unit with edge replaced
                                -- by link
  use win_unit (move = path)    -- instantiate win_unit with move replaced
                                -- by path
```

One could also define `edge` in place of `link` above, and then `path_unit` can be used without rebinding the name `edge`, as follows.

```
kunit win_path_unit:             -- as above
  edge = {(1,2), (1,3), ...}     -- as above but use edge in place of link
  use path_unit ()               -- as above but without replacing edge by
                                 -- link
  use win_unit (move = path)     -- as above
```                                                                    ∎

Defined from Undefined Positions. Sets and predicates can be defined using the set of values of arguments for which a given predicate is undefined. This is not possible in previous 3-valued logic like WFS, because anything depending on undefined can only be undefined.

Example. Using the win-not-win game, the predicates `move_to_draw` and `reach_from_draw` below define the set of positions that have a move to a draw position, and the set of positions that have a special move from a draw position, respectively.

```
kunit draw_unit:
  move = {(1,1), (2,3), (3,1)}
  use win_unit ()

  move_to_draw(x) ← move(x,y) ∧ win.U(y)

  special_move = {(1,4), (4,2)}
  use path_unit (edge = special_move)

  reach_from_draw(y) ← win.U(x) ∧ path(x,y)
```

In `draw_unit`, we have `win.U(1)`, that is, 1 is a draw position. Then we have `move_to_draw(3)`, and we have `reach_from_draw(4)` and `reach_from_draw(2)`.

Note that we could copy the single win rule here in place of `use win_unit ()` and obtain an equivalent `draw_unit`. We avoid copying when possible because this is a good principle, and in general, a kunit may contain many rules and facts.   ∎

Unique Undefined Positions. Among the most critical information is information that is true in all possible ways of satisfying given constraints but cannot be determined to be true by just following founded reasoning. Having both founded semantics and constraint semantics at the same time allows one to find such information.

Example. Predicate `unique` in `cmp_unit` below finds positions in the game in `win_unit1` that are U in the founded model but, if a constraint model exists, are winning in all possible models in constraint semantics.

```
kunit win_unit1:
  prolog ← ¬ asp
  asp ← ¬ prolog
  move(1,0) ← prolog
  move(1,0) ← asp
  move(1,1)
  use win_unit ()

kunit cmp_unit:
  use win_unit1 ()

  unique(x) ← win.U(x) ∧ ∃ m ∈ win_unit1.CS
                     ∧ ∀ m ∈ win_unit1.CS | m.win(x)
```

In `win_unit1`, founded semantics gives `move.T(1,1)`, `move.U(1,0)`, `win.U(0)`, and `win.U(1)`. `win_unit1.CS = {{move(1,1), move(1,0), win(1)}}`, that is, `win(1)` is true, and `win(0)` is false. So `win.U(1)` and `win.U(0)` are imprecise, and `unique(1)` is true in `cmp_unit`.   ∎

Multiple Uncertain Worlds. Given multiple worlds with different models, different uncertainties can arise from different worlds, yielding multiple uncertain worlds. It is simple to represent this using predicates that are possibly 3-valued and that are parameterized by a 2-valued model.

Example. The game in `win_unit2` uses `win_unit` on a set of moves. The game in `win_set_unit` has its own moves, but the moves are valid if and only if they start from a position that is a winning position in a model in the constraint semantics of `win_unit2`.

```
kunit win_unit2:
  move = {(1,4),(4,1)}
  use win_unit ()

kunit win_set_unit:
  move = {(1,2),(2,3),(3,1),(4,4),(5,6)}
  valid_move(x,y,m) ← move(x,y), win_unit2.CS(m), m.win(x)

  use win_unit (move = valid_move(m), win = valid_win(m))

  win_some(x) ← valid_win(x,m)
  win_each(x) ← win_some(x) ∧ ∀ m ∈ win_unit2.CS | valid_win(x,m)
```

In `win_unit2`, there is a 2-edge cycle of moves, so `win_unit2.CS = {m1,m2}`, where `m1.win = {1}` and `m2.win = {4}`. In `win_set_unit`, each m in `win_unit2` leads to a separately defined predicate `valid_move` under argument m, which is then used to define a separate predicate `valid_win` under argument m by instantiating `win_unit` with `move` and `win` parameterized by additional argument m.   ∎

6   Related Work and Conclusion

Many logic languages and semantics have been proposed. Several overview articles [2,11,20,21,24] give a good sense of the complications and challenges when there is unrestricted negation. Notable different semantics include Clark completion [5] and similar additions, e.g., [4,12,15,19,22,23], Fitting semantics or Kripke-Kleene semantics [10], supported model semantics [1], stratified semantics [1], WFS [25], and SMS [13]. Note that these semantics disagree, in contrast to different styles of semantics that agree [9].

There are also a variety of works on relating and unifying different semantics. These include Dung's study of relationships [8], partial stable models, also called stationary models [20], Loop fomulas [16], FO(ID) [7], and founded semantics and constraint semantics [18]. FO(ID) is more powerful than works prior to it, by supporting both first-order logic and inductive definitions while also being similar to SMS [3]. However, it does not support any 3-valued semantics. Founded semantics and constraint semantics uniquely unify different semantics, by capturing their different assumptions using predicates declared to be certain, complete, and closed, or not.

However, founded semantics and constraint semantics by themselves do not provide a way for different semantics to be used for solving different parts of a problem or even the same part of the problem. DA logic supports these, and supports everything completely declaratively, in a unified language.

Specifically, DA logic allows different assumptions under different semantics to be specified easily as meta-constraints, and allows the results of different semantics to be built upon, including defining predicates using undefined values in a 3-valued model and using models in a set of 2-valued models, and parameterizing predicates by a set of 2-valued models. More fundamentally, DA logic allows different parts of a problem to be solved with different knowledge units, where every predicate is a parameter that can be instantiated with new predicates, including new predicates with additional arguments. These are not supported in prior languages.

Among many directions for future work, one particularly important and intriguing problem is to study precise complexity guarantees for inference and queries for DA logic.

Acknowledgments. This work was supported in part by NSF under grants CCF-1414078, CNS-1421893, and IIS-1447549.

References

1. Apt, K.R., Blair, H.A., Walker, A.: Towards a theory of declarative knowledge. In: Foundations of Deductive Databases and Logic Programming, pp. 89–148. Morgan Kaufman (1988)
2. Apt, K.R., Bol, R.N.: Logic programming and negation: a survey. J. Log. Program. **19**, 9–71 (1994)

3. Bruynooghe, M., Denecker, M., Truszczynski, M.: First order logic with inductive definitions for model-based problem solving. AI Mag. **37**(3), 69–80 (2016)
4. Chan, D.: Constructive negation based on the completed database. In: Proceedings of the 5th International Conference and Symposium on Logic Programming, pp. 111–125. MIT Press (1988)
5. Clark, K.L.: Negation as failure. In: Gallaire, H., Minker, J. (eds.) Logic and Databases, pp. 293–322. Plenum Press (1978)
6. Dasseville, I., Van der Hallen, M., Janssens, G., Denecker, M.: Semantics of templates in a compositional framework for building logics. Theory Pract. Log. Program. **15**(4–5), 681–695 (2015)
7. Denecker, M., Ternovska, E.: A logic of nonmonotone inductive definitions. ACM Trans. Comput. Log. **9**(2), 14 (2008)
8. Dung, P.M.: On the relations between stable and well-founded semantics of logic programs. Theoret. Comput. Sci. **105**(1), 7–25 (1992)
9. Ershov, Y.L., Goncharov, S.S., Sviridenko, D.I.: Semantic foundations of programming. In: Budach, L., Bukharajev, R.G., Lupanov, O.B. (eds.) FCT 1987. LNCS, vol. 278, pp. 116–122. Springer, Heidelberg (1987). https://doi.org/10.1007/3-540-18740-5_28
10. Fitting, M.: A Kripke-Kleene semantics for logic programs. J. Log. Program. **2**(4), 295–312 (1985)
11. Fitting, M.: Fixpoint semantics for logic programming: a survey. Theoret. Comput. Sci. **278**(1), 25–51 (2002)
12. Foo, N.Y., Rao, A.S., Taylor, A., Walker, A.: Deduced relevant types and constructive negation. In: Proceedings of the 5th International Conference and Symposium on Logic Programming, pp. 126–139 (1988)
13. Gelfond, M., Lifschitz, V.: The stable model semantics for logic programming. In: Proceedings of the 5th International Conference and Symposium on Logic Programming, pp. 1070–1080. MIT Press (1988)
14. Hou, P., De Cat, B., Denecker, M.: FO(FD): extending classical logic with rule-based fixpoint definitions. Theory Pract. Log. Program. **10**(4–6), 581–596 (2010)
15. Jaffar, J., Lassez, J.-L., Maher, M.J.: Some issues and trends in the semantics of logic programming. In: Shapiro, E. (ed.) ICLP 1986. LNCS, vol. 225, pp. 223–241. Springer, Heidelberg (1986). https://doi.org/10.1007/3-540-16492-8_78
16. Lin, F., Zhao, Y.: ASSAT: computing answer sets of a logic program by sat solvers. Artif. Intell. **157**(1–2), 115–137 (2004)
17. Liu, Y.A., Stoller, S.D.: Founded semantics and constraint semantics of logic rules. Computing Research Repository arXiv:1606.06269 [cs.LO], June 2016 (Revised April 2017)
18. Liu, Y.A., Stoller, S.D.: Founded semantics and constraint semantics of logic rules. In: Artemov, S., Nerode, A. (eds.) LFCS 2018. LNCS, vol. 10703, pp. 221–241. Springer, Cham (2018). https://doi.org/10.1007/978-3-319-72056-2_14
19. Lloyd, J.W., Topor, R.W.: Making Prolog more expressive. J. Log. Program. **1**(3), 225–240 (1984)
20. Przymusinski, T.C.: Well-founded and stationary models of logic programs. Ann. Math. Artif. Intell. **12**(3), 141–187 (1994)
21. Ramakrishnan, R., Ullman, J.D.: A survey of deductive database systems. J. Log. Program. **23**(2), 125–149 (1995)
22. Sato, T., Tamaki, H.: Transformational logic program synthesis. In: Proceedings of the International Conference on Fifth Generation Computer Systems, pp. 195–201 (1984)

23. Stuckey, P.J.: Constructive negation for constraint logic programming. In: Proceedings of the 6th Annual IEEE Symposium on Logic in Computer Science, pp. 328–339 (1991)
24. Truszczynski, M.: An introduction to the stable and well-founded semantics of logic programs. In: Kifer, M., Liu, Y.A. (eds.) Declarative Logic Programming: Theory, Systems, and Applications, pp. 121–177. ACM and Morgan & Claypool (2018)
25. Van Gelder, A., Ross, K., Schlipf, J.S.: The well-founded semantics for general logic programs. J. ACM **38**(3), 620–650 (1991)

A Globally Sound Analytic Calculus
for Henkin Quantifiers

Matthias Baaz[1](✉)[iD] and Anela Lolic[2][iD]

[1] Institute of Discrete Mathematics and Geometry, TU Wien, Vienna, Austria
baaz@logic.at
[2] Institute of Logic and Computation, TU Wien, Vienna, Austria
anela@logic.at

Abstract. This paper presents a methodology to construct globally sound but possibly locally unsound analytic calculi for partial theories of Henkin quantifiers. It is demonstrated that locally sound analytic calculi do not exist for any reasonable fragment of the full theory of Henkin quantifiers. This is due to the combination of strong and weak quantifier inferences in one quantifier rule.

Keywords: Henkin quantifiers · Sequent calculus · Cut-elimination

1 Introduction

Henkin introduced the general idea of dependent quantifiers extending classical first-order logic [3], cf. [4] for an overview. This leads to the notion of a partially ordered quantifier with m universal quantifiers and n existential quantifiers, where F is a function that determines for each existential quantifier on which universal quantifiers it depends (m and n may be any finite number). The simplest Henkin quantifier that is not definable in ordinary first-order logic is the quantifier Q_H binding four variables in a formula. A formula A using Q_H can be written as $A_H = \begin{pmatrix} \forall x \, \exists u \\ \forall y \, \exists v \end{pmatrix} A(x,y,u,v)$. This is to be read "For every x there is a u and for every y there is a v (depending only on y)" s.t. $A(x,y,u,v)$. If the semantical meaning of this formula is given in second-order notation, the above formula is semantically equivalent to the second-order formula $\exists f \exists g \forall x \forall y A(x,y,f(x),g(y))$, where f and g are function variables (the investigation of this quantifier is generic for all Henkin quantifiers). Systems of partially ordered quantification are intermediate in strength between first-order logic and second-order logic. Similar to second-order logic, first-order logic extended by Q_H is incomplete [6]. In proof theory incomplete logics are represented by partial proof systems, c.f. the wealth of approaches dealing with

M. Baaz—This work is partially supported by FWF projects P 31063 and P 31955.
A. Lolic—Recipient of a DOC Fellowship of the Austrian Academy of Sciences at the Institute of Logic and Computation at TU Wien.

S. Artemov and A. Nerode (Eds.): LFCS 2020, LNCS 11972, pp. 128–143, 2020.
https://doi.org/10.1007/978-3-030-36755-8_9

partial proof systems for second-order logic. In an analytic setting, these partial systems allow the extraction of implicit information in proofs, i.e. proof mining. However, in contrast to second-order logic only a few results deal with the proof theoretic aspect of the use of branching quantifiers in partial systems.[1]

The first step is to establish an analytic function calculus with a suitable partial Henkin semantics. In this paper we choose a multiplicative function calculus based on pairs of multisets as sequents corresponding to term models. We refer to this calculus as **LF**. Besides the usual propositional inference rules of **LK** the quantifier inference rules of **LF** are

- ∀-introduction for second-order function variables

$$\frac{A(t(t_1^*, \ldots, t_n^*))\Gamma \to \Delta}{\forall f^* A(f^*(t_1^*, \ldots, t_n^*)), \Gamma \to \Delta} \ \forall_l^n$$

t is a term and $t_1^*, \ldots, t_n^*$ are semi-terms.

$$\frac{\Gamma \to \Delta, A(f(t_1^*, \ldots, t_n^*))}{\Gamma \to \Delta, \forall f^* A(f^*(t_1^*, \ldots, t_n^*))} \ \forall_r^n$$

f is a free function variable (eigenvariable) of arity n which does not occur in the lower sequent and $t_1^*, \ldots, t_n^*$ are semi-terms.

- ∃-introduction for second-order function variables

$$\frac{A(f(t_1^*, \ldots, t_n^*)), \Gamma \to \Delta}{\exists f^* A(f^*(t_1^*, \ldots, t_n^*)), \Gamma \to \Delta} \ \exists_l^n$$

f is a free function variable (eigenvariable) of arity n which does not occur in the lower sequent and $t_1^*, \ldots, t_n^*$ are semi-terms.

$$\frac{\Gamma \to \Delta, A(t(t_1^*, \ldots, t_n^*))}{\Gamma \to \Delta, \exists f^* A(f^*(t_1^*, \ldots, t_n^*))} \ \exists_r^n$$

t is a term and $t_1^*, \ldots, t_n^*$ are semi-terms.

LF is obviously cut-free complete w.r.t. term models by the usual Schütte argument and admits effective cut-elimination.

The question arises why not to be content with the second-order representation of Henkin quantifiers. The answer is twofold: First of all, a lot of information can be extracted from cut-free proofs but only on first-order level. This includes

- suitable variants of Herbrand's theorem with or without Skolemization,
- the construction of term-minimal cut-free proofs,
- the development of suitable tableaux provers.

---

[1] The most relevant paper is the work of Lopez-Escobar [5], describing a natural deduction system for Q_H. The setting is of course intuitionistic logic. The formulation of the introduction rule for Q_H corresponds to the introduction rule right in the sequent calculus developed in this paper. The system lacks an elimination rule.

The first item fails due to the failure of second-order Skolemization. The second and third item fail because of the undecidability of second-order unification and the impossibility to obtain most general solutions.

Moreover, this paper can be considered as a general study of the analytic first-order representation of second-order calculi and the necessary weakening of eigenvariable conditions.

2   The Derivation of First-Order Rules from Second-Order Rule Macros: A First Approach

The language $\mathcal{L}_H$ of the intended calculus **LH** is based on the usual language of first-order logic with exception that the quantifiers are replaced by the quantifier Q_H. With exception of the quantifier-rules, **LH** corresponds to the calculus **LK** in a multiplicative setting based on pairs of multisets as sequents. The idea is to abstract the eigenvariable conditions from the premises of the inference macros in **LF**.

Definition 1 (LH). LH *is* **LK**, *except that we replace the quantifier rules by:*

$$\frac{\Gamma \rightarrow \Delta, A(a, b, t_1, t_2)}{\Gamma \rightarrow \Delta, \begin{pmatrix} \forall x \, \exists u \\ \forall y \, \exists v \end{pmatrix} A(x, y, u, v)} \, Q_{Hr}$$

a and b are eigenvariables ($a \neq b$) not allowed to occur in the lower sequent and t_1 and t_2 are terms s.t. t_1 must not contain b and t_2 must not contain a.[2]

$$\frac{A(t_1', t_2', a, b), \Pi \rightarrow \Gamma}{\begin{pmatrix} \forall x \, \exists u \\ \forall y \, \exists v \end{pmatrix} A(x, y, u, v), \Pi \rightarrow \Gamma} \, Q_{Hl_1}$$

where a and b are eigenvariables ($a \neq b$) not allowed to occur in the lower sequent and t_1', t_2' are terms s.t. b does not occur in t_2' and a and b do not occur in t_1'.

$$\frac{A(t_1', t_2', a, b), \Pi \rightarrow \Gamma}{\begin{pmatrix} \forall x \, \exists u \\ \forall y \, \exists v \end{pmatrix} A(x, y, u, v), \Pi \rightarrow \Gamma} \, Q_{Hl_2}$$

where a and b are eigenvariables ($a \neq b$) not allowed to occur in the lower sequent and t_1', t_2' are terms s.t. a does not occur in t_1' and a and b do not occur in t_2'. The usual quantifier rules of **LK** *($\forall_l, \forall_r$ and $\exists_l$, $\exists_r$) can be obtained by partial dummy applications of Q_H:*

– *we define $\forall$ by Q_H, where $\begin{pmatrix} \forall x \, \exists u \\ \forall y \, \exists v \end{pmatrix} A(x, y, u, v)$ binds only on x.*

– *we define $\exists$ by Q_H, where $\begin{pmatrix} \forall x \, \exists u \\ \forall y \, \exists v \end{pmatrix} A(x, y, u, v)$ binds only on u.*

---

[2] Note that such a rule was already used by Lopez-Escobar in [5].

The defined quantifiers $\forall, \exists$ will be denoted as usual.

The rule Q_{H_r} originates from an analysis of a corresponding sequence of inferences in a suitable partial second-order calculus for functions:

$$
\cfrac{
  \cfrac{
    \cfrac{
      \cfrac{
        \Gamma \rightarrow \Delta, A(a, b, s(a), t(b))
      }{
        \Gamma \rightarrow \Delta, \forall y A(a, y, s(a), t(y))
      } \forall_r
    }{
      \Gamma \rightarrow \Delta, \forall x \forall y A(x, y, s(x), t(y))
    } \forall_r
  }{
    \Gamma \rightarrow \Delta, \exists g \forall x \forall y A(x, y, s(x), g(y))
  } \exists_r
}{
  \Gamma \rightarrow \Delta, \exists f \exists g \forall x \forall y A(x, y, f(x), g(y))
} \exists_r
$$

The rules $Q_{H_{l_1}}$ and $Q_{H_{l_2}}$ originate from

$$
\cfrac{
  \cfrac{
    \cfrac{
      \cfrac{
        A(t, t', f'(t), g'(t')), \Gamma \rightarrow \Delta
      }{
        \forall y A(t, y, f'(t), g'(y)), \Gamma \rightarrow \Delta
      } \forall_l
    }{
      \forall x \forall y A(x, y, f'(x), g'(y)), \Gamma \rightarrow \Delta
    } \forall_l
  }{
    \exists g \forall x \forall y A(x, y, f'(x), g(y)), \Gamma \rightarrow \Delta
  } \exists_l
}{
  \exists f \exists g \forall x \forall y A(x, y, f(x), g(y)), \Gamma \rightarrow \Delta
} \exists_l
$$

f', g' eigenvariables. $f'(t)$ can obviously not occur in t and $g'(t')$ can obviously not occur in t'. $f'(t)$ either does not occur in t' or $g'(t')$ does not occur in t.

Example 1. Consider the sequent

$$
\begin{pmatrix} \forall x \ \exists u \\ \forall y \ \exists v \end{pmatrix} A(x, y, u, v) \rightarrow \forall x \forall y \exists u \exists v A(x, y, u, v).
$$

Its **LH**-proof is:

$$
\cfrac{
  \cfrac{
    \cfrac{
      \cfrac{
        A(a, b, c, d) \rightarrow A(a, b, c, d)
      }{
        A(a, b, c, d) \rightarrow \exists v A(a, b, c, v)
      } \exists_r
    }{
      A(a, b, c, d) \rightarrow \exists u \exists v A(a, b, u, v)
    } \exists_r
  }{
    \begin{pmatrix} \forall x \ \exists u \\ \forall y \ \exists v \end{pmatrix} A(x, y, u, v) \rightarrow \exists u \exists v A(a, b, u, v)
  } Q_{H_l}
}{
  \cfrac{
    \begin{pmatrix} \forall x \ \exists u \\ \forall y \ \exists v \end{pmatrix} A(x, y, u, v) \rightarrow \forall y \exists u \exists v A(a, y, u, v)
  }{
    \begin{pmatrix} \forall x \ \exists u \\ \forall y \ \exists v \end{pmatrix} A(x, y, u, v) \rightarrow \forall x \forall y \exists u \exists v A(x, y, u, v)
  } \forall_r
} \forall_r
$$

Cuts in **LH** can be eliminated following Gentzen's procedure.

Theorem 1. LH *admits cut-elimination.*

Proof. We follow Gentzen's procedure and illustrate only the cases for the reduction of Q_H. There are two cases for the reduction of the quantifier Q_H, where $\begin{pmatrix} \forall x \ \exists u \\ \forall y \ \exists v \end{pmatrix} A(x, y, u, v)$ does not occur in Δ or Π.

1.

$$\frac{\Gamma \to \Delta, A(a,b,s(a),t(b))}{\Gamma \to \Delta, \left(\begin{smallmatrix}\forall x\ \exists u\\ \forall y\ \exists v\end{smallmatrix}\right) A(x,y,u,v)}Q_{H_r} \qquad \frac{A(s',t'(u),u,v),\Pi \to \Lambda}{\left(\begin{smallmatrix}\forall x\ \exists u\\ \forall y\ \exists v\end{smallmatrix}\right) A(x,y,u,v),\Pi \to \Lambda}Q_{H_{L_1}}}{\Gamma,\Pi \to \Delta, \Lambda}$$

$(a \neq b,\ u \neq v$, all occurrences of a,b,u,v are indicated). This is reduced to

$$\frac{\Gamma \to \Delta, F_1 \qquad F_1, \Pi \to \Lambda}{\Gamma, \Pi' \to \Delta', \Lambda}$$

where $F_1 = A(s', t'(s(s')), s(s'), t(t'(s(s'))))$.

2.

$$\frac{\Gamma \to \Delta, A(a,b,s(a),t(b))}{\Gamma \to \Delta, \left(\begin{smallmatrix}\forall x\ \exists u\\ \forall y\ \exists v\end{smallmatrix}\right) A(x,y,u,v)}Q_{H_r} \qquad \frac{A(s'(v),t',u,v),\Pi \to \Lambda}{\left(\begin{smallmatrix}\forall x\ \exists u\\ \forall y\ \exists v\end{smallmatrix}\right) A(x,y,u,v),\Pi \to \Lambda}Q_{H_{L_1}}}{\Gamma,\Pi \to \Delta, \Lambda}$$

$(a \neq b,\ u \neq v$, all occurrences of a,b,u,v are indicated). This is reduced to

$$\frac{\Gamma \to \Delta, F_2 \qquad F_2, \Pi \to \Lambda}{\Gamma, \Pi' \to \Delta', \Lambda}$$

where $F_2 = A(s'(t(t')), t', s(s'(t(t'))), t(t'))$.

Corollary 1 (midsequent theorem). *For every proof of a prenex sequent in* **LH** *there is a cut-free proof with a midsequent s.t. every inference above the midsequent is structural or propositional and every inference below the midsequent is structural or Q_{H_r}, $Q_{H_{l_1}}$, $Q_{H_{l_2}}$.*

Proof. As in **LK** all quantifier inferences can be postponed until all propositional inferences are completed.

Corollary 2. **LH** *admits interpolation. If the axioms contain Q_H only in the dummy forms of $\forall, \exists$ there is an interpolant in the usual first-order language.*

Proof. We adapt Maehara's lemma (c.f. [7]). For every partition $(\Gamma_1 \to \Delta_1, \Gamma_2 \to \Delta_2)$ of $\Gamma \to \Delta$ an interpolant I is constructed s.t. $\Gamma_1 \to \Delta_1, I$ and $I, \Gamma_2 \to \Delta_2$ are derivable. The relevant cases are:

1. case

$$\frac{\Gamma \to \Delta, A(a,b,t,t')}{\Gamma \to \Delta, \left(\begin{smallmatrix}\forall x\ \exists u\\ \forall y\ \exists v\end{smallmatrix}\right) A(x,y,u,v)}Q_{H_r}$$

The inference is in the right partition: Block a, b using $\forall_l$ in the interpolant of the right partition. Infer $\forall_r$ dually in the left partition. The inference is in the left partition: Block a, b using a $\exists_r$ in the left partition, infer $\exists_l$ dually in the right partition.

2. case

$$\frac{A(t, t', u, v), \Gamma \to \Delta}{\begin{pmatrix} \forall x\ \exists u \\ \forall y\ \exists v \end{pmatrix} A(x, y, u, v), \Gamma \to \Delta} \ Q_{H_{l_{1,2}}}$$

The inference is in the right partition: Block a, b using $\exists_l$ in the interpolant of the right partition. Infer $\exists_r$ dually in the left partition. The inference is in the left partition: Block a, b using a $\forall_r$ in the left partition, infer $\forall_l$ dually in the right partition.

Corollary 2 hints at the weakness of this calculus which turns out not to be adequate for our aims.

Theorem 2. LH *is incomplete for any reasonable partial semantics.*

Proof. Assume towards a contradiction the sequent $\begin{pmatrix} \forall x\ \exists u \\ \forall y\ \exists v \end{pmatrix} A(x, y, u, v) \to$ $\begin{pmatrix} \forall x\ \exists u \\ \forall y\ \exists v \end{pmatrix} (A(x, y, u, v) \vee C)$ is provable. Then it is provable without cuts. A cut-free derivation after deletion of weakenings and contractions has the form:

$$\frac{A(a, b, c, d) \to A(a, b, c, d)}{A(a, b, c, d) \to A(a, b, c, d) \vee C}$$

$$\vdots$$

Due to the mixture of strong and weak positions[3] in Q_H none of Q_{H_r}, $Q_{H_{l_1}}$, $Q_{H_{l_2}}$ can be applied.

Corollary 3. *Compound axioms $A \to A$ cannot be reduced to atomic ones.*

Remark 1. The provable sequents of **LH** are usually asymmetric w.r.t. Q_H if Q_H does not originate in an axiom or a weakening.

3   LK++: A Globally Sound Calculus, Cf. [1]

The inherent incompleteness of **LH** even for trivial statements is a consequence of the fact that Q_H represents a quantifier inference macro combining quantifiers in a strong and a weak position. This phenomenon occurs already on the level of usual first-order logic when quantifiers defined by macros of quantifiers such as $\forall x \exists y$ are considered [2].

The solution is to consider sequent calculi with concepts of proof which are globally but not locally sound. This means that all derived statements are true but that not every sub-derivation is meaningful.

---

[3] In **LK** strong quantifier inferences are $\forall_r$ and $\exists_l$ and weak quantifier inferences are $\forall_l$ and $\exists_r$.

Definition 2 (side variable relation $<_{\varphi,\mathbf{LK}}$). *Let φ be an* **LK***-derivation. We say b is a side variable of a in φ (written $a <_{\varphi,\mathbf{LK}} b$) if φ contains a strong quantifier inference of the form*

$$\frac{\Gamma \to \Delta, A(a, b, \overline{c})}{\Gamma \to \Delta, \forall x A(x, b, \overline{c})} \ \forall_r$$

or of the form

$$\frac{A(a, b, \overline{c}), \Gamma \to \Delta}{\exists x A(x, b, \overline{c}), \Gamma \to \Delta} \ \exists_l$$

In addition to strong and weak quantifier inferences we define $\mathbf{LK}^{++}$-suitable quantifier inferences.

Definition 3 ($\mathbf{LK}^{++}$-suitable quantifier inferences). *We say a quantifier inference is suitable for a proof φ if either it is a weak quantifier inference, or the following three conditions are satisfied:*

- *(substitutability) the eigenvariable does not appear in the conclusion of φ.*
- *(side variable condition) the relation $<_{\varphi,\mathbf{LK}}$ is acyclic.*
- *(very weak regularity) the eigenvariable of an inference with main formula A is different to the eigenvariable of an inference with main formula A' whenever $A \neq A'$.*

We obtain $\mathbf{LK}^{++}$ from $\mathbf{LK}$ by replacing the usual eigenvariable conditions by $\mathbf{LK}^{++}$-suitable ones[4].

Remark 2. Note that eigenvariables may occur outside of the scope of the intended quantifier.

Theorem 3. *If a sequent is $\mathbf{LK}^{++}$-derivable, then it is already $\mathbf{LK}$-derivable.*

Proof. (Sketch.) Let φ be an $\mathbf{LK}^{++}$-proof. Replace every unsound universal quantifier inference by an $\supset_l$ inference:

$$\frac{\Gamma \to \Delta, A(a) \qquad \forall x\, A(x) \to \forall x\, A(x)}{\Gamma, A(a) \supset \forall x\, A(x) \to \Delta, \forall x\, A(x)} \ \supset_l$$

Similarly replace every unsound existential quantifier by an $\supset_l$ inference

$$\frac{\exists x\, A(x) \to \exists x\, A(x) \qquad A(a), \Gamma \to \Delta}{\Gamma, \exists x\, A(x), \exists x\, A(x) \supset A(a) \to \Delta} \ \supset_l$$

---

[4] $\mathbf{LK}^+$ in [1] coincides with $\mathbf{LK}^{++}$ with exception to the notion of regularity, which is the usual one.

By doing this, we obtain a proof of the desired sequent, together with formulae of the form

$$A(a) \supset \forall x \, A(x) \quad \text{or} \quad \exists x \, A(x) \supset A(a)$$

on the left-hand side. However, we can eliminate each of them by adding an existential quantifier inference and cutting with formulae of the form

$$\rightarrow \exists y \, \big(A(y) \supset \forall x \, A(x)\big) \quad \text{or} \quad \rightarrow \exists y \, \big(\exists x \, A(x) \supset A(y)\big),$$

both of which are easily derivable. Note that the existential quantifier inferences can be carried out using contractions in a way that is permissible by **LK** because the $<_{\varphi,\mathbf{LK}}$ does not loop.

Example 2. Consider the following locally unsound but globally sound $\mathbf{LK}^{++}$-derivation φ:

$$\cfrac{\cfrac{\cfrac{A(a) \rightarrow A(a)}{A(a) \rightarrow \forall y A(y)} \, \forall_r}{\rightarrow A(a) \supset \forall y A(y)} \, \supset_r}{\rightarrow \exists x (A(x) \supset \forall y A(y))} \, \exists_r$$

As a is the only eigenvariable the side variable relation $<_{\varphi,\mathbf{LK}}$ is empty.

In [1] the focus has been on the strongly reduced complexity of cut-free $\mathbf{LK}^{++}$ proofs (Theorem 2.6 and Corollary 2.7). The focus of this paper is to provide a framework s.t. cut-free complete calculi for Henkin quantifiers can be established.

4   The Analytic Sequent Calculus LH^{++}

From the example in the Sect. 2 it becomes obvious that there will be no analytic calculus with local rules to represent any reasonable fragment of the full logic with Q_H. The reason is that the inference rules for Q_H need eigenvariables in both polarities. The solution is to keep global soundness but to give up local soundness. To this aim, the eigenvariable conditions will be weakened.

The weakened eigenvariable conditions will be obtained from the weakened eigenvariable conditions of the globally sound but possibly locally unsound sequent calculus $\mathbf{LF}^{++}$ which corresponds to **LF** as $\mathbf{LK}^{++}$ corresponds to **LK**.

Definition 4 (side variable relation $<_{\varphi,\mathbf{LF}}$). *Let $f_1, \ldots, f_k, g$ be all free variables occurring in the main formula of an inference of $\forall_r^n$ or $\exists_l^n$ with eigenvariable g. Then $f_1, \ldots, f_k$ are side variables of g ($g <_{\varphi,\mathbf{LF}} f_1, \ldots, g <_{\varphi,\mathbf{LF}} f_k$).*

Definition 5 ($\mathbf{LF}^{++}$-suitable quantifier inferences). *A quantifier inference is $\mathbf{LF}^{++}$-suitable for a proof φ if the following three conditions are satisfied:*

– *(substitutability) the eigenvariable does not appear in the conclusion of φ.*
– *(side-variable condition) the relation $<_{\varphi,\mathbf{LF}}$ is acyclic.*

– (very weak regularity) the eigenvariable of an inference with main formula A is different to the eigenvariable of an inference with main formula A' whenever $A \neq A'$.

The sequent calculus $\mathbf{LF}^{++}$ is $\mathbf{LF}$, except that we replace quantifier inferences with $\mathbf{LF}^{++}$-suitable quantifier inferences.

The calculus $\mathbf{LF}^{++}$ is possibly locally unsound, but globally sound:

Theorem 4. *If a sequent is derivable in $\mathbf{LF}^{++}$, then it is derivable in $\mathbf{LF}$.*

Proof. The proof corresponds to the proof of Theorem 3. Let φ be an $\mathbf{LF}^{++}$-proof. Replace every unsound universal quantifier inference by a $\supset_l$ inference

$$\frac{\Gamma \to \Delta, A(f(t_1^*,\ldots,t_n^*)) \qquad \forall f^* A(f^*(t_1^*,\ldots,t_n^*)) \to \forall f^* A(f^*(t_1^*,\ldots,t_n^*))}{A(f(t_1^*,\ldots,t_n^*)) \supset \forall f^* A(f^*(t_1^*,\ldots,t_n^*)), \Gamma \to \Delta, \forall f^* A(f^*(t_1^*,\ldots,t_n^*))} \supset_l$$

Similarly, replace every unsound existential quantifier inference by

$$\frac{\exists f^* A(f^*(t_1^*,\ldots,t_n^*)) \to \exists f^* A(f^*(t_1^*,\ldots,t_n^*)) \qquad A(f(t_1^*,\ldots,t_n^*)), \Gamma \to \Delta}{\exists f^* A(f^*(t_1^*,\ldots,t_n^*)), \exists f^* A(f^*(t_1^*,\ldots,t_n^*)) \supset A(f(t_1^*,\ldots,t_n^*)), \Gamma \to \Delta} \supset_l$$

We obtain a proof of the desired sequent together with formulae

$$A(f(t_1^*,\ldots,t_n^*)) \supset \forall f^* A(f^*(t_1^*,\ldots,t_n^*))$$

and

$$\exists f^* A(f^*(t_1^*,\ldots,t_n^*)) \supset A(f(t_1^*,\ldots,t_n^*))$$

on the left-hand side. However, we can eliminate each of them: To each of these formulae a corresponding free variable is associated which is covered by the quantifier. We order the associated variables in a total ordering $<$ extending $<_{\varphi,\mathbf{LF}}$. Let f be the associated variable to

$$A(f(t_1^*,\ldots,t_n^*)) \supset \forall f A(f(t_1^*,\ldots,t_n^*)),$$

which is a $<_{\varphi,\mathbf{LF}}$-least variable for the formulae still present in the end-sequent

$$A(f(t_1^*,\ldots,t_n^*)) \supset \forall f A(f(t_1^*,\ldots,t_n^*)), \Pi \to \Delta.$$

The eigenvariable condition for f is fulfilled. Therefore we can introduce an existential quantifier to obtain $\exists g^* (A(g^*(t_1^*,\ldots,t_n^*)) \supset \forall f^* A(f^*(t_1^*,\ldots,t_n^*)))$ and cut with the derivation

$$\frac{\dfrac{\dfrac{\dfrac{\dfrac{\dfrac{A(f(t_1^*,\ldots,t_n^*)) \to A(f(t_1^*,\ldots,t_n^*))}{A(f(t_1^*,\ldots,t_n^*)) \to A(f(t_1^*,\ldots,t_n^*)), \forall f A(f(t_1^*,\ldots,t_n^*))} w_r}{\to A(f(t_1^*,\ldots,t_n^*)), A(f(t_1^*,\ldots,t_n^*)) \supset \forall f^* A(f^*(t_1^*,\ldots,t_n^*))} \supset_r}{\to \forall f^* A(f^*(t_1^*,\ldots,t_n^*)), A(f(t_1^*,\ldots,t_n^*)) \supset \forall f^* A(f^*(t_1^*,\ldots,t_n^*))} \forall_r^n}{\dfrac{A(f(t_1^*,\ldots,t_n^*)) \to \forall f^* A(f^*(t_1^*,\ldots,t_n^*)), A(f(t_1^*,\ldots,t_n^*)) \supset \forall f^* A(f^*(t_1^*,\ldots,t_n^*))}{\to A(f(t_1^*,\ldots,t_n^*)) \supset \forall f^* A(f^*(t_1^*,\ldots,t_n^*)), A \supset \forall f^* A(f^*(t_1^*,\ldots,t_n^*))} \supset_r} w_l}{\dfrac{\to A(f(t_1^*,\ldots,t_n^*)) \supset \forall f^* A(f^*(t_1^*,\ldots,t_n^*))}{\to \exists g^* (A(g^*(t_1^*,\ldots,t_n^*)) \supset \forall f^* A(f^*(t_1^*,\ldots,t_n^*)))} \exists_r^n} c_r$$

If f is the associated variable to $\exists f^* A(f^*(t_1^*, \ldots, t_n^*)) \supset A(f(t_1^*, \ldots, t_n^*))$, we proceed analogously.

Note that every **LF** derivation is also an $\mathbf{LF}^{++}$ derivation.

In the construction of $\mathbf{LH}^{++}$ we use valuations of eigenvariables to guarantee the eigenvariable conditions of $\mathbf{LF}^{++}$ on the first-order level.

Definition 6 (valuation of eigenvariables). *We assign constants* $c_a, c_b, \ldots$ *to all eigenvariables* $a, b, \ldots$ *occurring in rule applications* $Q_{H_r}, Q_{H_{l_1}}, Q_{H_{l_2}}$ *in a derivation* φ. *Let* $C = \{c_a, c_b, \ldots\}$ *and* a *an eigenvariable in* φ. *Then* $\nu(a) \subseteq C$ *is a valuation, assigning a subset of* C *to* a. $\nu(t) = \bigcup \nu(x)$ *where* x *is an eigenvariable in the term* t.

The valuation of eigenvariables represents the generalized eigenvariable conditions of a suitable second-order function calculus together with the form of the premisses of quantifier inferences.

Definition 7 (side variable relation $<_{\varphi, \mathbf{LH}}$). *The side variable conditions are obtained from the inference rules* Q_{H_r} *and* $Q_{H_{l_{1,2}}}$. *Consider the inferences*

$$\frac{\Gamma \to \Delta, A(a, b, t_1, t_2, z_1, \ldots, z_k)}{\Gamma \to \Delta, \begin{pmatrix} \forall x \, \exists u \\ \forall y \, \exists v \end{pmatrix} A(x, y, u, v, z_1, \ldots, z_k)} \; Q_{H_r}$$

in an **LH**-*derivation* φ, *where* a *does not occur in* t_2, b *does not occur in* t_1 *and all eigenvariables of the derivation occurring in* A *are indicated. The variables* $z_1, \ldots, z_k$ *are side variables. Let* c_a *and* c_b *be the corresponding constants:*

$$\nu(a) = \{c_a\}, \quad \nu(b) = \{c_b\}, \quad c_a \notin \nu(t_2), \quad c_b \notin \nu(t_1).$$

The the side variable condition is given by

$$\{c_a <_{\varphi, \mathbf{LH}} c, c_b <_{\varphi, \mathbf{LH}} c \mid c \in (\nu(t_1) \cup \nu(t_2)) \backslash \{c_a, c_b\} \;\cup\; c \in \bigcup_{i=1}^{k} \nu(z_i)\}$$

$$\cup \; \{c_a <_{\varphi, \mathbf{LH}} c_b\}.$$

Similarly, consider

$$\frac{A(t_1, t_2, a, b, z_1, \ldots, z_k), \Pi \to \Gamma}{\begin{pmatrix} \forall x \, \exists u \\ \forall y \, \exists v \end{pmatrix} A(x, y, u, v, z_1, \ldots, z_k), \Pi \to \Gamma} \; Q_{H_{l_1}}$$

in an **LH**-*derivation* φ, *where* b *does not occur in* t_2, a *and* b *do not occur in* t_1 *and all eigenvariables of the derivation occurring in* A *are indicated. The variables* $z_1, \ldots, z_k$ *are side variables. Let* c_f *and* c_g *be the corresponding constants:*

$$\nu(a) = \nu(t_1) \cup \{c_f\}, \quad \nu(b) = \nu(t_2) \cup \{c_g\},$$

$$c_f \notin \nu(t_1), \quad c_g \notin \nu(t_2), \quad c_g \notin \nu(t_1),$$
$$c_f \in \nu(t_2) \Rightarrow \nu(t_1) \subseteq \nu(t_2).$$

The the side variable condition is given by

$$\{c_f <_{\varphi,\mathbf{LH}} c, c_g <_{\varphi,\mathbf{LH}} c \mid c \in \bigcup_{i=1}^k \nu(z_i)\} \cup \{c_g <_{\varphi,\mathbf{LH}} c_f\}.$$

For the second left inference

$$\frac{A(t_1, t_2, a, b, z_1, \ldots, z_k), \Pi \to \Gamma}{\begin{pmatrix} \forall x \, \exists u \\ \forall y \, \exists v \end{pmatrix} A(x, y, u, v, z_1, \ldots, z_k), \Pi \to \Gamma} \, Q_{Hl_2}$$

where a does not occur in t_1, a and b do not occur in t_2 and all eigenvariables of the derivation occurring in A are indicated. The variables $z_1, \ldots, z_k$ are side variables. Let c_f and c_g be the corresponding constants.

$$\nu(a) = \nu(t_1) \cup \{c_f\}, \quad \nu(b) = \nu(t_2) \cup \{c_g\},$$
$$c_f \notin \nu(t_1), \quad c_g \notin \nu(t_2), \quad c_f \notin \nu(t_2),$$
$$c_g \in \nu(t_1) \Rightarrow \nu(t_2) \subseteq \nu(t_1).$$

The the side variable condition is given by

$$\{c_f <_{\varphi,\mathbf{LH}} c, c_g <_{\varphi,\mathbf{LH}} c \mid c \in \bigcup_{i=1}^k \nu(z_i)\} \cup \{c_g <_{\varphi,\mathbf{LH}} c_f\}.$$

The aim of the valuations is to guarantee the embedding into $\mathbf{LF}^{++}$-derivations.

Definition 8 ($\mathbf{LH}^{++}$-suitable quantifier inferences). *A quantifier inference is $\mathbf{LH}^{++}$-suitable for a proof φ if the following three conditions are satisfied:*

- *(substitutability) the eigenvariable does not appear in the conclusion of φ.*
- *(side variable condition) the relation $<_{\varphi,\mathbf{LH}}$ is acyclic.*
- *(very weak regularity) the eigenvariables of an inference with main formula A are different to the eigenvariables of an inference with main formula A' whenever $A \neq A'$.*

The sequent calculus $\mathbf{LH}^{++}$ is $\mathbf{LH}$, except that we replace quantifier inferences with $\mathbf{LH}^{++}$-suitable quantifier inferences.

Example 3. Consider the cut-free derivation $\varphi =$

$$\frac{\dfrac{\dfrac{A(a,b,c,d) \to A(a,b,c,d)}{A(a,b,c,d) \to A(a,b,c,d), C} \, w_r}{A(a,b,c,d) \to A(a,b,c,d) \vee C} \, \vee_r}{\dfrac{A(a,b,c,d) \to \begin{pmatrix} \forall x \, \exists u \\ \forall y \, \exists v \end{pmatrix} (A(x,y,u,v) \vee C)}{\begin{pmatrix} \forall x \, \exists u \\ \forall y \, \exists v \end{pmatrix} A(x,y,u,v) \to \begin{pmatrix} \forall x \, \exists u \\ \forall y \, \exists v \end{pmatrix} (A(x,y,u,v) \vee C)} \, Q_{Hl_1}} \, Q_{Hr}$$

φ is regular and the eigenvariables a, b, c, d do not occur in the end-sequent. Let c_a, c_b the corresponding constants for the eigenvariable in the right inference and c_f, c_g the corresponding constants for the left inference. $\nu(a) = \{c_a\}$, $\nu(b) = \{c_b\}$, $\nu(c) = \{c_f, c_a\}$ and $\nu(d) = \{c_g, c_b\}$. Consequently we obtain $c_a <_{\varphi,\mathbf{LH}} c_f$, $c_b <_{\varphi,\mathbf{LH}} c_g$, $c_a <_{\varphi,\mathbf{LH}} c_b$, $c_a <_{\varphi,\mathbf{LH}} c_g$ and $c_b <_{\varphi,\mathbf{LH}} c_f$.

It is essential that the side variable order is acyclic. Consider the following example with an end-sequent which is already unsound in **LK**.

Example 4. Let φ be the following cut-free derivation

$$\frac{\dfrac{\dfrac{\dfrac{A(a,b) \to A(a,b)}{A(a,b) \to \forall y A(a,y)}\,\forall_r}{A(a,b) \to \exists x \forall y A(x,y)}\,\exists_r}{\exists y A(y,b) \to \exists x \forall y A(x,y)}\,\exists_l}{\forall x \exists y A(y,x) \to \exists x \forall y A(x,y)}\,\forall_l$$

We obtain side variable conditions $c_a <_{\varphi,\mathbf{LH}} c_b$ and $c_b <_{\varphi,\mathbf{LH}} c_a$, which loop.

5 Soundness, Completeness and Cut-Elimination for LH^{++}

It is impossible to embed **LH**$^{++}$ into **LH** with additional cuts, similar to the embedding of **LK**$^{++}$ to **LK**, because **LH** admits cut-elimination, which demonstrates that **LH** is strictly weaker than **LH**$^{++}$. As usual cut-elimination methods do not work for **LH**$^{++}$, we will embed **LH**$^{++}$ into **LF**$^{++}$.

Theorem 5. *There is no Gentzen-style cut-elimination for* **LH**$^{++}$.

Proof. Consider the following derivation $\varphi =$

$$\frac{\dfrac{\dfrac{A(a) \to A(a)}{A(a) \to \forall x A(x)}\,\forall_r \qquad \dfrac{A(f(a)) \to A(f(a))}{\forall x A(x) \to A(f(a))}\,\forall_l}{A(a) \to A(f(a))}\,cut}{\to A(a) \supset A(f(a))}\,\supset_r$$

$\nu(a) = \{c_a\}$. The side variable relation is empty, as there are no further eigenvariables. If the cut on $\forall x A(x)$ was eliminable according to the Gentzen procedure, the proof would be in **LJ**. But the proven formula is not intuitionistically valid.

Remark 3. This example shows that there will be no analytic intuitionistic companion of **LH**$^{++}$ and, a fortiori, no usual natural deduction system with normal forms.

Lemma 1. *An* **LH**$^{++}$-*derivation φ with cuts can be immediately transformed into an* **LF**$^{++}$-*derivation φ' with cuts.*

Proof. By induction on the depth of φ. The proof in $\mathbf{LH}^{++}$ is step-wise transformed into a proof in $\mathbf{LF}^{++}$. We also show that every function term replacing an eigenvariable is compatible with the valuation $\nu(a)$ of this derivation. This means that all functions h in the term $f(t)$ are represented by constants c_h in $\nu(a)$. The constructed side variable order for the derivation in $\mathbf{LF}^{++}$ is given by $c_f <_{\varphi,\mathbf{LH}} c_g \Leftrightarrow f <_{\varphi,\mathbf{LF}} g$. In

$$\frac{\Gamma \to \Delta, A(a,b,s,t)}{\Gamma \to \Delta, \begin{pmatrix} \forall x\, \exists u \\ \forall y\, \exists v \end{pmatrix} A(x,y,u,v)}\ Q_{Hr}$$

we take the premise and introduce the quantifiers to obtain

$$\frac{\dfrac{\dfrac{\dfrac{\Gamma \to \Delta, A(a,b,s,t)}{\Gamma \to \Delta, \forall y A(a,y,s,t)}\ \forall_r^n}{\Gamma \to \Delta, \forall x \forall y A(x,y,s,t)}\ \forall_r^n}{\Gamma \to \Delta, \exists g \forall x \forall y A(x,y,s,g(y))}\ \exists_r^n}{\Gamma \to \Delta, \exists f \exists g \forall x \forall y A(x,y,f(x),g(y))}\ \exists_r^n$$

In

$$\frac{A(s,t,a,b), \Pi \to \Gamma}{\begin{pmatrix} \forall x\, \exists u \\ \forall y\, \exists v \end{pmatrix} A(x,y,u,v), \Pi \to \Gamma}\ Q_{Hl_1}$$

we replace $A(s,t,a,b)$ in the premise with $A(s,t',f(s),g(t'))$, where $t' = t\{a \leftarrow f(s)\}$ to obtain

$$\frac{\dfrac{\dfrac{\dfrac{A(s,t',f(s),g(t')), \Gamma \to \Delta}{\forall y A(s,y,f(s),g(y)), \Gamma \to \Delta}\ \forall_l^n}{\forall x \forall y A(x,y,f(x),g(y)), \Gamma \to \Delta}\ \forall_l^n}{\exists g^* \forall x \forall y A(x,y,f(x),g^*(y)), \Gamma \to \Delta}\ \exists_l^n}{\exists f^* \exists g^* \forall x \forall y A(x,y,f^*(x),g^*(y)), \Gamma \to \Delta}\ \exists_{rl}^n$$

And in

$$\frac{A(s,t,a,b), \Pi \to \Gamma}{\begin{pmatrix} \forall x\, \exists u \\ \forall y\, \exists v \end{pmatrix} A(x,y,u,v), \Pi \to \Gamma}\ Q_{Hl_2}$$

we replace $A(s,t,a,b)$ in the premise with $A(s',t,f(s'),g(t))$, where $s' = s\{b \leftarrow g(t)\}$ to obtain

$$\frac{\dfrac{\dfrac{\dfrac{A(s',t,f(s'),g(t)), \Gamma \to \Delta}{\forall y A(s',y,f(s'),g(y)), \Gamma \to \Delta}\ \forall_l^n}{\forall x \forall y A(x,y,f(x),g(y)), \Gamma \to \Delta}\ \forall_l^n}{\exists g^* \forall x \forall y A(x,y,f(x),g^*(y)), \Gamma \to \Delta}\ \exists_l^n}{\exists f^* \exists g^* \forall x \forall y A(x,y,f^*(x),g^*(y)), \Gamma \to \Delta}\ \exists_{rl}^n$$

Note that by the suitability of the $\mathbf{LH}^{++}$-derivation the eigenvariable conditions for the constructed $\mathbf{LF}^{++}$-derivation are guaranteed.

Lemma 2. *An $\mathbf{LF}^{++}$-derivation φ where the end-sequent contains only quantifiers in blocked distinct sequences $\exists f \exists g \forall x \forall y$ can be transformed into a cut-free $\mathbf{LF}^{++}$-derivation φ' where the quantifiers in the sequence $\exists f \exists g \forall x \forall y$ belonging to a block in the end-sequent are inferred immediately one after the other.*

Proof. We transform the $\mathbf{LF}^{++}$-proof φ in an $\mathbf{LF}$-proof with cuts by Theorem 4 and eliminate the cuts to obtain an $\mathbf{LF}$-proof without cuts, where we replace compound axioms with atomic ones. Then we use the weakened eigenvariable condition of $\mathbf{LF}^{++}$ to infer the quantifier in the block one after the other to obtain φ'. The eigenvariable conditions are fulfilled, as the proof is regular the eigenvariables do not occur in the end-sequent and the side variable conditions are respected.

Lemma 3. *A cut-free $\mathbf{LF}^{++}$-proof φ with blocked quantifier inferences $\exists f \exists g \forall x \forall y$ from atomic axioms and only such blocks of quantifiers in the end-sequent can be transformed into a cut-free $\mathbf{LH}^{++}$-proof φ' from atomic axioms. Whenever $c_f <_{\varphi',\mathbf{LF}} c_g$ then $f <_{\varphi,\mathbf{LH}} g$.*

Proof. By induction on the proof-depth of φ. We step-wise replace function terms $f(t)$ by variables $x_{f(t)}$, where $\nu(x_{f(t)}) = \{c_h \mid h \text{ is a function occurring in } f(t)\}$. For eigenvariables a (free 0-placed functions) $\nu(a) = \{c_a\}$. Every sequence of quantifier introductions in φ of the form

$$
\frac{
\dfrac{
\dfrac{
\dfrac{\Gamma \to \Delta, A(a, b, s, t)}{\Gamma \to \Delta, \forall y A(a, y, s, t)} \forall_r^n
}{\Gamma \to \Delta, \forall x \forall y A(x, y, s, t)} \forall_r^n
}{\Gamma \to \Delta, \exists g \forall x \forall y A(x, y, s, g(y))} \exists_r^n
}{\Gamma \to \Delta, \exists f \exists g \forall x \forall y A(x, y, f(x), g(y))} \exists_r^n
$$

is replaced by the quantifier introduction for Q_H:

$$
\frac{\Gamma \to \Delta, A(a, b, s, t)}{\Gamma \to \Delta, \begin{pmatrix} \forall x \ \exists u \\ \forall y \ \exists v \end{pmatrix} A(x, y, u, v)} Q_{Hr}
$$

If a occurred in t or b occurred in s an inference of this form would be impossible (the eigenvariable conditions are fulfilled). In case we have a sequence of quantifier introductions in φ of the form

$$
\frac{
\dfrac{
\dfrac{
\dfrac{A(s, t, f(s), g(t)), \Gamma \to \Delta}{\forall y A(s, y, f(s), g(y)), \Gamma \to \Delta} \forall_l^n
}{\forall x \forall y A(x, y, f(x), g(y)), \Gamma \to \Delta} \forall_l^n
}{\exists g^* \forall x \forall y A(x, y, f(x), g^*(y)), \Gamma \to \Delta} \exists_l^n
}{\exists f^* \exists g^* \forall x \forall y A(x, y, f^*(x), g^*(y)), \Gamma \to \Delta} \exists_{rl}^n
$$

we replace $f(s)$ everywhere in the proof with $a_{f(s)}$ and $g(t)$ with $b_{g(t)}$. This replacement is possible because the original $\mathbf{LF}^{++}$-derivation inferred only these blocks of quantifiers one after the other. $f(s)$ can obviously not be contained in s and $g(t)$ can obviously not be contained in t. However, it is impossible that both $f(s)$ is contained in t and $g(t)$ is contained in s. Therefore the eigenvariable conditions for one variant of the left rule for Q_H are fulfilled.

Now we are ready to state the main result of this section.

Theorem 6. $\mathbf{LH}^{++}$ *is sound, cut-free complete w.r.t. the intended semantics and admits an effective cut-elimination[5].*

Proof. Follows immediately by Lemmas 1, 2 and 3.

Corollary 4 (midsequent theorem). *For every proof of a prenex sequent in $\mathbf{LH}^{++}$ there is a cut-free proof with a midsequent s.t. every inference above the midsequent is structural or propositional and every inference below the midsequent is structural or a quantifier inference.*

Proof. All quantifier inferences can be postponed until all propositional inferences are completed.

Corollary 5. *Assume the language contains constants c, d, e. If $\mathbf{LH}^{++}$ derives*

$$\rightarrow \begin{pmatrix} \forall x\ \exists u \\ \forall y\ \exists v \end{pmatrix} A(x, y, u, v),$$ *where A is quantifier-free and does not contain free variables, then*

$$\rightarrow \bigvee_{i=1}^{n} A(c_i, d_i, s_i(c_i, \overline{h_i}), t_i(d_i, \overline{j_i})),$$

where $\overline{h_i}, \overline{j_i} \subseteq \{c_k, \ldots, c_n \mid k = i+1\} \cup \{e\}$.

Proof. From the midsequent theorem, the argument is similar to the derivation of Σ_2 forms of the theorem of Herbrand in classical logic: the order of components of the disjunctions is induced by the side variable order $\leq_{\varphi.\mathbf{LH}}$.

Corollary 6. *If $\mathbf{LH}^{++}$ derives*

$$\rightarrow \begin{pmatrix} \forall x\ \exists u \\ \forall y\ \exists v \end{pmatrix} A(x, y, u, v),$$

where A is quantifier-free and does not contain free variables, then $\mathbf{LH}$ derives

$$\rightarrow \begin{pmatrix} \forall x\ \exists u \\ \forall y\ \exists v \end{pmatrix} A(x, y, u, v).$$

Remark 4. Note that it is easy to construct an automated deduction calculus in tableaux-format from $\mathbf{LH}^{++}$ as contrary to $\mathbf{LF}$ or $\mathbf{LF}^{++}$ the usual unification algorithm is applicable.

---

[5] Note that usual regularity is not sufficient as the formation of quantifier blocks of inferences in Lemma 2 in cut-free $\mathbf{LF}$-derivations might distribute $\exists f \exists g$ inferences by contractions to more than one branch of the proof.

6   Conclusion

It is obvious that the methodology developed in this paper can be extended to arbitrary Henkin quantifiers, it is however an open question whether this approach can be extended to arbitrary macros of function quantifiers.

The most important question is however which types of quantifiers in general can be represented with such global conditions on eigenvariables whose order is recorded as external information.

References

1. Aguilera, J.P., Baaz, M.: Unsound inferences make proofs shorter. J. Symb. Log. **84**(1), 102–122 (2019). https://doi.org/10.1017/jsl.2018.51
2. Baaz, M., Lolic, A.: Note on globally sound analytic calculi for quantifier macros. In: Iemhoff, R., Moortgat, M., de Queiroz, R. (eds.) WoLLIC 2019. LNCS, vol. 11541, pp. 486–497. Springer, Heidelberg (2019). https://doi.org/10.1007/978-3-662-59533-6_29
3. Henkin, L.: Some remarks on infinitely long formulas. J. Symb. Log. **30**, 167–183 (1961). https://doi.org/10.2307/2270594
4. Krynicki, M., Mostowski, M.: Henkin quantifiers. In: Krynicki, M., Mostowski, M., Szczerba, L.W. (eds.) Quantifiers: Logics, Models and Computation. Synthese Library (Studies in Epistemology, Logic, Methodology, and Philosophy of Science), vol. 248, pp. 193–262. Springer, Dordrecht (1995). https://doi.org/10.1007/978-94-017-0522-6_7
5. Lopez-Escobar, E.G.K.: Formalizing a non-linear Henkin quantifier. Fundamenta Mathematicae **138**(2), 93–101 (1991)
6. Mostowski, M., Zdanowski, K.: Degrees of logics with Henkin quantifiers in poor vocabularies. Arch. Math. Log. **43**(5), 691–702 (2004). https://doi.org/10.1007/s00153-004-0220-8
7. Takeuti, G.: Proof Theory, 2nd edn. North-Holland, Amsterdam (1987)

Feedback Hyperjump

Robert S. Lubarsky[✉]

Department of Mathematical Sciences, Florida Atlantic University,
Boca Raton, FL 33431, USA
`Robert.Lubarsky@alum.mit.edu`

Abstract. Feedback is oracle computability when the oracle consists exactly of the con- and divergence information about computability relative to that same oracle. Here we study the feedback version of the hyperjump.

Keywords: Hyperjump · Feedback · Kleene's $\mathcal{O}$

1 Introduction

Imagine a notion of computability which allows for an oracle. A natural choice of oracle is the halting problem, the set of halting programs. What if the programs in the oracle were exactly the halting programs relative to that same oracle?

That is the essence of feedback. There is not (yet) a general definition of feedback, one which is based on an unspecified notion of computation, perhaps a notion with some properties given axiomatically. What we do have are particular examples of feedback, including feedback Turing machines [1,2], feedback infinite time Turing machines [8], and feedback primitive recursion [1,2]. Experience has shown that the way feedback is defined has to be adapted to each different setting. Hence it is useful and interesting to examine more instances of feedback. The purpose of this work is to introduce feedback hyperjump.

There are several aspects of this that could be of interest. There are naturally not one but two different kinds of feedback for the hyperjump. Called below strict and loose, the difference between them is that the loose version has a kind of built-in parallelism. If there were a general definition of feedback then there should be only one kind of feedback hyperjump, but as it turns out we see no prima facie reason to choose one of strict and loose over the other. Another aspect that bears mention is that, in both cases, the central concept is arguably that of well-foundedness, but it plays a two-faced role. In some more detail, with feedback the escape from paradox threatened by diagonalization is provided by the possibility of computations freezing. For instance, if a computation asks the oracle about itself, then, in deciding what to answer, that same computation must be run, which will eventually ask the oracle about itself, ad infinitum. More generally, if a computation involves an infinite nested chain of oracle calls, then the oracle (depending on the setting) may not have a good answer and the computation could freeze. So the ill-foundedness of the tree of oracle calls

© Springer Nature Switzerland AG 2020
S. Artemov and A. Nerode (Eds.): LFCS 2020, LNCS 11972, pp. 144–155, 2020.
https://doi.org/10.1007/978-3-030-36755-8_10

(typically) leads to the computation freezing. On the other hand, considering $\mathcal{O}$ as a simple example of a hyperjump (the hyperjump of $\emptyset$), membership of n in $\mathcal{O}$ is given by the well-foundedness of the tree of ordinal notations less than n (less than in the sense of $<_\mathcal{O}$). Conversely, non-membership in $\mathcal{O}$ (in the interesting cases) is witnessed by the ill-foundedness of the induced tree of potential ordinal notations. With feedback, this is just the kind of information we want to capture. So ill-foundedness here will give us positive information. In the end, we will need to distinguish between two different kinds of trees, the tree of sub-computations and the tree of ordinal notations, the well- or ill-foundedness of each having very different consequences. Actually, this description is clean only for strict feedback hyperjump; for the loose version, as a kind of parallelism, even an ill-founded subcomputation tree can lead to a non-freezing computation. Ultimately the point for the moment is that we will be taking a very close look at the well-foundedness of trees associated with these computations. Finally, the results themselves might be of interest, as providing alternate descriptions of some ordinals which have already appeared in the literature.

To simplify the exposition, instead of defining the feedback hyperjump of an arbitrary real X, this will be done for only the empty set; the relativization to an arbitrary X is straightforward. The next section will review some of the basics of the regular hyperjump of 0, a.k.a. $\mathcal{O}$; all of this material is standard for the field, and serves only as a refresher and to introduce some of the notation and terminology we will use. The sections after that will study strict and loose feedback hyperjumps respectively.

To provide some historical context, feedback was clearly identified in [12] (pp. 406–407), although the topic was not pursued at that time. It was re-introduced in [8]. For an overview, see [2].

It should be mentioned that, as of this writing, not all of the proofs are complete. That notwithstanding, the author believes that they are far enough along to be convincing, and that the notions introduced are interesting enough to warrant public exposition.

2   Background on $\mathcal{O}$

For a much more thorough introduction to admissibility, $\mathcal{O}$, and such like, we refer the reader to [4] or [13]. Recall the mutual inductive definitions of Kleene's $\mathcal{O}$ and of the partial order $<_\mathcal{O}$ on $\mathcal{O}$. Regarding the former, $\mathcal{O}$ is the least set such that

1. $1 \in \mathcal{O}$,
2. if $n \in \mathcal{O}$ then $2^n \in \mathcal{O}$, and
3. if $\{e\}$ is total, and $\forall n \; \{e\}(n) <_\mathcal{O} \{e\}(n+1)$, then $3 \cdot 5^e \in \mathcal{O}$.

For the latter, $<_\mathcal{O}$ is the least transitive relation on $\mathcal{O}$ such that $n <_\mathcal{O} 2^n$ and $\{e\}(n) <_\mathcal{O} 3 \cdot 5^e$. (Notice that this is finer than the ordering on the ordinals represented by the members of $\mathcal{O}$.) Numbers not in $\mathcal{O}$ are incomparable with everything.

Given $n \in \mathcal{O}$, $\{k \mid k <_{\mathcal{O}} n\}$ is naturally ordered as a tree, with root n. The children of n are given by the primitive relations just mentioned, i.e. n is the unique child of 2^n, and if $n = 3 \cdot 5^e$ then the numbers $\{e\}(k)$ are the children of n. This tree is well-founded. In fact, that essentially characterizes the members of $\mathcal{O}$. That is, every $n \in \mathbb{N}$, whether in $\mathcal{O}$ or not, induces such a tree T_n, defined recursively, the well-foundedness of which, or not, determines membership in $\mathcal{O}$. We think of T_n, and related trees to be defined later, as the tree of ordinal notations, although it would be more accurate to call it the tree of potential ordinal notations, since some entries may not actually be ordinal notations (making of course n also not an ordinal notation); the latter name being more cumbersome, we stick with the former.

Definition 1. *1. T_1 is the tree consisting of the single node 1.*
2. For $n \neq 0$, T_{2^n} has root 2^n, which has a unique child n, which is the root of the subtree T_n.
3. $T_{3 \cdot 5^e}$ has root $3 \cdot 5^e$, with children each $\{e\}(k)$ which is defined, which is the root of the subtree $T_{\{e\}(k)}$.
4. In all other cases, T_n consists of the single node n.
   *We say that the tree T_n is **ill-formed** if*

1. either T_n contains a node not of the form 2^m or $3 \cdot 5^e$,
2. or T_n contains a node $3 \cdot 5^e$, and either $\{e\}$ is partial, or, for some k, $\{e\}(k) \notin T_{\{e\}(k+1)}$.

*If T_n is not ill-formed then we say it is **well-formed**.*

Proposition 1. $n \in \mathcal{O}$ iff T_n is well-formed and well-founded.

It is easy to see that there are n's with T_n well-formed and ill-founded: work in some non-standard model of some kind of set theory (say KP or anything stronger) with standard part ω_1^{CK}, and let n be a notation (as interpreted in that model) for some non-standard ordinal.

3 Strict Feedback Hyperjump

We will ultimately define the feedback oracle $\mathcal{SO}$, or **strict feedback** $\mathcal{O}$. With regular (as opposed to feedback) oracle computation, an oracle can be taken to be a set, and an oracle query returns YES or NO depending upon whether the number queried is in the oracle or not. With feedback oracle computation, this will not work. One cannot avoid freezing, the possibility that the oracle just doesn't answer; this is how diagonalization is avoided. So a feedback oracle is taken to be a partial function, which on its domain returns either Y or N. A query which is not in the domain of an oracle is said to be a freezing query (relative to that oracle); if during the course of a computation the oracle is asked a freezing oracle query, it does not answer, and that computation freezes. In addition, our oracle will have to answer two different types of questions: not only "$n \in \mathcal{SO}$?", but also "$m <_{\mathcal{SO}} n$?". So a **feedback oracle** is a partial function from the set of queries of the form "$n \in \mathcal{SO}$?" and "$m <_{\mathcal{SO}} n$?" to $\{Y, N\}$.

Let P be a feedback oracle. By way of notation, $\{e\}^P$ is Turing computability relative to P.

The trees T_n defined in the previous section are sufficient to witness membership in $\mathcal{O}$; more crucially, they are necessary to witness non-membership in $\mathcal{O}$. Since non-membership in $\mathcal{SO}$ needs to be witnessed positively, we have need of the analogue U_n of T_n, also called the tree of ordinal notations, appropriate for the current setting. The definition of U_n is identical to that of T_n, except that the computations involved are feedback computations, with notation $\langle e \rangle$ and $\langle e \rangle(k)$, depending on whether the program e calls for an input. (Whether angle brackets $\langle \rangle$ are meant as feedback computation, as in $\langle e \rangle(k)$, or as forming a tuple, as in the ordered pair $\langle a, b \rangle$, should be clear from the context. Typically one argument $\langle e \rangle$ means feedback, and more than one a tuple.) Since we are in the midst of defining these very computations, in order to avoid circularity we must first define U_n^P, where P is a feedback oracle. We will then use this to define a one-step procedure from the set of feedback oracles to itself, and observe that this procedure is positive in P (i.e. $P \subseteq Q$ implies $U_n^P \subseteq U_n^Q$). Then on general principles there will be a least fixed point of that procedure, which we will call $\mathcal{SO}$. With $\mathcal{SO}$ in hand, the ultimate tree of interest U_n can be taken to be $U_n^{\mathcal{SO}}$.

Terminology. When building a tree T_a recursively from a parameter a, we will typically give the children b of the root, and then want to continue defining the descendants of b in T_a as essentially the members of T_b. We will give some precise definitions here, to fall back on as need be, although we may abuse notation for convenience (for instance sometimes identifying a piece of code, when convergent, with its output, or identifying a tuple of length 1 with its only entry). A tree is a set of tuples of natural numbers of positive length, closed under truncation. The label of a node of a tree is the last entry of the node as a tuple. For instance, the tree T_a will have root $\langle a \rangle$, which is labeled a. If σ is a node in T, **to append a tree U beneath σ in T** means to include in T all tuples of the form $\sigma^\frown \tau$, where τ is a node in U (and $^\frown$ is concatenation).

Definition 2. 1. U_1^P is the tree consisting of only the root, labeled 1.

2. For $m \neq 0$, $U_{2^m}^P$ has root labeled 2^m, which has a unique child labeled m, and U_m^P is appended to $U_{2^m}^P$ beneath the root.

3. $U_{3 \cdot 5^e}^P$ has root labeled $3 \cdot 5^e$, with a child labeled by the pair $\langle e, k \rangle$ for each $k \in \mathbb{N}^1$; if $\{e\}^P(k)$ is defined, then append $U_{\{e\}^P(k)}^P$ beneath the root, by abuse of notation identifying the label $\langle e, k \rangle$ with the label $\{e\}^P(k)$ of the root of $U_{\{e\}^P(k)}^P$; if $\{e\}^P(k)$ is not defined (either freezing or divergent) then the node labeled $\langle e, k \rangle$ has no children.

4. In all other cases, U_n^P consists of a single node labeled n.

We say that U_n^P **freezes** if there is a node in U_n^P labeled $\langle e, k \rangle$ such that $\{e\}^P(k)$ freezes.

---

[1] This child is to be thought of as a piece of syntax acting as a placeholder, and not, for instance, as feedback computation, for which angle brackets $\langle \rangle$ are also used.

U_n^P *is* **ill-formed** *if*

1. *either U_n^P contains a node not of the form 2^m or $3 \cdot 5^e$,*
2. *or U_n^P contains a node $3 \cdot 5^e$, and either $\{e\}^P$ is partial, or, for some k, the oracle call "$\{e\}^P(k) <_P \{e\}^P(k+1)$?" returns N.*

If U_n^P does not freeze and is not ill-formed then we say it is **well-formed**.

Proposition 2. *If U_n^P does not freeze and $Q \supseteq P$, then $U_n^Q = U_n^P$.*

Viewing a feedback oracle P as giving partial information about a fixed point feedback oracle, P induces its own version of answers to oracle queries, which we want to view as the successor feedback oracle to P, hence the notation P^+.

Definition 3. 1. *$P^+(n \in \mathcal{SO}?) = Y$ if U_n^P is well-formed and well-founded.*
2. *$P^+(n \in \mathcal{SO}?) = N$ if U_n^P either is not freezing and ill-formed, or is well-formed and ill-founded.*
3. *$P^+(m <_{\mathcal{SO}} n?) = Y$ if $P^+(n \in \mathcal{SO}?) = Y$ and $m \neq n$ is a node in U_n^P.*
4. *$P^+(m <_{\mathcal{SO}} n?) = N$ if $P^+(n \in \mathcal{SO}?) = N$, or if $P^+(n \in \mathcal{SO}?) = Y$ and either $m = n$ or m is not a node in U_n^P.*

Proposition 3. *$P^+(m <_{\mathcal{SO}} n?)$ returns a value iff $P^+(n \in \mathcal{SO}?)$ returns a value. Also, $P^+(n \in \mathcal{SO}?)$ returns a value iff U_n^P does not freeze.*

Proposition 4. *The definition of P^+ is positive in P. Hence if $P \subseteq Q$ then $P^+ \subseteq Q^+$.*

The preceding proposition justifies the following.

Definition 4. $\mathcal{SO}$ *or* **strict** $\mathcal{O}$ *is the least fixed point of the operation $P \mapsto P^+$. U_n is $U_n^{\mathcal{SO}}$.*

Sometimes we think of $\mathcal{SO}$ not as a feedback oracle but rather as a partial set of numbers, as captured by the following convention; which way to think about $\mathcal{SO}$ should always be clear from the context.
Notation:

- "$n \in \mathcal{SO}$" is an abbreviation for $\mathcal{SO}(n \in \mathcal{SO}?) = Y$.
- "$n \notin \mathcal{SO}$" is an abbreviation for $\mathcal{SO}(n \in \mathcal{SO}?) = N$.
- "$n ? \mathcal{SO}$" is an abbreviation for $n \in \mathcal{SO}?$ not being in the domain of $\mathcal{SO}$.
- "$m <_{\mathcal{SO}} n$" is an abbreviation for $\mathcal{SO}(m <_{\mathcal{SO}} n?) = Y$.
- "$m \not<_{\mathcal{SO}} n$" is an abbreviation for $\mathcal{SO}(m <_{\mathcal{SO}} n?) = N$.
- "$m ?_{\mathcal{SO}} n$" is an abbreviation for $m <_{\mathcal{SO}} n?$ not being in the domain of $\mathcal{SO}$.
- $\langle e \rangle$ is the computation $\{e\}^{\mathcal{SO}}$.
- Q is said to be a freezing query if it is freezing relative to $\mathcal{SO}$, i.e. not in the domain of $\mathcal{SO}$.

Theorem 1. $\mathcal{SO}$ *is a system of notation for ordinals through the least recursively inaccessible ordinal α, and is a complete Σ_1 set over L_α.*

In order to prove this, we will need witnesses within L_α for assertions like $n \in \mathcal{SO}$ and $n \notin \mathcal{SO}$. Of course this will involve the trees of ordinal notation U_n. (Whether they are well- or ill-founded will determine whether n is in or out of $\mathcal{SO}$.) It will also involve the definition of $\mathcal{SO}$ as a least fixed point. Least fixed points of positive inductive definitions can be viewed as developed in stages indexed by the ordinals. In our case, elements enter $\mathcal{SO}$ at later stages because of computations based on fragments of $\mathcal{SO}$ from earlier stages, which are themselves based on sub-computations from yet earlier stages. It turns out to be useful to organize these sub-computations into a tree, the well-foundedness of which, or not, will be crucial. Please note that the trees of sub-computations are not to be confused with the trees of ordinal notations.

Definition 5. *For e a natural number and Q an oracle query, the trees S_e and S_Q (S for sub-computations) are defined recursively.*

For S_e, the root is e. Start running the oracle Turing computation $\{e\}^2$. If it makes an oracle query Q, then append S_Q beneath the root in S_e. If the oracle $\mathcal{SO}$ returns an answer to Q, then the computation $\{e\}$ continues. Similarly if at any time the run of $\{e\}$ makes an oracle call R, then S_R is appended beneath the root in S_e, with the root of S_R being a child of e to the right of all earlier oracle calls Q.

For S_Q, the root is Q. Whether Q is $n \in \mathcal{SO}$? or $m <_{\mathcal{SO}} n$?, let n_Q be n. To answer Q one would need to consider U_{n_Q}. The only nodes in U_{n_Q} that require any computation are nodes labeled $\langle e, k \rangle$. Append $S_{\langle e,k \rangle}$ for each such $\langle e, k \rangle$ beneath the root in S_Q.

Proposition 5. $\langle e \rangle$ *is non-freezing iff S_e is well-founded, and Q is non-freezing iff S_Q is well-founded.*

Proof. First we show inductively on ranks that if S_e resp. S_Q is well-founded then $\langle e \rangle$ resp. Q is non-freezing.

The immediate sub-trees of S_e are the S_Q's, where the Q's are e's oracle queries. If S_e is well-founded then each such S_Q has smaller rank, so Q is non-freezing, hence the run of $\langle e \rangle$ does not freeze.

To say that Q is non-freezing means that Q is in the domain of $\mathcal{SO}$. Since $\mathcal{SO}$ is a fixed point of the operation $P \mapsto P^+$, it suffices to show that Q is in the domain of $\mathcal{SO}^+$. By an earlier proposition, that holds iff $U_{n_Q}^{\mathcal{SO}}$ does not freeze. Toward that end, we must consider only nodes of $U_{n_Q}^{\mathcal{SO}}$ labeled $\langle e, k \rangle$. For any such e and k, $S_{\langle e,k \rangle}$ is an immediate sub-tree of S_Q. Hence it has lower rank, so inductively $\langle e \rangle(k)$ is non-freezing.

The converse hinges on $\mathcal{SO}$ being the least fixed point: there is no need to go beyond the realm in which the trees of sub-computations are well-founded. Toward this end, let P be $\mathcal{SO} \upharpoonright \{Q \mid S_Q \text{ is well-founded}\}$. We will show that P is a fixed point, which suffices.

---

[2] Sometimes we will have occasion to consider the computation $\{\bar{e}\}(k)$ instead. Then implicitly $e = \langle \bar{e}, k \rangle$.

Let Q be a query. We need to show that if $P^+(Q)$ returns an answer then so does $P(Q)$. Because $P \subseteq \mathcal{SO}$ and $\mathcal{SO}$ is a fixed point, $P^+ \subseteq \mathcal{SO}$. That means that Q is in the domain of $\mathcal{SO}$. To get Q to be in the domain of P, we need only show that S_Q is well-founded. The immediate sub-trees of S_Q are all of the form $S_{\langle e, k \rangle}$, for a node labeled $\langle e, k \rangle$ from U_{n_Q}. Because $P^+(Q)$ is defined, by the proposition $U_{n_Q}^P$ does not freeze. Because $\mathcal{SO} \supseteq P$, $U_{n_Q} = U_{n_Q}^{\mathcal{SO}} = U_{n_Q}^P$. Hence all of the nodes $\langle e, k \rangle$ we must consider from U_{n_Q} are already in $U_{n_Q}^P$. Because $U_{n_Q}^P$ does not freeze, every such $\{e\}^P(k)$ does not freeze. That means that when running $\{e\}(k)$, every time an oracle query R is made, the oracle P responds. By the definition of P, S_R is well-founded. Since the immediate sub-trees of $S_{\langle e, k \rangle}$ are all of that form, $S_{\langle e, k \rangle}$ is well-founded. Hence S_Q is well-founded.

Corollary 1. U_n is not freezing iff $S_{n \in \mathcal{SO}?}$ is well-founded.

We are now ready to prove the main theorem, that $\mathcal{SO}$ is a complete Σ_1 set over the least recursively inaccessible L_α.

Proof. Sketch of proof: Let P be $\mathcal{SO}$ restricted to those queries Q with $S_Q \in L_\alpha$. We will show that P is a fixed point. It is then immediate that $\mathcal{SO}$ is Σ_1 definable over L_α, as $\mathcal{SO}(Q) = Y$ resp. N iff there are a tree S_Q which is well-founded and a computation witnessing the answer Y resp. N.

All of the statements of interest, when true, have witnesses. For instance, that $n \in \mathcal{SO}$ is witnessed by $S_{n \in \mathcal{SO}}$ (as well as a ranking function to the ordinals), to show that U_n is not freezing, and U_n itself (as well as witnesses that U_n is well-formed and well-founded). That $\langle e \rangle$ does not freeze is witnessed by S_e (along with its ranking function). Of course, the constructions of the objects U_n, S_e, S_Q serve at witnesses that those objects are what they are purported to be. Hence we can think of these objects as being generated as we ascend the L-hierarchy. This justifies the notation U_n^β, S_e^β, etc., as U_n resp. S_e as defined over L_β, using only the witnesses within L_β. We will show that over L_α no new computations are defined over L_α, and that $U_n = U_n^\alpha, S_e = S_e^\alpha, S_Q = S_Q^\alpha$.

If $\langle e \rangle$ converges then that is witnessed by a finite run of a Turing machine, along with the witnesses to finitely many oracle calls. If all of the oracle call witnesses are in L_α, so is this finite sequence.

If $\langle e \rangle$ diverges then that is witnessed by an ω-sequence which is the divergent run of $\langle e \rangle$, along with a sequence of witnesses to oracle calls of length at most ω. If each such witness is in L_α, then by the admissibility of L_α so is the ω-sequence. Then the run of $\langle e \rangle$ is definable over that latter sequence.

In building U_n, work on one level at a time. (The root is level 0, its children level 1, etc.) It is immediate to determine if a node has any children and what those children are, except for nodes labeled $3 \cdot 5^e$. Since U_n is (by assumption) not freezing, each child $\langle e, k \rangle$ has a witness as to whether $\langle e \rangle(k)$ converges or diverges. Since each level can be arranged in an ω-sequence, this induces a total function from ω to these witnesses. By admissibility this function is in L_α. Furthermore, we have to repeat this construction on all ω-many levels of U_n, which again by admissibility is in L_α.

S_e is essentially the witnesses to the oracle calls from above that $\langle e \rangle$ does not freeze.

For S_Q well-founded, U_{n_Q} was already seen to be in L_α. The immediate subtrees of S_Q are the $S_{\langle e,k \rangle}$'s for $\langle e,k \rangle$ a node in U_{n_Q}. In L_α there is a sequence of such nodes of length at most ω. If each $S_{\langle e,k \rangle}$ were in L_α, then again by admissibility the sequence of such is in L_α, which puts S_Q into L_α.

Finally, in order to answer an oracle question Q, the only time Q has an answer is when U_{n_Q} is not freezing. So then U_{n_Q} is in L_α, as above. Whether it is ill-formed or not is witnessed within L_α, again using the admissibility of L_α. When U_{n_Q} is well-formed, whether it is well-founded is also witnessed within L_α, this time using the fact that α is a limit of admissibles: if a tree is in L_α, then a ranking function for the tree's well-founded part is definable over the least admissible set containing the tree, and hence is in L_α.

The harder direction is to show that L_α is a lower bound, in that every real in L_α is $\mathcal{SO}$-computable and moreover that Σ_1 questions about L_α can be converted uniformly to questions about membership in $\mathcal{SO}$. It should be clear from the presence of the oracle calls that the $\mathcal{SO}$-computable reals are closed under hyperjump. For instance, $\mathcal{O}$ is computable: for n to be a candidate for membership in $\mathcal{O}$, when building U_n no computations along the way may consult with an oracle, so there is no possibility for U_n to freeze. Hence the oracle answer to $n \in \mathcal{SO}$? is the correct information for whether n is in $\mathcal{O}$. Since this argument relativizes, the computable reals are closed under the hyperjump. In particular, $\mathcal{O}$ exists, as do $\mathcal{O}', \mathcal{O}''$, etc. It is not hard to define the join of the $\mathcal{O}^{(n)}$'s: split the work tape up into ω-many infinite tapes, and dedicate the nth tape to $\mathcal{O}^{(n)}$. So the computable reals go beyond the first limit of admissibles.

To show however that we can go past any inadmissible limit of admissibles, this needs in some form the Gandy Selection Theorem, that the computable predicates are closed under a search through ω. This turns a fact of inadmissibility – $\forall i \in \omega \, \exists A_i \; \phi(i, A_i)$ – into a computable sequence $\langle A_i \rangle_{i \in \omega}$. Gandy Selection can be proved via a Stage Comparison Theorem.

4   Loose Feedback Hyperjump

In the inductive generation of $\mathcal{O}$ and $\mathcal{SO}$, there is really no difference in the ways numbers get put into those sets, the ways numbers get accepted as ordinal notations. The difference between them is that $\mathcal{SO}$ contains negative information too, that $\mathcal{SO}$ will tell you when something is not a member of $\mathcal{SO}$. This negative information is essentially the ill-foundedness of a certain tree (T_n resp. U_n). For $\mathcal{O}$, there was no possibility of T_n being freezing, whereas some U_n's most certainly are. It is part of the definition of $\mathcal{SO}$ that for an oracle call "$n \in \mathcal{SO}$?" to be non-freezing the tree U_n must be non-freezing. The requirement that all nodes in U_n be non-freezing can be explained or justified by thinking of U_n being generated by a (transfinite) breadth-first search. That is, first evaluate all the computations on the first level of U_n, then all those on the second level, and so on. After all ω-many levels, one can see whether the tree is well-founded. If any of those computations freezes, then this procedure cannot be completed.

This outcome, that a single freezing node in U_n freezes the question $n \in \mathcal{SO}$?, is necessary if the answer is to be "yes", because a "yes" answer is to be taken as a guarantor of U_n's well-foundedness. Imagine by way of contrast that some node of an otherwise well-founded tree were freezing. That freezing happens when a particular oracle call is made. You can think of the machine at that point as waiting for a response. This waiting can be taken to be measured along the ordinals as indexing L, but it does not have to be. As an alternative, while we are studying in this paper the semantics of the least fixed point, we don't have to. Perhaps a larger fixed point is generated by some random (albeit appropriate) computation all of a sudden no longer freezing. Perhaps such sudden removal of blockages can be organized in a partial order, like a kind of (or actual!) Kripke model. Then time could be taken as movement along this partial order. There could be other interpretations of time. A conservative way of thinking about freezing is that one should use no information about a freezing node, not even that it is freezing. So back to our pseudo well-founded tree with one freezing node. If that node ever gets unstuck, depending on what happens after that, the tree could become ill-founded. This issue does not come up if instead U_n is not freezing: U_n cannot change at all, even as or if the oracle changes; hence a well-founded U_n will remain well-founded.

In contrast, such prudence is not necessary for ill-foundedness. Once a tree is ill-founded, even as the tree grows it will remain ill-founded. Allowing an ill-founded tree, even when freezing, to qualify as a witness that a number is not ordinal notation is what we are calling **loose feedback** $\mathcal{O}$, or $\mathcal{LO}$. This allowance can be explained or justified by thinking of U_n as being generated while ascending through L. As one climbs through the L_α's, α increasing, more computations become completed (i.e. converge or diverge), so more nodes get placed into U_n. If at any stage U_n is seen to be ill-founded, even if it still contains freezing nodes, then we can take that as a witness that $n \notin \mathcal{LO}$.

In comparing the negative information in $\mathcal{LO}$ with that of $\mathcal{SO}$, it comes down to a kind of parallelism. Normally in mathematical logic parallelism plays no role, since it can be simulated by sequential computation via dovetailing. This does not work with feedback around: once a freezing oracle call is made, the entire computation stops. This was used in [9] to define parallel feedback (Turing) computability, by which an ω-sequence of machines was run in parallel, which was shown to be stronger than (sequential) feedback Turing computability. Parallelism was also defined for infinite time Turing machines [8] (and ultimately analyzed in [16], even if the framework there is Kleene's higher types [6], as the results are translatable to feedback ITTMs). In the current setting, it's as though we're searching for an infinite branch in a tree, even if another part of the tree is freezing. Instead of running ω-many machines in parallel, what we have here can be called **tree parallelism**.

We will make use of the same trees U_n^P as in the previous section. Given an oracle, there is no change from before about the induced tree of ordinal notations. The difference from before is the inductive step on feedback oracles, there called P^+, here $P^{\&}$.

Definition 6. *1. $P^\&(n \in \mathcal{SO}?) = Y$ if U_n^P is well-formed and well-founded.*
2. $P^\&(n \in \mathcal{SO}?) = N$ if U_n^P is either ill-formed or ill-founded.
3. $P^\&(m <_{\mathcal{SO}} n?) = Y$ if $P^\&(n \in \mathcal{SO}?) = Y$ and $m \neq n$ is a node in U_n^P.
4. $P^\&(m <_{\mathcal{SO}} n?) = N$ if $P^\&(n \in \mathcal{SO}?) = N$, or if $P^\&(n \in \mathcal{SO}?) = Y$ and either $m = n$ or m is not a node in U_n^P.

Note that the only difference between P^+ and $P^\&$ is in clause 2.[3]

Proposition 6. *The definition of $P^\&$ is positive in P. Hence if $P \subseteq Q$ then $P^\& \subseteq Q^\&$.*

The preceding proposition justifies the following.

Definition 7. $\mathcal{LO}$ *or* **loose** $\mathcal{O}$ *is the least fixed point of the operation $P \mapsto P^\&$. V_n is $U_n^{\mathcal{LO}}$.*

While we're at it, we will also define the trees of sub-computations $S_e^\&$ and $S_Q^\&$. Formally speaking, they are defined the same way S_e and S_Q were in the previous section, only with reference to $\mathcal{SO}$ replaced by $\mathcal{LO}$. This affects $S_e^\&$ directly: if $\{e\}$ makes an oracle call, it is more likely to get an answer from $\mathcal{LO}$ than $\mathcal{SO}$. Then this affects $S_Q^\&$, which is defined in terms of $S_e^\&$ the way S_Q is defined in terms of S_e.

Definition 8. *Let Γ be a collection of formulas, X a class of ordinals, and ν^{+X} the least member of X greater than ν. We say that α is Γ-**reflecting** on X if, for all $\phi \in \Gamma$, if $L_{\alpha+X} \models \phi(\alpha)$, then for some $\beta < \alpha$, $L_{\beta+X} \models \phi(\beta)$.*

We are interested in the case $\Gamma = \Pi_1$ and $X =$ the collection of admissible ordinals. For this choice of X, we abbreviate ν^{+X} by ν^+, which is standard notation for the next admissible anyway. This is called Π_1 **gap-reflection on admissibles**. Let γ be the least such ordinal.

It may seem like a strange notion. But this is not the first time it has come up. Extending work in [11], it was shown in [7] that such ordinals are exactly the Σ_1^1 reflecting ordinals. (In this context, the superscript 1 refers not to reals

---

[3] It bears mention that there are several options for dealing with this clause. In all cases, the evidence that n is not an ordinal notation is that its tree U_n^P of smaller ordinal notations is bad somehow, either ill-formed or ill-founded. For P^+, we took this in the strictest possible sense: U_n^P had to be non-freezing in order to qualify as evidence. For $P^\&$, there is no such requirement on U_n^P ever; once we have any evidence that U_n^P will not be acceptable, we take it. In contrast with both of these, one could work in the middle. That is, the reasons that U_n^P activate clause 2 are that it has a node not of the right form, or that the function named by a node is partial, or that the function named by a node is not increasing (in the sense of $<_P$), or that the tree has an infinite descending path; the requirement that U_n^P be non-freezing could, in principle, be levied on some and not all of these conditions. We find the two extreme cases isolated here to be the most natural ones; we believe that the only condition of any real importance is the well-foundedness of U_n^P, and that varying the others will make no difference; determining this is left for future work.

but to subsets of the structure over which the formula is being evaluated.) The reason this topic came up in the latter paper is that a particular case of its main theorem is that γ is the closure point of Σ_2-definable sets of integers in the μ-calculus. (The μ-calculus is first-order logic augmented with least and greatest fixed-point operators; see [3]. In this context, Σ_2 refers to the complexity of the fixed points in the formula, namely, in normal form, a least fixed point in front, followed by a greatest fixed point, followed by a fixed-point-free matrix.) In [11] it was also shown that the least Σ_1^1 reflecting ordinal is also the closure point of Σ_1^1 monotone inductive definitions. (Here the superscript does refer to reals.) Furthermore, that is the same least ordinal which provides winning strategies for all Σ_2^0 games (Solovay, see [10] 7C.10 or [15]). (If Player I has a winning strategy, then there is one in L_γ; if II does, then there is one in $L_{\gamma+}$.) As though that weren't enough, [14] shows the equivalence of closure under Σ_1^1 monotone inductive definitions with the Σ_1^1 Ramsey property. (For all Σ_1^1 partitions P of ω there is an infinite set $H \subseteq \omega$ such that the infinite subsets of H are either all in P or all not in P.) An ordinal α is Gandy if the α-computable well-orderings are cofinal through α^+; γ is the least non-Gandy ordinal [5]. Closest of all to the work being discussed here, γ is also the closure ordinal of context-dependent deterministic parallel feedback Turing computability [2,9]. With all of these applications, this definition counts as natural.

Theorem 2. $\mathcal{LO}$ *is a system of notation for ordinals through the least ordinal* γ *which is* Π_1 *gap-reflecting on admissibles, and is a complete* Σ_1 *set over* L_γ.

Proof. It is easier to show that γ is an upper bound. The notations $V_n^\beta, S_e^{\&\beta}, S_Q^{\&\beta}$ mean $V_n, S_e^\&, S_Q^\&$ as interpreted in L_β. In the definition of $P^\&$, clauses 1, 3, and 4 are the same as for P^+. So by the arguments for the previous section $\mathcal{LO}^\gamma$ is closed under those clauses by the admissibility of γ. Similarly for the ill-formedness condition of clause 2: if V_n^γ is ill-formed, then so is some V_n^β ($\beta < \gamma$). Now suppose V_n^γ were ill-founded. Because V_n^γ is definable over L_γ, its ill-foundedness is a Π_1 statement over $L_{\gamma+}$ with parameter γ. Therefore, by the choice of γ, there is a smaller β with $L_{\beta+} \models$ "V_n^β is ill-founded." So there is already a witness to n not being in $\mathcal{LO}$ in L_γ.

To show that γ is a lower bound, we will interpret, or simulate, parallel feedback Turing computability [2,9] within (computability relative to) $\mathcal{LO}$. Since the former has already been shown to compute everything in L_γ, this suffices. The reason this reduction would hold is that the same structures are involved with both of them. In more detail, the $\mathcal{LO}$ computations are run by the trees V_n of ordinal notations and $S_e^\&, S_Q^\&$ of sub-computations. For a computation not to freeze, it is not necessary that V_n not freeze (as opposed to U_n), much less be well-founded. For $S_e^\&$ and $S_Q^\&$, it is not necessary that they be well-founded (as opposed to S_e and S_Q), well-founded in the right way: an infinite path through V_n determines infinitely many sub-trees (rooted on the first level) of S_Q, where Q is $n \in \mathcal{LO}?$, and they all must be well-founded. Now consider the trees that come up in parallel feedback. Most prominent is $C^{(e,n)}$, the tree of runs. In order for the parallel feedback computation $\langle e \rangle(n)$ not to freeze, it is not necessary that $C^{(e,n)}$ not freeze; rather, $C^{(e,n)}$ could have a terminal node, which is the uninteresting case in all the proofs, or it is ill-founded. This is the same behavior as the V_n's. The work on parallel feedback did not discuss the tree of sub-computations, because it no longer had to be well-founded; in fact, the only well-foundedness that matters is that of an infinite set of sub-trees as determined by some infinite path through $C^{(e,n)}$. The stopping conditions are the same in both cases. That is why ultimately each can code the other.

References

1. Ackerman, N., Freer, C., Lubarsky, R.: Feedback turing computability, and turing computability as feedback. In: Proceedings of LICS 2015, Kyoto, Japan (2015). http://math.fau.edu/lubarsky/pubs.html
2. Ackerman, N., Freer, C., Lubarsky, R.: An introduction to feedback turing computability. J. Log. Comput., special issue on LFCS '16 (2020, submitted). http://math.fau.edu/lubarsky/pubs.html
3. Arnold, A., Niwinski, D.: Rudiments of μ-Calculus. Studies in Logic and the Foundations of Mathematics, vol. 146. North Holland, Amsterdam (2001)
4. Barwise, J.: Admissible Sets and Structures. Perspectives in Mathematical Logic. Springer, Berlin (1975)
5. Gostanian, R.: The next admissible ordinal. Ann. Math. Log. **17**, 171–203 (1979)
6. Kleene, S.C.: Recursive functionals and quantifiers of finite types I. Trans. Am. Math. Soc. **91**, 1–53 (1959)
7. Lubarsky, R.: μ-definable sets of integers. J. Symb. Log. **58**(1), 291–313 (1993)
8. Lubarsky, R.: ITTMs with feedback. In: Schindler, R. (ed.) Ways of Proof Theory, pp. 341–354. Ontos, Frankfurt (2010). http://math.fau.edu/lubarsky/pubs.html
9. Lubarsky, R.S.: Parallel feedback turing computability. In: Artemov, S., Nerode, A. (eds.) LFCS 2016. LNCS, vol. 9537, pp. 236–250. Springer, Cham (2016). https://doi.org/10.1007/978-3-319-27683-0_17
10. Moschovakis, Y.: Descriptive Set Theory, 1st edn. North Holland, Amsterdam (1987). 2nd edn. AMS (2009)
11. Richter, W., Aczel, P.: Inductive definitions and reflecting properties of admissible ordinals. In: Fenstad, J.E., Hinman, P.G. (eds.) Generalized Recursion Theory, pp. 301–381. North-Holland, Amsterdam (1974)
12. Rogers, H.: Theory of Recursive Functions and Effective Computability. McGraw-Hill, New York (1967)
13. Sacks, G.: Higher Recursion Theory. Perspectives in Mathematical Logic. Springer, Berlin (1990)
14. Tanaka, K.: The Galvin-Prikry theorem and set existence axioms. Ann. Pure Appl. Log. **42**(1), 81–104 (1989)
15. Tanaka, K.: Weak axioms of determinacy and subsystems of analysis II (Σ_2^0 games). Ann. Pure Appl. Log. **52**(1–2), 181–193 (1991)
16. Welch, P.: $G_{\delta\sigma}$-games and generalized computation (to appear)

Syntactic Cut-Elimination
for Intuitionistic Fuzzy Logic
via Linear Nested Sequents

Tim Lyon[✉]

Institut für Logic and Computation, Technische Universität Wien,
1040 Vienna, Austria
lyon@logic.at

Abstract. This paper employs the *linear nested sequent* framework to design a new cut-free calculus (LNIF) for intuitionistic fuzzy logic—the first-order Gödel logic characterized by linear relational frames with constant domains. Linear nested sequents—which are nested sequents restricted to linear structures—prove to be a well-suited proof-theoretic formalism for intuitionistic fuzzy logic. We show that the calculus LNIF possesses highly desirable proof-theoretic properties such as invertibility of all rules, admissibility of structural rules, and syntactic cut-elimination.

Keywords: Cut-elimination · Fuzzy logic · Gödel logic · Intermediate logic · Intuitionistic logic · Linear nested sequents · Proof theory

1 Introduction

Intuitionistic fuzzy logic (IF) has attracted considerable attention due to its unique nature as a logic blending fuzzy reasoning and constructive reasoning [1,3,4,13,25]. The logic, which was initially defined by Takeuti and Titani in [25], has its roots in the work of Kurt Gödel. Gödel introduced extensions of propositional intuitionistic logic (now called, "Gödel logics") in order to prove that propositional intuitionistic logic does not possess a finite characteristic matrix [11]. These logics were later studied by Dummett who extended Gödel's finite-valued semantics to include an infinite number of truth-values [7]. Dummett additionally provided an axiomatization for the propositional fragment of IF [7]. The first-order logic IF also admits a finite axiomatization, obtained by extending an axiomatization of first-order intuitionistic logic with the *linearity axiom* $(A \supset B) \vee (B \supset A)$ and the *quantifier shift axiom* $(\forall x)(A(x) \vee C) \supset \forall x A(x) \vee C$ (where x does not occur free in C) [14].

Over the last few decades, propositional and first-order Gödel logics (including the prominent logic IF) have been applied in various areas of logic and computer science [3,4,6,13,17,18,27]. For example, Visser [27] applied the propositional fragment of IF while analyzing the provability logic of Heyting arithmetic,

© Springer Nature Switzerland AG 2020
S. Artemov and A. Nerode (Eds.): LFCS 2020, LNCS 11972, pp. 156–176, 2020.
https://doi.org/10.1007/978-3-030-36755-8_11

Lifschitz et al. [18] employed a Gödel logic to model the strong equivalence of logic programs, and Borgwardt et al. [6] studied standard reasoning problems of first-order Gödel logics in the context of fuzzy description logics. Additionally— and quite significantly—the logic IF has been recognized as one of the fundamental formalizations of fuzzy logic [13].

The question of whether or not a logic possesses an *analytic* proof calculus— that is, a calculus which stepwise (de)composes the formula to be proved—is of critical importance. Such calculi are effective tools for designing automated reasoning procedures and for proving certain (meta-)logical properties of a logic. For example, analytic calculi have been leveraged to provide decidability procedures for logics [10], to prove that logics interpolate [17], for counter-model extraction [20], and to understand the computational content of proofs [23].

In his seminal work [10], Gentzen proposed the *sequent calculus* framework for classical and intuitionistic logic, and subsequently, proved his celebrated *Hauptsatz* (i.e. cut-elimination theorem), which ultimately provided analytic calculi for the two logics. Gentzen's sequent calculus formalism has become one of the preferred proof-theoretic frameworks for providing analytic calculi, and indeed, many logics of interest have been equipped with such calculi. Nevertheless, one of the alluring features of the formalism—namely, its simplicity—has also proven to be one of the formalism's drawbacks; there remain many logics for which no cut-free, or analytic, sequent calculus (à la Gentzen) is known [12,24]. In response to this, the sequent calculus formalism has been extended in various ways over the last 30 years to include additional structure, allowing for numerous logics to be supplied with cut-free, analytic calculi. Some of the most prominent extensions of Gentzen's formalism include *display calculi* [5], *labelled calculi* [20,26], *hypersequent calculi* [24], and *nested calculi* [8,12].

In this paper, we employ the *linear nested sequent* formalism, introduced by Lellmann in [15]. Linear nested sequents fall within the *nested calculus paradigm*, but where sequents are restricted to linear, instead of treelike, structures. Linear nested sequents are based off of Masini's 2-sequent framework [22,23] that was used to provide cut-free calculi for the modal logic KD as well as various other constructive logics. The linear nested formalism proves to be highly compatible with the well-known first-order Gödel logic IF (i.e. intuitionistic fuzzy logic), due to the fact that IF can be semantically characterized by *linear* relational frames (see Sect. 2). The present work provides the linear nested calculus LNIF for IF, which enjoys a variety of fruitful properties, such as:[1]

▶ *Separation*: Each logical rule exhibits no other logical connectives than the one to be introduced.
▶ *Symmetry*: Each logical connective has a left and right introduction rule.
▶ *Internality*: Each sequent translates into a formula of the logical language.
▶ *Cut-eliminability*: There exists an algorithm allowing the permutation of a (*cut*) rule (encoding reasoning with lemmata) upwards in a derivation until the rule is completely eliminated from the derivation.

---

[1] We refer to [28] for a detailed discussion of fundamental proof-theoretic properties.

▶ *Subformula property*: Every formula occurring in a derivation is a subformula of some formula in the end sequent.
▶ *Admissibility of structural rules*: Everything derivable with a structural rule (cf. (iw) and (mrg) in Sect. 4) is derivable without the structural rule.
▶ *Invertibility of rules*: If the conclusion of an inference rule is derivable, then so is the premise.

In [4], a cut-free hypersequent calculus HIF for IF was introduced to overcome the shortcomings of previously introduced systems [14,25] that violated fundamental proof-theoretic properties such as cut-elimination. In contrast to HIF, the current approach of exploiting linear nested sequents has two main benefits. First, the admissibility of structural rules has not been shown in HIF, and as such, the calculus does not offer a purely formula-driven approach to proof search. Therefore, the calculus LNIF serves as a better basis for automated reasoning in IF—bottom-up applications of the rules in LNIF simply decompose or propagate formulae, and so, the question of if/when structural rules need to be applied does not arise. Second, the calculus HIF cannot be leveraged to prove interpolation for the logic IF (see [17]) via the so-called *proof-theoretic method* (cf. [17,21]) due to the presence of the communication structural rule [1]. In [17], it was shown that the propositional fragment of LNIF can be harnessed to prove Lyndon interpolation for the propositional fragment of IF. This result suggests that LNIF, in conjunction with the aforementioned proof-theoretic method, may potentially be harnessed to study and determine interpolable fragments of IF, or to solve the longstanding open problem of if the entire logic IF interpolates or not.

The contributions and organization of this paper can be summarized as follows: In Sect. 2, we introduce the semantics and axiomatization for intuitionistic fuzzy logic (IF). Section 3 introduces linear nested sequents and the calculus LNIF, as well as proves the calculus sound and complete relative to IF. In Sect. 4, we provide invertibility, structural rule admissibility, and cut-elimination results. Last, Sect. 5 concludes and discusses future work.

2   Logical Preliminaries

Our language consists of denumerably many *variables* $\{x, y, z, \ldots\}$, denumerably many n-ary *predicates* $\{p, q, r, \ldots\}$ (with $n \in \mathbb{N}$), the *connectives* $\bot, \wedge, \vee, \supset$, the *quantifiers* $\forall, \exists$, and *parentheses* '(' and ')'. We define the language $\mathcal{L}$ via the BNF grammar below, and will use A, B, C, etc. to represent formulae from $\mathcal{L}$.

$$A ::= p(x_1, \ldots, x_n) \mid \bot \mid (A \vee A) \mid (A \wedge A) \mid (A \supset A) \mid (\forall x)A \mid (\exists x)A$$

In the above grammar, p is any n-ary predicate symbol and $x_1, \ldots, x_n, x$ are variables. We refer to formulae of the form $p(x_1, \ldots, x_n)$ as *atomic formulae*, and (more specifically) refer to formulae of the form p as *propositional variables* (i.e. a 0-ary predicate p is a propositional variable). The *free variables* of a formula A are defined in the usual way as variables unbound by a quantifier, and *bound variables* as those bounded by a quantifier.

We opt for the relational semantics of IF—as opposed to the fuzzy semantics (cf. [4])—since the structure of linear nested sequents is well-suited for interpretation via linear relational frames.

Definition 1 (Relational Frames, Models [9]). *A relational frame is a triple $F = (W, R, D)$ such that: (i) W is a non-empty set of worlds $w, u, v, \ldots$, (ii) R is a reflexive, transitive, antisymmetric, and connected binary relation on W, and (iii) D is a function that maps a world $w \in W$ to a non-empty set of* parameters D_w *called the* domain of w *such that the following condition is met:*

(CD) *If Rwu, then $D_w = D_u$.*

A model M is a tuple (F, V) where F is a relational frame and V is a valuation function such that $V(p, w) \subseteq (D_w)^n$ for each n-ary predicate p and

(TP) *If Rwu, then $V(p, w) \subseteq V(p, u)$ (if p is of arity $n > 0$);*
      If Rwu and $w \in V(p, w)$, then $u \in V(p, v)$ (if p is of arity 0).

We uphold the convention in [9] and assume that for each world $w \in W$, $(D_w)^0 = \{w\}$, so $V(p, w) = \{w\}$ or $V(p, w) = \emptyset$, for a propositional variable p.

The distinctive feature of relational frames for IF is the *connected* property, which states that for any $w, u, v \in W$ of a frame $F = (W, R, D)$, if Rwu and Rwv, then either Ruv or Rvu. Imposing this property on reflexive, transitive, and antisymmetric (i.e. intuitionistic) frames causes the set of worlds to become linearly ordered, thus validating the linearity axiom $(A \supset B) \vee (B \supset A)$ (shown in Fig. 1). Furthermore, the constant domain condition **(CD)** validates the quantifier shift axiom $\forall x(A(x) \vee B) \supset \forall x A(x) \vee B$ (also shown in Fig. 1).

Rather than interpret formulae from $\mathcal{L}$ in relational models, we follow [9] and introduce D_w-sentences to be interpreted in relational models. This gives rise to a notion of validity for formulae in $\mathcal{L}$ (see Definition 3). The definition of validity also depends on the *universal closure* of a formula: if a formula A contains only $x_0, \ldots, x_m$ as free variables, then the universal closure $\overline{\forall} A$ is taken to be the formula $\forall x_0 \ldots \forall x_m A$.

Definition 2 (D_w-Sentence). *Let M be a relational model with $w \in W$ of M. We define $\mathcal{L}_{D_w}$ to be the language $\mathcal{L}$ expanded with parameters from the set D_w. We define a D_w-formula to be a formula in $\mathcal{L}_{D_w}$, and we define a D_w-sentence to be a D_w-formula that does not contain any free variables. Last, we use $a, b, c, \ldots$ to denote parameters in a set D_w.*

Definition 3 (Semantic Clauses [9]). *Let $M = (W, R, D, V)$ be a relational model with $w \in W$ and $R(w) := \{v \in W \mid (w, v) \in R\}$. The satisfaction relation $M, w \Vdash A$ between $w \in W$ and a D_w-sentence A is inductively defined as follows:*

- *$M, w \nVdash \bot$*
- *If p is a propositional variable, then $M, w \Vdash p$ iff $w \in V(p, w)$;*

- *If p is an n-ary predicate symbol (with $n > 0$), then $M, w \Vdash p(a_1, \cdots, a_n)$ iff $(a_1, \cdots, a_n) \in V(p, w)$;*
- *$M, w \Vdash A \vee B$ iff $M, w \Vdash A$ or $M, w \Vdash B$;*
- *$M, w \Vdash A \wedge B$ iff $M, w \Vdash A$ and $M, w \Vdash B$;*
- *$M, w \Vdash A \supset B$ iff for all $u \in R(w)$, if $M, u \Vdash A$, then $M, u \Vdash B$;*
- *$M, w \Vdash \forall x A(x)$ iff for all $u \in R(w)$ and all $a \in D_u$, $M, u \Vdash A(a)$;*
- *$M, w \Vdash \exists x A(x)$ iff there exists an $a \in D_w$ such that $M, w \Vdash A(a)$.*

We say that a formula A is globally true *on M, written $M \Vdash A$, iff $M, u \Vdash \overline{\forall} A$ for all worlds $u \in W$. A formula A is* valid, *written $\Vdash A$, iff it is globally true on all relational models.*

Lemma 1 (Persistence). *Let M be a relational model with $w, u \in W$ of M. For any D_w-sentence A, if $M, w \Vdash A$ and Rwu, then $M, u \Vdash A$.*

Proof. See [9, Lem. 3.2.16] for details.   □

A sound and complete axiomatization for the logic IF is provided in Fig. 1. We define the *substitution* $[y/x]$ of the variable y for the free variable x on a formula A in the standard way as the replacement of all free occurrences of x in A with y. The substitution $[a/x]$ of the parameter a for the free variable x is defined similarly. Last, the side condition y *is free for* x (see Fig. 1) is taken to mean that y does not become bound by a quantifier if substituted for x.

$A \supset (B \supset A) \qquad (A \supset (B \supset C)) \supset ((A \supset B) \supset (A \supset C)) \qquad A \supset (B \supset (A \wedge B))$

$$\frac{A \quad A \supset B}{B} \; mp$$

$(A \wedge B) \supset A \qquad (A \wedge B) \supset B \qquad A \supset (A \vee B) \qquad B \supset (A \vee B)$

$$\frac{A}{\forall x A} \; gen$$

$(A \supset C) \supset ((B \supset C) \supset ((A \vee B) \supset C)) \qquad \bot \supset A \qquad (A \supset B) \vee (B \supset A)$

$\forall x A \supset A[y/x] \; y \text{ free for } x \quad A[y/x] \supset \exists x A \; y \text{ free for } x \quad \forall x (B \supset A(x)) \supset (B \supset \forall x A(x))$

$\forall x (A(x) \supset B) \supset (\exists x A(x) \supset B) \qquad \forall x (A(x) \vee B) \supset \forall x A(x) \vee B \; \text{ with } x \notin B$

Fig. 1. Axiomatization for the logic IF [9]. The logic IF is the smallest set of formulae from $\mathcal{L}$ closed under substitutions of the axioms and applications of the inference rules *mp* and *gen*. We write $\vdash_{\mathsf{IF}} A$ to denote that A is an element, or *theorem*, of IF.

Theorem 1 (Adequacy of IF). *For any $A \in \mathcal{L}$, $\Vdash A$ iff $\vdash_{\mathsf{IF}} A$.*

Proof. The forward direction follows from [9, Prop. 7.2.9] and [9, Prop. 7.3.6], and the backwards direction follows from [9, Lem. 3.2.31].   □

3   Soundness and Completeness of **LNIF**

Let us define *linear nested sequents* (which we will refer to as *sequents*) to be syntactic objects $\mathcal{G}$ given by the BNF grammar shown below:

$$\mathcal{G} ::= \Gamma \vdash \Gamma \mid \mathcal{G} \mathbin{/\!/} \mathcal{G} \text{ where } \Gamma ::= A \mid \Gamma, \Gamma \text{ with } A \in \mathcal{L}.$$

Each sequent $\mathcal{G}$ is of the form $\Gamma_1 \vdash \Delta_1 \mathbin{/\!/} \cdots \mathbin{/\!/} \Gamma_n \vdash \Delta_n$ with $n \in \mathbb{N}$. We refer to each $\Gamma_i \vdash \Delta_i$ (for $1 \leq i \leq n$) as a *component* of $\mathcal{G}$ and use $\|\mathcal{G}\|$ to denote the number of components in $\mathcal{G}$.

We often use $\mathcal{G}$, $\mathcal{H}$, $\mathcal{F}$, and $\mathcal{K}$ to denote sequents, and will use Γ and Δ to denote antecedents and consequents of components. Last, we take the comma operator to be commutative and associative; for example, we identify the sequent $p(x) \vdash q(x), r(y), p(x)$ with $p(x) \vdash r(y), p(x), q(x)$. This interpretation lets us view an antecedent Γ or consequent Δ as a multiset of formulae.

To ease the proof of cut-elimination (Theorem 4), we follow [8] and syntactically distinguish between *bound variables* $\{x, y, z, \ldots\}$ and *parameters* $\{a, b, c, \ldots\}$, which will take the place of free variables occurring in formulae. Thus, our sequents make use of formulae from $\mathcal{L}$ where each free variable has been replaced by a unique parameter. For example, we would use the sequent $p(a) \vdash \forall x q(x, b) \mathbin{/\!/} \bot \vdash r(a)$ instead of the sequent $p(x) \vdash \forall x q(x, y) \mathbin{/\!/} \bot \vdash r(x)$ in a derivation (where the parameter a has been substituted for the free variable x and b has been substituted for y). We also use the notation $A(a_0, \ldots, a_n)$ to denote that the parameters $a_0, \ldots, a_n$ occur in the formula A, and write $A(\vec{a})$ as shorthand for $A(a_0, \ldots, a_n)$. This notation extends straightforwardly to sequents as well.

The linear nested calculus **LNIF** for **IF** is given in Fig. 2. (NB. The linear nested calculus **LNG** introduced in [17] is the propositional fragment of **LNIF**, i.e. **LNG** is the calculus **LNIF** without the quantifier rules and where propositional variables are used in place of atomic formulae.) The $(\supset_{r2})$ and $(\forall_{r2})$ rules in **LNIF** are particularly noteworthy; as will be seen in the next section, the rules play a vital role in ensuring the invertibility and admissibility of certain rules, ultimately permitting the elimination of (cut) (see Theorem 4).

To obtain soundness, we interpret each sequent as a formula in $\mathcal{L}$ and utilize the notion of validity in Definition 3. The following definition specifies how each sequent is interpreted.

Definition 4 (Interpretation ι).   *The interpretation of a sequent is defined inductively as follows:*

$$\iota(\Gamma \vdash \Delta) := \bigwedge \Gamma \supset \bigvee \Delta \qquad \iota(\Gamma \vdash \Delta \mathbin{/\!/} \mathcal{G}) := \bigwedge \Gamma \supset \left(\bigvee \Delta \vee \iota(\mathcal{G}) \right)$$

We interpret a sequent $\mathcal{G}$ as a formula in $\mathcal{L}$ by taking the universal closure $\overline{\forall}\iota(\mathcal{G})$ of $\iota(\mathcal{G})$ and we say that $\mathcal{G}$ is valid if and only if $\Vdash \overline{\forall}\iota(\mathcal{G})$.

$$\dfrac{}{\mathcal{G} \mathbin{/\mkern-5mu/} \Gamma, p(\vec{a}) \vdash p(\vec{a}), \Delta \mathbin{/\mkern-5mu/} \mathcal{H}} \; (id_1) \qquad \dfrac{}{\mathcal{G} \mathbin{/\mkern-5mu/} \Gamma_1, p(\vec{a}) \vdash \Delta_1 \mathbin{/\mkern-5mu/} \mathcal{H} \mathbin{/\mkern-5mu/} \Gamma_2 \vdash p(\vec{a}), \Delta_2 \mathbin{/\mkern-5mu/} \mathcal{F}} \; (id_2)$$

$$\dfrac{}{\mathcal{G} \mathbin{/\mkern-5mu/} \Gamma, \bot \vdash \Delta \mathbin{/\mkern-5mu/} \mathcal{H}} \; (\bot_l) \qquad \dfrac{\mathcal{G} \mathbin{/\mkern-5mu/} \Gamma, A, B \vdash \Delta \mathbin{/\mkern-5mu/} \mathcal{H}}{\mathcal{G} \mathbin{/\mkern-5mu/} \Gamma, A \wedge B \vdash \Delta \mathbin{/\mkern-5mu/} \mathcal{H}} \; (\wedge_l) \qquad \dfrac{\mathcal{G} \mathbin{/\mkern-5mu/} \Gamma \vdash \Delta, A, B \mathbin{/\mkern-5mu/} \mathcal{H}}{\mathcal{G} \mathbin{/\mkern-5mu/} \Gamma \vdash \Delta, A \vee B \mathbin{/\mkern-5mu/} \mathcal{H}} \; (\vee_r)$$

$$\dfrac{\mathcal{G} \mathbin{/\mkern-5mu/} \Gamma \vdash \Delta, A \mathbin{/\mkern-5mu/} \mathcal{H} \quad \mathcal{G} \mathbin{/\mkern-5mu/} \Gamma \vdash \Delta, B \mathbin{/\mkern-5mu/} \mathcal{H}}{\mathcal{G} \mathbin{/\mkern-5mu/} \Gamma \vdash \Delta, A \wedge B \mathbin{/\mkern-5mu/} \mathcal{H}} \; (\wedge_r) \qquad \dfrac{\mathcal{G} \mathbin{/\mkern-5mu/} \Gamma, A \vdash \Delta \mathbin{/\mkern-5mu/} \mathcal{H} \quad \mathcal{G} \mathbin{/\mkern-5mu/} \Gamma, B \vdash \Delta \mathbin{/\mkern-5mu/} \mathcal{H}}{\mathcal{G} \mathbin{/\mkern-5mu/} \Gamma, A \vee B \vdash \Delta \mathbin{/\mkern-5mu/} \mathcal{H}} \; (\vee_l)$$

$$\dfrac{\mathcal{G} \mathbin{/\mkern-5mu/} \Gamma \vdash \Delta \mathbin{/\mkern-5mu/} A \vdash B}{\mathcal{G} \mathbin{/\mkern-5mu/} \Gamma \vdash \Delta, A \supset B} \; (\supset_{r1}) \qquad \dfrac{\mathcal{G} \mathbin{/\mkern-5mu/} \Gamma, B \vdash \Delta \mathbin{/\mkern-5mu/} \mathcal{H} \quad \mathcal{G} \mathbin{/\mkern-5mu/} \Gamma, A \supset B \vdash A, \Delta \mathbin{/\mkern-5mu/} \mathcal{H}}{\mathcal{G} \mathbin{/\mkern-5mu/} \Gamma, A \supset B \vdash \Delta \mathbin{/\mkern-5mu/} \mathcal{H}} \; (\supset_l)$$

$$\dfrac{\mathcal{G} \mathbin{/\mkern-5mu/} \Gamma_1, A \vdash \Delta_1 \mathbin{/\mkern-5mu/} \Gamma_2, A \vdash \Delta_2 \mathbin{/\mkern-5mu/} \mathcal{H}}{\mathcal{G} \mathbin{/\mkern-5mu/} \Gamma_1, A \vdash \Delta_1 \mathbin{/\mkern-5mu/} \Gamma_2 \vdash \Delta_2 \mathbin{/\mkern-5mu/} \mathcal{H}} \; (lift) \qquad \dfrac{\mathcal{G} \mathbin{/\mkern-5mu/} \Gamma, A[a/x], \forall x A \vdash \Delta \mathbin{/\mkern-5mu/} \mathcal{H}}{\mathcal{G} \mathbin{/\mkern-5mu/} \Gamma, \forall x A \vdash \Delta \mathbin{/\mkern-5mu/} \mathcal{H}} \; (\forall_l)$$

$$\dfrac{\mathcal{G} \mathbin{/\mkern-5mu/} \Gamma \vdash \Delta \mathbin{/\mkern-5mu/} \vdash A[a/x]}{\mathcal{G} \mathbin{/\mkern-5mu/} \Gamma \vdash \Delta, \forall x A} \; (\forall_{r1})^\dagger \qquad \dfrac{\mathcal{G} \mathbin{/\mkern-5mu/} \Gamma, A[a/x] \vdash \Delta \mathbin{/\mkern-5mu/} \mathcal{H}}{\mathcal{G} \mathbin{/\mkern-5mu/} \Gamma, \exists x A \vdash \Delta \mathbin{/\mkern-5mu/} \mathcal{H}} \; (\exists_l)^\dagger \qquad \dfrac{\mathcal{G} \mathbin{/\mkern-5mu/} \Gamma \vdash A[a/x], \Delta \mathbin{/\mkern-5mu/} \mathcal{H}}{\mathcal{G} \mathbin{/\mkern-5mu/} \Gamma \vdash \exists x A, \Delta \mathbin{/\mkern-5mu/} \mathcal{H}} \; (\exists_r)$$

$$\dfrac{\mathcal{G} \mathbin{/\mkern-5mu/} \Gamma_1 \vdash \Delta_1 \mathbin{/\mkern-5mu/} A \vdash B \mathbin{/\mkern-5mu/} \Gamma_2 \vdash \Delta_2 \mathbin{/\mkern-5mu/} \mathcal{H} \quad \mathcal{G} \mathbin{/\mkern-5mu/} \Gamma_1 \vdash \Delta_1 \mathbin{/\mkern-5mu/} \Gamma_2 \vdash \Delta_2, A \supset B \mathbin{/\mkern-5mu/} \mathcal{H}}{\mathcal{G} \mathbin{/\mkern-5mu/} \Gamma_1 \vdash \Delta_1, A \supset B \mathbin{/\mkern-5mu/} \Gamma_2 \vdash \Delta_2 \mathbin{/\mkern-5mu/} \mathcal{H}} \; (\supset_{r2})$$

$$\dfrac{\mathcal{G} \mathbin{/\mkern-5mu/} \Gamma_1 \vdash \Delta_1 \mathbin{/\mkern-5mu/} \vdash A[a/x] \mathbin{/\mkern-5mu/} \Gamma_2 \vdash \Delta_2 \mathbin{/\mkern-5mu/} \mathcal{H} \quad \mathcal{G} \mathbin{/\mkern-5mu/} \Gamma_1 \vdash \Delta_1 \mathbin{/\mkern-5mu/} \Gamma_2 \vdash \Delta_2, \forall x A \mathbin{/\mkern-5mu/} \mathcal{H}}{\mathcal{G} \mathbin{/\mkern-5mu/} \Gamma_1 \vdash \Delta_1, \forall x A \mathbin{/\mkern-5mu/} \Gamma_2 \vdash \Delta_2 \mathbin{/\mkern-5mu/} \mathcal{H}} \; (\forall_{r2})^\dagger$$

Fig. 2. The Calculus LNIF. The side condition † stipulates that the parameter a is an *eigenvariable*, i.e. it does not occur in the conclusion. Occasionally, we write $\vdash_{\mathsf{LNIF}} \mathcal{G}$ to mean that the sequent $\mathcal{G}$ is derivable in LNIF.

Theorem 2 (Soundness of LNIF). *For any linear nested sequent $\mathcal{G}$, if $\mathcal{G}$ is provable in* LNIF, *then* $\Vdash \overline{\forall} \iota(\mathcal{G})$.

Proof. We prove the result by induction on the height of the derivation of

$$\mathcal{G} = \Gamma_1 \vdash \Delta_1 \mathbin{/\mkern-5mu/} \cdots \mathbin{/\mkern-5mu/} \Gamma_n \vdash \Delta_n \mathbin{/\mkern-5mu/} \Gamma_{n+1} \vdash \Delta_{n+1} \mathbin{/\mkern-5mu/} \cdots \mathbin{/\mkern-5mu/} \Gamma_m \vdash \Delta_m$$

and only present the more interesting $\forall$ quantifier cases in the inductive step. All remaining cases can be found in the online appended version [19]. Each inference rule considered is of one of the following two forms.

$$\dfrac{\mathcal{G}'}{\mathcal{G}} \; (r_1) \qquad\qquad \dfrac{\mathcal{G}_1 \quad \mathcal{G}_2}{\mathcal{G}} \; (r_2)$$

We argue by contraposition and prove that if $\mathcal{G}$ is invalid, then at least one premise is invalid. Assuming $\mathcal{G}$ is invalid (i.e. $\not\Vdash \overline{\forall} \iota(\mathcal{G})$) implies the existence of a model $M = (W, R, D, V)$ with world $v \in W$ such that Rvw_0, $\vec{a} \in D_{w_0}$, and $M, w_0 \not\Vdash \iota(\mathcal{G})(\vec{a})$, where $\vec{a}$ represents all parameters in $\iota(\mathcal{G})$. Hence, there is a sequence of worlds $w_1, \cdots, w_m \in W$ such that $Rw_j w_{j+1}$ (for $0 \le j \le m-1$), $M, w_i \Vdash \bigwedge \Gamma_i$, and $M, w_i \not\Vdash \bigvee \Delta_i$, for each $1 \le i \le m$. We assume all parameters in $\bigwedge \Gamma_i$ and $\bigvee \Delta_i$ are interpreted as elements of the associated domain D_{w_i} (for $1 \le i \le m$).

$(\forall_{r1})$**-rule:** By our assumption $M, w_m \Vdash \bigwedge \Gamma_m$ and $M, w_m \not\Vdash \bigvee \Delta_m \vee \forall x A$. The latter implies that $M, w_m \not\Vdash \forall x A$, meaning there exists a world $w_{m+1} \in W$

such that $Rw_m w_{m+1}$ and $M, w_{m+1} \not\Vdash A[b/x]$ for some $b \in D_{w_{m+1}}$. If we interpret the eigenvariable of the premise as b, then the premise is shown invalid.

$(\forall_{r2})$-**rule:** It follows from our assumption that $M, w_n \Vdash \bigwedge \Gamma_n$, $M, w_n \not\Vdash \bigvee \Delta_n \vee \forall x A$, $M, w_{n+1} \Vdash \bigwedge \Gamma_{n+1}$, and $M, w_{n+1} \not\Vdash \bigvee \Delta_{n+1}$. The fact that $M, w_n \not\Vdash \bigvee \Delta_n \vee \forall x A$ implies that there exists a world $w \in W$ such that $Rw_n w$ and for some $b \in D_w$, $M, w \not\Vdash A[b/x]$. Since our frames are connected, there are two cases to consider: (i) Rww_{n+1}, or (ii) $Rw_{n+1}w$. Case (i) falsifies the left premise, and case (ii) falsifies the right premise.

$(\forall_l)$-**rule:** We know that $M, w_n \Vdash \bigwedge \Gamma_n \wedge \forall x A$ and $M, w_n \not\Vdash \bigvee \Delta_n$. Hence, for any world $w \in W$, if $Rw_n w$, then $M, w \Vdash A[b/x]$ for all $b \in D_w$. Since $Rw_n w_n$, it follows that $M, w_n \Vdash A[b/x]$ for any $b \in D_{w_n}$. If a occurs in the conclusion $\mathcal{G}$, then by the constant domain condition (**CD**), we know that $a \in D_{w_n}$, so we may falsify the premise of the rule. If a does not occur in $\mathcal{G}$, then it is an eigenvariable, and assigning a to any element of D_{w_n} will falsify the premise. $\square$

Theorem 3 (Completeness of LNIF). *If $\vdash_{\mathsf{IF}} A$, then A is provable in* LNIF.

Proof. It is not difficult to show that LNIF can derive each axiom of IF and can simulate each inference rule. We refer the reader to the online appended version for details [19]. $\square$

4   Proof-Theoretic Properties of LNIF

In this section, we present the fundamental proof-theoretic properties of LNIF, thus extending the results in [17] from the propositional setting to the first-order setting. (NB. We often leverage results from [17] to simplify our proofs.) Most results are proved by induction on the *height* of a given derivation Π, i.e. on the length (number of sequents) of the longest branch from the end sequent to an initial sequent in Π. Proofs of Lemmas 14, 16, and Theorem 4 are given by induction on the lexicographic ordering of pairs $(|A|, h)$, where $|A|$ is the *complexity* of a certain formula A (defined in the usual way as the number of logical operators in A) and h is the height of the derivation. Lemmata whose proofs are omitted can be found in the online appended version [19].

$$\frac{\mathcal{G} \parallel \Gamma_1 \vdash \Delta_1 \parallel \mathcal{H}}{\mathcal{G} \parallel \Gamma_1, \Gamma_2 \vdash \Delta_1, \Delta_2 \parallel \mathcal{H}} \ (iw) \qquad \frac{\mathcal{G} \parallel \Gamma, A, A \vdash \Delta \parallel \mathcal{H}}{\mathcal{G} \parallel \Gamma, A \vdash \Delta \parallel \mathcal{H}} \ (ic_l) \qquad \frac{\mathcal{G} \parallel \mathcal{H}}{\mathcal{G} \parallel \ \vdash \ \parallel \mathcal{H}} \ (ew)$$

$$\frac{\mathcal{G}}{\mathcal{G}[b/a]} \ (sub) \qquad \frac{\mathcal{G} \parallel \Gamma_1 \vdash A, \Delta_1 \parallel \Gamma_2 \vdash \Delta_2 \parallel \mathcal{H}}{\mathcal{G} \parallel \Gamma_1 \vdash \Delta_1 \parallel \Gamma_2 \vdash A, \Delta_2 \parallel \mathcal{H}} \ (lwr) \qquad \frac{\mathcal{G} \parallel \Gamma \vdash \Delta, \bot \parallel \mathcal{H}}{\mathcal{G} \parallel \Gamma \vdash \Delta \parallel \mathcal{H}} \ (\bot_r)$$

$$\frac{\mathcal{G} \parallel \Gamma \vdash A, A, \Delta \parallel \mathcal{H}}{\mathcal{G} \parallel \Gamma \vdash A, \Delta \parallel \mathcal{H}} \ (ic_r) \qquad \frac{\mathcal{G} \parallel \Gamma_1 \vdash \Delta_1 \parallel \Gamma_2 \vdash \Delta_2 \parallel \mathcal{H}}{\mathcal{G} \parallel \Gamma_1, \Gamma_2 \vdash \Delta_1, \Delta_2 \parallel \mathcal{H}} \ (mrg)$$

Fig. 3. Admissible rules in LNIF.

We say that a rule is *admissible* in LNIF iff derivability of the premise(s) implies derivability of the conclusion in LNIF. Additionally, a rule is *height preserving (hp-)admissible* in LNIF iff if the premise of the rule has a derivation of a certain height in LNIF, then the conclusion of the rule has a derivation of the same height or less in LNIF. Last, a rule is *invertible (hp-invertible)* iff derivability of the conclusion implies derivability of the premise(s) (with a derivation of the same height or less). Admissible rules of LNIF are given in Fig. 3.

Lemma 2. *For any* A, Γ, Δ, $\mathcal{G}$, *and* $\mathcal{H}$, $\vdash_{\mathsf{LNIF}} \mathcal{G} /\!/ \Gamma, A \vdash A, \Delta /\!/ \mathcal{H}$.

Lemma 3. *The* $(\bot_r)$ *rule is hp-admissible in* LNIF.

Proof. By induction on the height of the given derivation. In the base case, applying $(\bot_r)$ to (id_1), (id_2), or $(\bot_l)$ gives an initial sequent, and for each case of the inductive step we apply IH followed by the corresponding rule.   □

Lemma 4. *The* (sub) *rule is hp-admissible in* LNIF.

Lemma 5. *The* (iw) *rule is hp-admissible in* LNIF.

Lemma 6. *The* (ew) *rule is admissible in* LNIF.

Proof. By [17, Lem. 5.6] we know that (ew) is admissible in LNG, thus leaving us to prove the $(\forall_{r1})$, $(\exists_l)$, $(\forall_{r2})$, $(\forall_l)$, and $(\exists_r)$ cases. The $(\exists_l)$, $(\forall_l)$, and $(\exists_r)$ cases are easily shown by applying IH and then the rule. We therefore prove the $(\forall_{r1})$ and $(\forall_{r2})$ cases, beginning with the former, which is split into the two subcases, shown below:

$$\dfrac{\dfrac{\mathcal{G} /\!/ \Gamma \vdash \Delta /\!/ \vdash A[a/x]}{\mathcal{G} /\!/ \Gamma \vdash \Delta, \forall x A}(\forall_{r1})}{\mathcal{G}' /\!/ \Gamma \vdash \Delta, \forall x A}(ew) \qquad \dfrac{\dfrac{\mathcal{G} /\!/ \Gamma \vdash \Delta /\!/ \vdash A[a/x]}{\mathcal{G} /\!/ \Gamma \vdash \Delta, \forall x A}(\forall_{r1})}{\mathcal{G} /\!/ \Gamma \vdash \Delta, \forall x A /\!/ \vdash}(ew)$$

In the top left case, where we weaken in a component prior to the component $\Gamma \vdash \Delta, \forall x A$, we may freely permute the two rule instances. The top right case is resolved as shown below.

$$\dfrac{\dfrac{}{\mathcal{G} /\!/ \Gamma \vdash \Delta /\!/ \vdash A[a/x] /\!/ \vdash}\text{ IH} \qquad \dfrac{\dfrac{}{\mathcal{G} /\!/ \Gamma \vdash \Delta /\!/ \vdash /\!/ \vdash A[a/x]}\text{ IH}}{\dfrac{\mathcal{G} /\!/ \Gamma \vdash \Delta /\!/ \vdash \forall x A}{}(\forall_{r1})}(\forall_{r2})}{\mathcal{G} /\!/ \Gamma \vdash \Delta, \forall x A /\!/ \vdash}$$

Suppose now that we have an $(\forall_{r2})$ inference (as in Fig. 2) followed by an (ew) inference. The only nontrivial case (which is resolved as shown below) occurs when a component is weakened in directly after the component $\Gamma_1 \vdash \Delta_1, \forall x A$. All other cases follow by an application of IH followed by an application of the $(\forall_{r2})$ rule.

$$\dfrac{\dfrac{}{\mathcal{G} /\!/ \Gamma_1 \vdash \Delta_1 /\!/ \vdash A[a/x] /\!/ \vdash /\!/ \Gamma_2 \vdash \Delta_2 /\!/ \mathcal{H}}\text{ IH} \qquad \Pi}{\mathcal{G} /\!/ \Gamma_1 \vdash \Delta_1, \forall x A /\!/ \vdash /\!/ \Gamma_2 \vdash \Delta_2 /\!/ \mathcal{H}}(\forall_{r2})$$

$$\Pi = \left\{ \dfrac{\dfrac{\mathcal{G} \,/\!/\, \Gamma_1 \vdash \Delta_1 /\!/ \vdash /\!/ \vdash A[a/x] /\!/ \Gamma_2 \vdash \Delta_2 /\!/ \mathcal{H}}{} \ \text{IH} \qquad \dfrac{\mathcal{G} \,/\!/\, \Gamma_1 \vdash \Delta_1 /\!/ \vdash /\!/ \Gamma_2 \vdash \Delta_2, \forall x A /\!/ \mathcal{H}}{} \ \text{IH}}{\mathcal{G} \,/\!/\, \Gamma_1 \vdash \Delta_1 /\!/ \vdash \forall x A /\!/ \Gamma_2 \vdash \Delta_2 /\!/ \mathcal{H}} \ (\forall_{r2}) \right.$$

$\square$

Lemma 7. *The rule* (lwr) *is hp-admissible in* LNIF.

Proof. By [17, Lem. 5.7] we know that (lwr) is admissible in LNG, and so, we may prove the claim by extending it to include the quantifier rules. We have two cases to consider: either (i) the lower-formula is a side formula in the quantifier inference, or (ii) it is principal. In case (i), the $(\forall_{r1})$, $(\forall_l)$, $(\exists_l)$, and $(\exists_r)$ cases can be resolved by applying IH followed by an application of the rule. Concerning the $(\forall_{r2})$ rule, all cases follow by applying IH and then the rule, with the exception of the following:

$$\dfrac{\dfrac{\mathcal{G} \,/\!/\, \Gamma_1 \vdash \Delta_1, A /\!/ \vdash B[a/x] /\!/ \Gamma_2 \vdash \Delta_2 /\!/ \mathcal{H} \qquad \mathcal{G} \,/\!/\, \Gamma_1 \vdash \Delta_1, A /\!/ \Gamma_2 \vdash \Delta_2, \forall x B /\!/ \mathcal{H}}{\mathcal{G} \,/\!/\, \Gamma_1 \vdash \Delta_1, A, \forall x B /\!/ \Gamma_2 \vdash \Delta_2 /\!/ \mathcal{H}} \ (\forall_{r2})}{\mathcal{G} \,/\!/\, \Gamma_1 \vdash \Delta_1, \forall x B /\!/ \Gamma_2 \vdash \Delta_2, A /\!/ \mathcal{H}} \ (lwr)$$

In the above case, we apply IH twice to the top left premise and apply IH once to the top right premise. A single application of $(\forall_{r2})$ gives the desired result.

Let us now consider case (ii). Observe that the principal formulae in $(\forall_{r1})$, $(\forall_l)$, and $(\exists_l)$ cannot be principal in the use of (lwr), so we need only consider the $(\exists_r)$ and $(\forall_{r2})$ cases. The $(\exists_r)$ case is shown below top-left and the case is resolved as shown below top-right. In the $(\forall_{r2})$ case (shown below bottom), we take the derivation of the top right premise as the proof of the desired conclusion.

$$\dfrac{\dfrac{\mathcal{G} \,/\!/\, \Gamma_1 \vdash \Delta_1, A[a/x], \exists x A /\!/ \Gamma_2 \vdash \Delta_2 /\!/ \mathcal{H}}{\mathcal{G} \,/\!/\, \Gamma_1 \vdash \Delta_1, \exists x A /\!/ \Gamma_2 \vdash \Delta_2 /\!/ \mathcal{H}} \ (\exists_r)}{\mathcal{G} \,/\!/\, \Gamma_1 \vdash \Delta_1 /\!/ \Gamma_2 \vdash \Delta_2, \exists x A /\!/ \mathcal{H}} \ (lwr) \qquad \dfrac{\mathcal{G} \,/\!/\, \Gamma_1 \vdash \Delta_1 /\!/ \Gamma_2 \vdash \Delta_2, A[a/x], \exists x A /\!/ \mathcal{H}}{\mathcal{G} \,/\!/\, \Gamma_1 \vdash \Delta_1 /\!/ \Gamma_2 \vdash \Delta_2, \exists x A /\!/ \mathcal{H}} \ \text{IH} \times 2 \ (\exists_r)$$

$$\dfrac{\dfrac{\mathcal{G} \,/\!/\, \Gamma_1 \vdash \Delta_1 /\!/ \vdash A[a/x] /\!/ \Gamma_2 \vdash \Delta_2 /\!/ \mathcal{H} \qquad \mathcal{G} \,/\!/\, \Gamma_1 \vdash \Delta_1 /\!/ \Gamma_2 \vdash \Delta_2, \forall x A /\!/ \mathcal{H}}{\mathcal{G} \,/\!/\, \Gamma_1 \vdash \Delta_1, \forall x A /\!/ \Gamma_2 \vdash \Delta_2 /\!/ \mathcal{H}} \ (\forall_{r2})}{\mathcal{G} \,/\!/\, \Gamma_1 \vdash \Delta_1 /\!/ \Gamma_2 \vdash \Delta_2, \forall x A /\!/ \mathcal{H}} \ (lwr)$$

$\square$

Our version of the $(lift)$ rule necessitates a stronger form of invertibility, called *m-invertibility*, for the $(\wedge_l)$, $(\vee_l)$, $(\supset_l)$, $(\forall_l)$, and $(\exists_l)$ rules (cf. [17]). We use A^{k_i} to represent k_i copies of a formula A, with $i \in \mathbb{N}$.

Lemma 8. *If* $\sum_{i=1}^{n} k_n \geq 1$, *then*

(i) (1) *implies* (2)   (iii) (6) *implies* (7) *and* (8)   (v) (11) *implies* (12)

(ii) (3) *implies* (4) *and* (5)   (iv) (9) *implies* (10)

$$\vdash_{\mathsf{LNIF}}\ \Gamma_1, (A \wedge B)^{k_1} \vdash \Delta_1 \parallel \cdots \parallel \Gamma_n, (A \wedge B)^{k_n} \vdash \Delta_n \tag{1}$$

$$\vdash_{\mathsf{LNIF}}\ \Gamma_1, A^{k_1}, B^{k_1} \vdash \Delta_1 \parallel \cdots \parallel \Gamma_n, A^{k_n}, B^{k_n} \vdash \Delta_n \tag{2}$$

$$\vdash_{\mathsf{LNIF}}\ \Gamma_1, (A \vee B)^{k_1} \vdash \Delta_1 \parallel \cdots \parallel \Gamma_n, (A \vee B)^{k_n} \vdash \Delta_n \tag{3}$$

$$\vdash_{\mathsf{LNIF}}\ \Gamma_1, A^{k_1} \vdash \Delta_1 \parallel \cdots \parallel \Gamma_n, A^{k_n} \vdash \Delta_n \tag{4}$$

$$\vdash_{\mathsf{LNIF}}\ \Gamma_1, B^{k_1} \vdash \Delta_1 \parallel \cdots \parallel \Gamma_n, B^{k_n} \vdash \Delta_n \tag{5}$$

$$\vdash_{\mathsf{LNIF}}\ \Gamma_1, (A \supset B)^{k_1} \vdash \Delta_1 \parallel \cdots \parallel \Gamma_n, (A \supset B)^{k_n} \vdash \Delta_n \tag{6}$$

$$\vdash_{\mathsf{LNIF}}\ \Gamma_1, B^{k_1} \vdash \Delta_1 \parallel \cdots \parallel \Gamma_n, B^{k_n} \vdash \Delta_n \tag{7}$$

$$\vdash_{\mathsf{LNIF}}\ \Gamma_1, (A \supset B)^{k_1} \vdash \Delta_1, A^{k_1} \parallel \cdots \parallel \Gamma_n, (A \supset B)^{k_n} \vdash \Delta_n, A^{k_n} \tag{8}$$

$$\vdash_{\mathsf{LNIF}}\ \Gamma_1, (\forall x A)^{k_1} \vdash \Delta_1 \parallel \cdots \parallel \Gamma_n, (\forall x A)^{k_n} \vdash \Delta_n \tag{9}$$

$$\vdash_{\mathsf{LNIF}}\ \Gamma_1, A[a/x]^{k_1}, (\forall x A)^{k_1} \vdash \Delta_1 \parallel \cdots \parallel \Gamma_n, A[a/x]^{k_n}, (\forall x A)^{k_n} \vdash \Delta_n \tag{10}$$

$$\vdash_{\mathsf{LNIF}}\ \Gamma_1, (\exists x A)^{k_1} \vdash \Delta_1 \parallel \cdots \parallel \Gamma_n, (\exists x A)^{k_n} \vdash \Delta_n \tag{11}$$

$$\vdash_{\mathsf{LNIF}}\ \Gamma_1, A[a/x]^{k_1} \vdash \Delta_1 \parallel \cdots \parallel \Gamma_n, A[a/x]^{k_n} \vdash \Delta_n \tag{12}$$

Lemma 9. *The $(\wedge_r)$, $(\vee_r)$, and $(\exists_r)$ rules are hp-invertible in* LNIF.

Proof. By [17, Lem. 5.8] we know that the claim holds for the $(\wedge_r)$ and $(\vee_r)$ rules relative to LNG. The proof may be extended to LNIF by considering the quantifier rules in the inductive step; however, it is quick to verify the claim for the quantifier rules by applying IH and then the corresponding rule. Proving invertibility of the $(\exists_r)$ rule is straightforward, and is shown by induction on the height of the given derivation. □

Lemma 10. *The $(\supset_{r2})$ rule is invertible in* LNIF.

Proof. We extend the proof of [17, Lem. 5.10] to include the quantifier rules, and prove the result by induction on the height of the given derivation of $\mathcal{G} \parallel \Gamma_1 \vdash \Delta_1, A \supset B \parallel \Gamma_2 \vdash \Delta_2 \parallel \mathcal{H}$. Derivability of the right premise $\mathcal{G} \parallel \Gamma_1 \vdash \Delta_1 \parallel \Gamma_2 \vdash \Delta_2, A \supset B \parallel \mathcal{H}$ follows from Lemma 7, so we focus on showing that the left premise $\mathcal{G} \parallel \Gamma_1 \vdash \Delta_1 \parallel A \vdash B \parallel \Gamma_2 \vdash \Delta_2 \parallel \mathcal{H}$ is derivable. For the $(\forall_{r1})$, $(\forall_l)$, $(\exists_l)$, and $(\exists_r)$ rules the desired conclusion is obtained by applying IH, followed by an application of the corresponding rule. The nontrivial $(\forall_{r2})$ case is shown below top and is resolved as shown below bottom. In all other $(\forall_{r2})$ cases, we apply IH followed by the $(\forall_{r2})$ rule.

$$\dfrac{\mathcal{G} \parallel \Gamma_1 \vdash \Delta_1, A \supset B \parallel \vdash C[a/x] \parallel \Gamma_2 \vdash \Delta_2 \parallel \mathcal{H} \qquad \mathcal{G} \parallel \Gamma_1 \vdash \Delta_1, A \supset B \parallel \Gamma_2 \vdash \Delta_2, \forall x C \parallel \mathcal{H}}{\mathcal{G} \parallel \Gamma_1 \vdash \Delta_1, \forall x C, A \supset B \parallel \Gamma_2 \vdash \Delta_2 \parallel \mathcal{H}} \ (\forall_{r2})$$

$$\cfrac{\Pi_1 \qquad \Pi_2}{\mathcal{G} /\!/ \Gamma_1 \vdash \Delta_1, \forall x C /\!/ A \vdash B /\!/ \Gamma_2 \vdash \Delta_2 /\!/ \mathcal{H}} \ (\forall_{r2}) \quad \Pi_1 = \begin{cases} \cfrac{\mathcal{G} /\!/ \Gamma_1 \vdash \Delta_1, A \supset B /\!/ \vdash C[a/x] /\!/ \Gamma_2 \vdash \Delta_2 /\!/ \mathcal{H}}{\cfrac{\mathcal{G} /\!/ \Gamma_1 \vdash \Delta_1 /\!/ \vdash C[a/x], A \supset B /\!/ \Gamma_2 \vdash \Delta_2 /\!/ \mathcal{H}}{\mathcal{G} /\!/ \Gamma_1 \vdash \Delta_1 /\!/ \vdash C[a/x] /\!/ A \vdash B /\!/ \Gamma_2 \vdash \Delta_2 /\!/ \mathcal{H}} \ \text{IH}} \ \text{Lem. 7} \end{cases}$$

$$\Pi_2 = \begin{cases} \cfrac{\cfrac{\mathcal{G} /\!/ \Gamma_1 \vdash \Delta_1, A \supset B /\!/ \vdash C[a/x] /\!/ \Gamma_2 \vdash \Delta_2 /\!/ \mathcal{H}}{\mathcal{G} /\!/ \Gamma_1 \vdash \Delta_1 /\!/ A \vdash B /\!/ \vdash C[a/x] /\!/ \Gamma_2 \vdash \Delta_2 /\!/ \mathcal{H}} \ \text{IH} \qquad \cfrac{\mathcal{G} /\!/ \Gamma_1 \vdash \Delta_1, A \supset B /\!/ \Gamma_2 \vdash \Delta_2, \forall x C /\!/ \mathcal{H}}{\mathcal{G} /\!/ \Gamma_1 \vdash \Delta_1 /\!/ A \vdash B /\!/ \Gamma_2 \vdash \Delta_2, \forall x C /\!/ \mathcal{H}} \ \text{IH}}{\mathcal{G} /\!/ \Gamma_1 \vdash \Delta_1 /\!/ A \vdash B, \forall x C /\!/ \Gamma_2 \vdash \Delta_2 /\!/ \mathcal{H}} \ (\forall_{r2}) \end{cases}$$

$\square$

Lemma 11. *The* $(\forall_{r2})$ *rule is invertible in* LNIF.

Proof. Let the sequent $\mathcal{G} /\!/ \Gamma_1 \vdash \Delta_1, \forall x A /\!/ \Gamma_2 \vdash \Delta_2 /\!/ \mathcal{H}$ be derivable in LNIF. Derivability of the right premise $\mathcal{G} /\!/ \Gamma_1 \vdash \Delta_1 /\!/ \Gamma_2 \vdash \Delta_2, \forall x A /\!/ \mathcal{H}$ follows from the hp-admissibility of (lwr) (Lemma 7). We prove that the left premise $\mathcal{G} /\!/ \Gamma_1 \vdash \Delta_1 /\!/ \vdash A[a/x] /\!/ \Gamma_2 \vdash \Delta_2 /\!/ \mathcal{H}$ is derivable by induction on the height of the given derivation.

Base Case. Regardless of if $\mathcal{G} /\!/ \Gamma_1 \vdash \Delta_1, \forall x A /\!/ \Gamma_2 \vdash \Delta_2 /\!/ \mathcal{H}$ is derived by an application of (id_1), (id_2), or $(\bot_l)$, $\mathcal{G} /\!/ \Gamma_1 \vdash \Delta_1 /\!/ \vdash A[a/x] /\!/ \Gamma_2 \vdash \Delta_2 /\!/ \mathcal{H}$ is an initial sequent as well.

Inductive Step. For all rules, with the exception of $(lift)$, $(\supset_{r2})$, $(\forall_{r1})$, $(\exists_l)$, and $(\forall_{r2})$, we apply IH to the premise(s) followed by the corresponding rule. We consider the aforementioned nontrivial cases below.

If the $(lift)$ rule is applied as shown below left, then the desired conclusion may be derived as shown below right. In all other cases, we apply IH and then $(lift)$ to achieve the desired result.

$$\cfrac{\mathcal{G} /\!/ \Gamma_1, B \vdash \Delta_1, \forall x A /\!/ \Gamma_2, B \vdash \Delta_2 /\!/ \mathcal{H}}{\mathcal{G} /\!/ \Gamma_1, B \vdash \Delta_1, \forall x A /\!/ \Gamma_2 \vdash \Delta_2 /\!/ \mathcal{H}} \ (lift) \qquad \cfrac{\cfrac{\cfrac{\cfrac{\mathcal{G} /\!/ \Gamma_1, B \vdash \Delta_1, \forall x A /\!/ \Gamma_2, B \vdash \Delta_2 /\!/ \mathcal{H}}{\mathcal{G} /\!/ \Gamma_1, B \vdash \Delta_1 /\!/ \vdash A[a/x] /\!/ \Gamma_2, B \vdash \Delta_2 /\!/ \mathcal{H}} \ \text{IH}}{\mathcal{G} /\!/ \Gamma_1, B \vdash \Delta_1 /\!/ B \vdash A[a/x] /\!/ \Gamma_2, B \vdash \Delta_2 /\!/ \mathcal{H}} \ \text{Lem. 5}}{\mathcal{G} /\!/ \Gamma_1, B \vdash \Delta_1 /\!/ B \vdash A[a/x] /\!/ \Gamma_2 \vdash \Delta_2 /\!/ \mathcal{H}} \ (lift)}{\mathcal{G} /\!/ \Gamma_1, B \vdash \Delta_1 /\!/ \vdash A[a/x] /\!/ \Gamma_2 \vdash \Delta_2 /\!/ \mathcal{H}} \ (lift)$$

If the $(\supset_{r2})$ rule is applied as shown below top, then the desired conclusion may be derived as shown below bottom. In all other cases, we apply IH and then the $(\supset_{r2})$ rule to obtain the desired result.

$$\cfrac{\mathcal{G} /\!/ \Gamma_1 \vdash \Delta_1, \forall x A /\!/ B \vdash C /\!/ \Gamma_2 \vdash \Delta_2 /\!/ \mathcal{H} \qquad \mathcal{G} /\!/ \Gamma_1 \vdash \Delta_1, \forall x A /\!/ \Gamma_2 \vdash \Delta_2, B \supset C /\!/ \mathcal{H}}{\mathcal{G} /\!/ \Gamma_1 \vdash \Delta_1, \forall x A, B \supset C /\!/ \Gamma_2 \vdash \Delta_2 /\!/ \mathcal{H}} \ (\supset_{r2})$$

$$\cfrac{\Pi_1 \qquad \Pi_2}{\mathcal{G} /\!/ \Gamma_1 \vdash \Delta_1, B \supset C /\!/ \vdash A[a/x] /\!/ \Gamma_2 \vdash \Delta_2 /\!/ \mathcal{H}} \ (\supset_{r2}) \quad \Pi_1 = \begin{cases} \cfrac{\cfrac{\mathcal{G} /\!/ \Gamma_1 \vdash \Delta_1, \forall x A /\!/ B \vdash C /\!/ \Gamma_2 \vdash \Delta_2 /\!/ \mathcal{H}}{\mathcal{G} /\!/ \Gamma_1 \vdash \Delta_1 /\!/ B \vdash C, \forall x A /\!/ \Gamma_2 \vdash \Delta_2 /\!/ \mathcal{H}} \ \text{Lem. 7}}{\mathcal{G} /\!/ \Gamma_1 \vdash \Delta_1 /\!/ B \vdash C /\!/ \vdash A[a/x] /\!/ \Gamma_2 \vdash \Delta_2 /\!/ \mathcal{H}} \ \text{IH} \end{cases}$$

$$\Pi_2 = \begin{cases} \cfrac{\cfrac{\mathcal{G} /\!/ \Gamma_1 \vdash \Delta_1, \forall x A /\!/ B \vdash C /\!/ \Gamma_2 \vdash \Delta_2 /\!/ \mathcal{H}}{\mathcal{G} /\!/ \Gamma_1 \vdash \Delta_1 /\!/ \vdash A[a/x] /\!/ B \vdash C /\!/ \Gamma_2 \vdash \Delta_2 /\!/ \mathcal{H}} \ \text{IH} \qquad \cfrac{\mathcal{G} /\!/ \Gamma_1 \vdash \Delta_1, \forall x A /\!/ \Gamma_2 \vdash \Delta_2, B \supset C /\!/ \mathcal{H}}{\mathcal{G} /\!/ \Gamma_1 \vdash \Delta_1 /\!/ \vdash A[a/x] /\!/ \Gamma_2 \vdash \Delta_2, B \supset C /\!/ \mathcal{H}} \ \text{IH}}{\mathcal{G} /\!/ \Gamma_1 \vdash \Delta_1 /\!/ \vdash A[a/x], B \supset C /\!/ \Gamma_2 \vdash \Delta_2 /\!/ \mathcal{H}} \ (\supset_{r2}) \end{cases}$$

In the $(\forall_{r}1)$ and $(\exists_{l})$ cases, we must ensure that the eigenvariable of the inference is not identical to the parameter a in $A[a/x]$ introduced by IH. However, this can always be ensured by Lemma 4. Therefore, we move onto the last nontrivial case, which concerns the $(\forall_{r}2)$ rule. The only nontrivial case occurs as shown below top and is resolved as shown below bottom. In all other cases, we apply IH followed by the $(\forall_{r}2)$ rule (invoking Lemma 4 if necessary).

$$\frac{\mathcal{G} \,/\!/\, \Gamma_1 \vdash \Delta_1, A \supset B \,/\!/\, \vdash C[a/x] \,/\!/\, \Gamma_2 \vdash \Delta_2 \,/\!/\, \mathcal{H} \qquad \mathcal{G} \,/\!/\, \Gamma_1 \vdash \Delta_1, A \supset B \,/\!/\, \Gamma_2 \vdash \Delta_2, \forall x C \,/\!/\, \mathcal{H}}{\mathcal{G} \,/\!/\, \Gamma_1 \vdash \Delta_1, \forall x C, A \supset B \,/\!/\, \Gamma_2 \vdash \Delta_2 \,/\!/\, \mathcal{H}} \,(\forall_{r}2)$$

$$\frac{\Pi_1 \qquad \Pi_2}{\mathcal{G} \,/\!/\, \Gamma_1 \vdash \Delta_1, \forall x C \,/\!/\, A \vdash B \,/\!/\, \Gamma_2 \vdash \Delta_2 \,/\!/\, \mathcal{H}} \,(\forall_{r}2) \quad \Pi_1 = \left\{ \begin{array}{c} \cfrac{\mathcal{G} \,/\!/\, \Gamma_1 \vdash \Delta_1, A \supset B \,/\!/\, \vdash C[a/x] \,/\!/\, \Gamma_2 \vdash \Delta_2 \,/\!/\, \mathcal{H}}{\cfrac{\mathcal{G} \,/\!/\, \Gamma_1 \vdash \Delta_1 \,/\!/\, \vdash C[a/x], A \supset B \,/\!/\, \Gamma_2 \vdash \Delta_2 \,/\!/\, \mathcal{H}}{\mathcal{G} \,/\!/\, \Gamma_1 \vdash \Delta_1 \,/\!/\, \vdash C[a/x] \,/\!/\, A \vdash B \,/\!/\, \Gamma_2 \vdash \Delta_2 \,/\!/\, \mathcal{H}} \text{ IH}} \text{ Lem. 4} \end{array} \right.$$

$$\Pi_2 = \left\{ \cfrac{\cfrac{\mathcal{G} \,/\!/\, \Gamma_1 \vdash \Delta_1, A \supset B \,/\!/\, \vdash C[a/x] \,/\!/\, \Gamma_2 \vdash \Delta_2 \,/\!/\, \mathcal{H}}{\mathcal{G} \,/\!/\, \Gamma_1 \vdash \Delta_1 \,/\!/\, A \vdash B \,/\!/\, \vdash C[a/x] \,/\!/\, \Gamma_2 \vdash \Delta_2 \,/\!/\, \mathcal{H}} \text{ IH} \qquad \cfrac{\mathcal{G} \,/\!/\, \Gamma_1 \vdash \Delta_1, A \supset B \,/\!/\, \Gamma_2 \vdash \Delta_2, \forall x C \,/\!/\, \mathcal{H}}{\mathcal{G} \,/\!/\, \Gamma_1 \vdash \Delta_1 \,/\!/\, A \vdash B \,/\!/\, \Gamma_2 \vdash \Delta_2, \forall x C \,/\!/\, \mathcal{H}} \text{ IH}}{\mathcal{G} \,/\!/\, \Gamma_1 \vdash \Delta_1 \,/\!/\, A \vdash B, \forall x C \,/\!/\, \Gamma_2 \vdash \Delta_2 \,/\!/\, \mathcal{H}} \,(\forall_{r}2) \right.$$

$\square$

Lemma 12. *The $(\supset_{r}1)$ rule is invertible in* LNIF.

Proof. We extend the proof of [17, Lem. 5.11] to include the quantifier cases. The claim is shown by induction on the height of the given derivation. When the last rule of the derivation is $(\forall_{l})$, $(\exists_{l})$, $(\exists_{r})$, or $(\forall_{r}2)$ in the inductive step, we apply IH to the premise(s) of the inference followed by an application of the corresponding rule. If the last inference of the derivation is an application of the $(\forall_{r}1)$ rule (as shown below left), then the case is resolved as shown below right.

$$\frac{\mathcal{G} \,/\!/\, \Gamma \vdash \Delta, A \supset B \,/\!/\, \vdash C[a/x]}{\mathcal{G} \,/\!/\, \Gamma \vdash \Delta, A \supset B, \forall x C} \,(\forall_{r}1) \qquad \frac{\cfrac{\cfrac{\mathcal{G} \,/\!/\, \Gamma \vdash \Delta, A \supset B \,/\!/\, \vdash C[a/x]}{\mathcal{G} \,/\!/\, \Gamma \vdash \Delta \,/\!/\, \vdash C[a/x], A \supset B} \text{ Lem. 7}}{\mathcal{G} \,/\!/\, \Gamma \vdash \Delta \,/\!/\, \vdash C[a/x] \,/\!/\, A \vdash B} \text{ IH} \qquad \cfrac{\cfrac{\mathcal{G} \,/\!/\, \Gamma \vdash \Delta, A \supset B \,/\!/\, \vdash C[a/x]}{\mathcal{G} \,/\!/\, \Gamma \vdash \Delta \,/\!/\, A \vdash B \,/\!/\, \vdash C[a/x]} \text{ Lem. 10}}{\mathcal{G} \,/\!/\, \Gamma \vdash \Delta \,/\!/\, A \vdash B, \forall x C} \,(\forall_{r}1)}{\mathcal{G} \,/\!/\, \Gamma \vdash \Delta, \forall x C \,/\!/\, A \vdash B}$$

$\square$

Lemma 13. *The $(\forall_{r}1)$ rule is invertible in* LNIF.

Proof. We prove the result by induction on the height of the given derivation of $\mathcal{G} \,/\!/\, \Gamma \vdash \Delta, \forall x A$ and show that $\mathcal{G} \,/\!/\, \Gamma \vdash \Delta \,/\!/\, \vdash A[a/x]$ is derivable.

Base case. If $\mathcal{G} \,/\!/\, \Gamma \vdash \Delta, \forall x A$ is obtained via (id_1), (id_2), or $(\perp_l)$, then $\mathcal{G} \,/\!/\, \Gamma \vdash \Delta \,/\!/\, \vdash A[a/x]$ is an instance of the corresponding rule as well.

Inductive Step. All cases, with the exception of the $(\supset_{r}1)$, $(\forall_{r}1)$, $(\exists_{l})$, and $(\forall_{r}2)$ rules, are resolved by applying IH to the premise(s) and then applying the relevant rule. Let us consider each of the additional cases in turn.

The $(\supset_{r}1)$ case is shown below left and is resolved as shown below right.

$$\frac{\mathcal{G} \,/\!/\, \Gamma \vdash \Delta, \forall x A \,/\!/\, B \vdash C}{\mathcal{G} \,/\!/\, \Gamma \vdash \Delta, \forall x A, B \supset C} \,(\supset_{r}1) \qquad \frac{\cfrac{\cfrac{\mathcal{G} \,/\!/\, \Gamma \vdash \Delta, \forall x A \,/\!/\, B \vdash C}{\mathcal{G} \,/\!/\, \Gamma \vdash \Delta \,/\!/\, B \vdash C, \forall x A} \text{ Lem. 7}}{\mathcal{G} \,/\!/\, \Gamma \vdash \Delta \,/\!/\, B \vdash C \,/\!/\, \vdash A[a/x]} \text{ IH} \qquad \cfrac{\cfrac{\mathcal{G} \,/\!/\, \Gamma \vdash \Delta, \forall x A \,/\!/\, B \vdash C}{\mathcal{G} \,/\!/\, \Gamma \vdash \Delta \,/\!/\, \vdash A[a/x] \,/\!/\, B \vdash C} \text{ Lem. 10}}{\mathcal{G} \,/\!/\, \Gamma \vdash \Delta \,/\!/\, \vdash A[a/x], B \supset C} \,(\supset_{r}1)}{\mathcal{G} \,/\!/\, \Gamma \vdash \Delta, B \supset C \,/\!/\, \vdash A[a/x]}$$

In the $(\forall_{r1})$ case where the relevant formula $\forall x A$ is principal, the premise of the inference is the desired conclusion. If the relevant formula $\forall x A$ is not principal, then the $(\forall_{r1})$ inference is of the form shown below left and is resolved as shown below right.

$$\dfrac{\mathcal{G} /\!/ \Gamma \vdash \Delta, \forall x A /\!/ B[b/y]}{\mathcal{G} /\!/ \Gamma \vdash \Delta, \forall x A, \forall y B} \ (\forall_{r1})$$

$$\dfrac{\dfrac{\dfrac{\mathcal{G} /\!/ \Gamma \vdash \Delta, \forall x A /\!/ \vdash B[b/y]}{\mathcal{G} /\!/ \Gamma \vdash \Delta /\!/ B[b/y], \forall x A} \ \text{Lem. 7}}{\mathcal{G} /\!/ \Gamma \vdash \Delta /\!/ \vdash B[b/y] /\!/ \vdash A[a/x]} \ \text{IH}}{\mathcal{G} /\!/ \Gamma \vdash \Delta, \forall y B /\!/ \vdash A[a/x]}$$

$$\dfrac{\dfrac{\dfrac{\mathcal{G} /\!/ \Gamma \vdash \Delta, \forall x A /\!/ \vdash B[b/y]}{\mathcal{G} /\!/ \Gamma \vdash \Delta /\!/ \vdash A[a/x] /\!/ \vdash B[b/y]} \ \text{Lem. 10}}{\mathcal{G} /\!/ \Gamma \vdash \Delta /\!/ \vdash A[a/x], \forall y B} \ (\forall_{r1})}{} \ (\forall_{r2})$$

If the last inference is an instance of the $(\exists_l)$ or $(\forall_{r2})$ rule, then we must ensure that the eigenvariable of the inference is not identical to the parameter a in $A[a/x]$ introduced by IH, but this can always be ensured due to Lemma 4. $\square$

Lemma 14. *The (ic_l) rule is admissible in* LNIF.

Proof. We extend the proof of [17, Lem. 5.12] and prove the result by induction on the lexicographic ordering of pairs $(|A|, h)$, where $|A|$ is the complexity of the contraction formula A and h is the height of the derivation. We know the result holds for LNG, and so, we argue the inductive step for the quantifier rules.

With the exception of the $(\exists_l)$ case shown below left, all quantifier cases are settled by applying IH followed by an application of the corresponding rule. The only nontrivial case occurs when a contraction is performed on a formula $\exists x A$ with one of the contraction formulae principal in the (ic_l) inference. The situation is resolved as shown below right.

$$\dfrac{\dfrac{\mathcal{G} /\!/ \Gamma, A[a/x], \exists x A \vdash \Delta /\!/ \mathcal{H}}{\mathcal{G} /\!/ \Gamma, \exists x A, \exists x A \vdash \Delta /\!/ \mathcal{H}} \ (\exists_l)}{\mathcal{G} /\!/ \Gamma, \exists x A \vdash \Delta /\!/ \mathcal{H}} \ (ic_l)$$

$$\dfrac{\dfrac{\dfrac{\mathcal{G} /\!/ \Gamma, A[a/x], \exists x A \vdash \Delta /\!/ \mathcal{H}}{\mathcal{G} /\!/ \Gamma, A[a/x], A[a/x] \vdash \Delta /\!/ \mathcal{H}} \ \text{Lem. 8}}{\mathcal{G} /\!/ \Gamma, A[a/x] \vdash \Delta /\!/ \mathcal{H}} \ \text{IH}}{\mathcal{G} /\!/ \Gamma, \exists x A \vdash \Delta /\!/ \mathcal{H}} \ (\exists_l)$$

Notice that IH is applicable since we are contracting on a formula of smaller complexity. $\square$

Lemma 15. *The (mrg) rule is admissible in* LNIF.

Proof. We extend the proof of [17, Lem. 5.13], which proves that (mrg) is admissible in LNG, and prove the admissibility of (mrg) in LNIF by induction on the height of the given derivation. We need only consider the quantifier rules due to [17, Lem. 5.13]. The $(\forall_{r1})$, $(\forall_l)$, $(\exists_l)$, and $(\exists_r)$ cases are all resolved by applying IH to the premise of the rule followed by an application of the rule. If (mrg) is applied to the principal components of the $(\forall_{r2})$ rule as follows:

$$\dfrac{\dfrac{\mathcal{G} /\!/ \Gamma_1 \vdash \Delta_1 /\!/ \vdash A[a/x] /\!/ \Gamma_2 \vdash \Delta_2 /\!/ \mathcal{H} \quad \mathcal{G} /\!/ \Gamma_1 \vdash \Delta_1 /\!/ \Gamma_2 \vdash \Delta_2, \forall x A /\!/ \mathcal{H}}{\mathcal{G} /\!/ \Gamma_1 \vdash \Delta_1, \forall x A /\!/ \Gamma_2 \vdash \Delta_2 /\!/ \mathcal{H}} \ (\forall_{r2})}{\mathcal{G} /\!/ \Gamma_1, \Gamma_2 \vdash \Delta_1, \Delta_2, \forall x A /\!/ \mathcal{H}} \ (mrg)$$

then the desired conclusion is obtained by applying IH to the top right premise. In all other cases, we apply IH to the premises of $(\forall_{r2})$ followed by an application of the rule. $\square$

Lemma 16. *The (ic_r) rule is admissible in* LNIF.

Proof. We extend the proof of [17, Lem. 5.14] to include the quantifier rules and argue the claim by induction on the lexicographic ordering of pairs $(|A|, h)$, where $|A|$ is the complexity of the contraction formula A and h is the height of the derivation. The $(\forall_l)$ and $(\exists_l)$ cases are settled by applying IH to the premise of the inference followed by an application of the rule. For the $(\exists_r)$ case, we invoke Lemma 9, apply IH, and then apply the corresponding rule. The nontrivial case (occurring when the principal formula is contracted) for the $(\forall_{r1})$ rule is shown below left, and the desired conclusion is derived as shown below right (where IH is applicable due to the decreased complexity of the contraction formula).

$$\frac{\mathcal{G} \ /\!/\ \Gamma \vdash \Delta, \forall x A \ /\!/ \vdash A[a/x]}{\dfrac{\mathcal{G} \ /\!/\ \Gamma \vdash \Delta, \forall x A, \forall x A}{\mathcal{G} \ /\!/\ \Gamma \vdash \Delta, \forall x A} \ (ic_r)} \ (\forall_{r1})$$

$$\frac{\dfrac{\dfrac{\dfrac{\dfrac{\mathcal{G} \ /\!/\ \Gamma \vdash \Delta, \forall x A \ /\!/ \vdash A[a/x]}{\mathcal{G} \ /\!/\ \Gamma \vdash \Delta /\!/ \vdash A[a/x] /\!/ \vdash A[a/x]} \ \text{Lem. 11}}{\mathcal{G} \ /\!/\ \Gamma \vdash \Delta /\!/ \vdash A[a/x], A[a/x]} \ \text{Lem. 15}}{\mathcal{G} \ /\!/\ \Gamma \vdash \Delta /\!/ \vdash A[a/x]} \ \text{IH}}{\mathcal{G} \ /\!/\ \Gamma \vdash \Delta, \forall x A} \ (\forall_{r1})$$

When the contracted formulae are both non-principal in an $(\forall_{r1})$ inference, we apply IH to the premise followed by an application of the $(\forall_{r1})$ rule. If the contracted formulae are both non-principal in an $(\forall_{r2})$ inference, then we apply IH to the premises followed by an application of the rule. If one of the contracted formulae is principal in an $(\forall_{r2})$ inference (as shown below top), then the case is settled as shown below bottom.

$$\frac{\mathcal{G} \ /\!/\ \Gamma_1 \vdash \Delta_1, \forall x A /\!/ \vdash A[a/x] /\!/ \Gamma_2 \vdash \Delta_2 /\!/ \mathcal{H} \qquad \mathcal{G} \ /\!/\ \Gamma_1 \vdash \Delta_1, \forall x A /\!/ \Gamma_2 \vdash \Delta_2, \forall x A /\!/ \mathcal{H}}{\mathcal{G} \ /\!/\ \Gamma_1 \vdash \Delta_1, \forall x A, \forall x A /\!/ \Gamma_2 \vdash \Delta_2 /\!/ \mathcal{H}} \ (\forall_{r2})$$

$$\frac{\dfrac{\dfrac{\dfrac{\mathcal{G} \ /\!/\ \Gamma_1 \vdash \Delta_1, \forall x A /\!/ \vdash A[a/x] /\!/ \Gamma_2 \vdash \Delta_2 /\!/ \mathcal{H}}{\mathcal{G} \ /\!/\ \Gamma_1 \vdash \Delta_1 /\!/ \vdash A[a/x] /\!/ \vdash A[a/x] /\!/ \Gamma_2 \vdash \Delta_2 /\!/ \mathcal{H}} \ \text{Lem. 11}}{\dfrac{\mathcal{G} \ /\!/\ \Gamma_1 \vdash \Delta_1 /\!/ \vdash A[a/x], A[a/x] /\!/ \Gamma_2 \vdash \Delta_2 /\!/ \mathcal{H}}{\mathcal{G} \ /\!/\ \Gamma_1 \vdash \Delta_1 /\!/ \vdash A[a/x] /\!/ \Gamma_2 \vdash \Delta_2 /\!/ \mathcal{H}} \ \text{IH}} \qquad \dfrac{\dfrac{\mathcal{G} \ /\!/\ \Gamma_1 \vdash \Delta_1, \forall x A /\!/ \Gamma_2 \vdash \Delta_2, \forall x A /\!/ \mathcal{H}}{\mathcal{G} \ /\!/\ \Gamma_1 \vdash \Delta_1 /\!/ \Gamma_2 \vdash \Delta_2, \forall x A, \forall x A /\!/ \mathcal{H}} \ \text{Lem. 7}}{\mathcal{G} \ /\!/\ \Gamma_1 \vdash \Delta_1 /\!/ \Gamma_2 \vdash \Delta_2, \forall x A /\!/ \mathcal{H}} \ \text{IH}}{\mathcal{G} \ /\!/\ \Gamma_1 \vdash \Delta_1, \forall x A /\!/ \Gamma_2 \vdash \Delta_2 /\!/ \mathcal{H}} \ (\forall_{r2})$$

Note that we may apply IH in the left branch of the derivation since the complexity of the contraction formula is less than $\forall x A$, and we may apply IH in the right branch since the height of the derivation is less than the original.   □

Before moving on to the cut-elimination theorem, we present the definition of the *splice* operation [17, 22]. The operation is used to formulate the *(cut)* rule.

Definition 5 (Splice [17]). *The splice $\mathcal{G} \oplus \mathcal{H}$ of two linear nested sequents $\mathcal{G}$ and $\mathcal{H}$ is defined as follows:*

$$(\Gamma_1 \vdash \Delta_1) \oplus (\Gamma_2 \vdash \Delta_2) := \Gamma_1, \Gamma_2 \vdash \Delta_1, \Delta_2$$
$$(\Gamma_1 \vdash \Delta_1) \oplus (\Gamma_2 \vdash \Delta_2 /\!/ \mathcal{F}) := \Gamma_1, \Gamma_2 \vdash \Delta_1, \Delta_2 /\!/ \mathcal{F}$$
$$(\Gamma_1 \vdash \Delta_1 /\!/ \mathcal{F}) \oplus (\Gamma_2 \vdash \Delta_2) := \Gamma_1, \Gamma_2 \vdash \Delta_1, \Delta_2 /\!/ \mathcal{F}$$
$$(\Gamma_1 \vdash \Delta_1 /\!/ \mathcal{F}) \oplus (\Gamma_2 \vdash \Delta_2 /\!/ \mathcal{K}) := \Gamma_1, \Gamma_2 \vdash \Delta_1, \Delta_2 /\!/ (\mathcal{F} \oplus \mathcal{K})$$

Theorem 4 (Cut-Elimination). *The rule*

$$\frac{\mathcal{G} \ /\!/ \ \Gamma \vdash \Delta, A \ /\!/ \ \mathcal{H} \qquad \mathcal{F} \ /\!/ \ A^{k_1}, \Gamma_1 \vdash \Delta_1 \ /\!/ \cdots /\!/ \ A^{k_n}, \Gamma_n \vdash \Delta_n}{(\mathcal{G} \oplus \mathcal{F}) \ /\!/ \ \Gamma, \Gamma_1 \vdash \Delta, \Delta_1 \ /\!/ \ (\mathcal{H} \oplus (\Gamma_2 \vdash \Delta_2 \ /\!/ \cdots /\!/ \ \Gamma_n \vdash \Delta_n))} \ (cut)$$

where $\| \mathcal{G} \| = \| \mathcal{F} \|$, $\| \mathcal{H} \| = n - 1$, *and* $\sum_{i=1}^{n} k_i \geq 1$, *is eliminable in* LNIF.

Proof. We extend the proof of [17, Thm. 5.16] and prove the result by induction on the lexicographic ordering of pairs $(|A|, h)$, where $|A|$ is the complexity of the cut formula A and h is the height of the derivation of the right premise of the (cut) rule. Moreover, we assume w.l.o.g. that (cut) is used once as the last inference of the derivation (given a derivation with multiple applications of (cut), we may repeatedly apply the elimination algorithm described here to the topmost occurrence of (cut), ultimately resulting in a cut-free derivation). By [17, Thm. 5.16], we know that (cut) is eliminable from any derivation in LNG, and therefore, we need only consider cases which incorporate quantifier rules.

If $h = 0$, then the right premise of (cut) is an instance of (id_1), (id_2), or $(\bot_l)$. If none of the cut formulae A are principal in the right premise, then the conclusion of (cut) is an instance of (id_1), (id_2), or $(\bot_l)$. If, however, one of the cut formulae A is principal in the right premise and is an atomic formula $p(\vec{a})$, then the top right premise of (cut) is of the form

$$\mathcal{F} \ /\!/ \ p(\vec{a})^{k_1}, \Gamma_1 \vdash \Delta_1 \ /\!/ \cdots /\!/ \ p(\vec{a})^{k_i}, \Gamma_i \vdash p(\vec{a}), \Delta_i' \ /\!/ \cdots /\!/ \ p(\vec{a})^{k_n}, \Gamma_n \vdash \Delta_n$$

where $\Delta_i = p(\vec{a}), \Delta_i'$. Observe that since Δ_i occurs in the conclusion of (cut), so does $p(\vec{a})$. To construct a cut-free derivation of the conclusion of (cut), we apply (lwr) to the left premise $\mathcal{G} \ /\!/ \ \Gamma \vdash \Delta, p(\vec{a}) \ /\!/ \ \mathcal{H}$ until $p(\vec{a})$ is in the ith component, and then apply hp-admissibility of (iw) (Lemma 5) to add in the missing formulae. Last, if the cut formula A is principal in the right premise and is equal to $\bot$, then the left premise of (cut) is of the form $\mathcal{G} \ /\!/ \ \Gamma \vdash \Delta, \bot \ /\!/ \ \mathcal{H}$. We obtain a cut-free derivation of the conclusion of (cut) by first applying hp-admissibility of $(\bot_r)$ (Lemma 3), followed by hp-admissibility of (iw) (Lemma 5) to add in the missing formulae.

Suppose that $h > 0$. If none of the cut formulae A are principal in the inference (r) of the right premise of (cut), then for all cases (with the exception of the $(\forall_{r1})$, $(\supset_{r1})$, $(\exists_l)$, $(\forall_{r2})$, and $(\supset_{r2})$ cases) we apply IH to the premise(s) of (r), followed by an application of (r). Let us now consider the $(\forall_{r1})$, $(\exists_l)$, $(\forall_{r2})$, $(\supset_{r1})$, and $(\supset_{r2})$ cases when none of the cut formulae A are principal. First, assume that $(\forall_{r1})$ is the rule used to derive the right premise of (cut):

$$\frac{\mathcal{G} \ /\!/ \ \Gamma \vdash \Delta, A \ /\!/ \ \mathcal{H} \qquad \dfrac{\mathcal{F} \ /\!/ \ A^{k_1}, \Gamma_1 \vdash \Delta_1 \ /\!/ \cdots /\!/ \ A^{k_n}, \Gamma_n \vdash \Delta_n /\!/ \vdash B[a/x]}{\mathcal{F} \ /\!/ \ A^{k_1}, \Gamma_1 \vdash \Delta_1 \ /\!/ \cdots /\!/ \ A^{k_n}, \Gamma_n \vdash \Delta_n, \forall x B} \ (\forall_{r1})}{(\mathcal{G} \oplus \mathcal{F}) \ /\!/ \ \Gamma, \Gamma_1 \vdash \Delta, \Delta_1 \ /\!/ \ (\mathcal{H} \oplus (\Gamma_2 \vdash \Delta_2 \ /\!/ \cdots /\!/ \ \Gamma_n \vdash \Delta_n, \forall x B))} \ (cut)$$

We invoke hp-admissibility of (sub) (Lemma 4) to substitute the eigenvariable a of $(\forall_{r1})$ with a fresh variable b that does not occur in either premise of (cut). We then apply admissibility of (ew) (Lemma 6) to the left premise of (cut), apply IH to the resulting derivations, and last apply the $(\forall_{r1})$ rule, as shown below:

$$\cfrac{\cfrac{\mathcal{G} \parallel \Gamma \vdash \Delta, A \parallel \mathcal{H}}{\mathcal{G} \parallel \Gamma \vdash \Delta, A \parallel \mathcal{H} \parallel \vdash} \text{Lem. 6} \qquad \cfrac{\cfrac{\mathcal{F} \parallel A^{k_1}, \Gamma_1 \vdash \Delta_1 \parallel \cdots \parallel A^{k_n}, \Gamma_n \vdash \Delta_n \parallel \vdash B[a/x]}{\mathcal{F} \parallel A^{k_1}, \Gamma_1 \vdash \Delta_1 \parallel \cdots \parallel A^{k_n}, \Gamma_n \vdash \Delta_n \parallel \vdash B[b/x]} \text{Lem. 4}}{} }{\cfrac{(\mathcal{G} \oplus \mathcal{F}) \parallel \Gamma, \Gamma_1 \vdash \Delta, \Delta_1 \parallel (\mathcal{H} \oplus (\Gamma_2 \vdash \Delta_2 \parallel \cdots \parallel \Gamma_n \vdash \Delta_n)) \parallel \vdash B[b/x]}{(\mathcal{G} \oplus \mathcal{F}) \parallel \Gamma, \Gamma_1 \vdash \Delta, \Delta_1 \parallel (\mathcal{H} \oplus (\Gamma_2 \vdash \Delta_2 \parallel \cdots \parallel \Gamma_n \vdash \Delta_n, \forall x B))} (\forall_{r1})} \text{IH}$$

In the $(\exists_l)$ case below

$$\cfrac{\mathcal{G} \parallel \Gamma \vdash \Delta, A \parallel \mathcal{H} \qquad \cfrac{\mathcal{F} \parallel A^{k_1}, \Gamma_1 \vdash \Delta_1 \parallel \cdots \parallel A^{k_i}, B[a/x], \Gamma_i \vdash \Delta_i \parallel \cdots \parallel A^{k_n}, \Gamma_n \vdash \Delta_n}{\mathcal{F} \parallel A^{k_1}, \Gamma_1 \vdash \Delta_1 \parallel \cdots \parallel A^{k_i}, \exists x B, \Gamma_i \vdash \Delta_i \parallel \cdots \parallel A^{k_n}, \Gamma_n \vdash \Delta_n} (\exists_l)}{(\mathcal{G} \oplus \mathcal{F}) \parallel \Gamma, \Gamma_1 \vdash \Delta, \Delta_1 \parallel (\mathcal{H} \oplus (\Gamma_2 \vdash \Delta_2 \parallel \cdots \parallel \exists x B, \Gamma_i \vdash \Delta_i \parallel \cdots \parallel \Gamma_n \vdash \Delta_n))} (cut)$$

we also make use of the hp-admissibility of (sub) to ensure that the $(\exists_l)$ rule can be applied after invoking the inductive hypothesis:

$$\cfrac{\mathcal{G} \parallel \Gamma \vdash \Delta, A \parallel \mathcal{H} \qquad \cfrac{\cfrac{\mathcal{F} \parallel A^{k_1}, \Gamma_1 \vdash \Delta_1 \parallel \cdots \parallel A^{k_i}, B[a/x], \Gamma_i \vdash \Delta_i \parallel \cdots \parallel A^{k_n}, \Gamma_n \vdash \Delta_n}{\mathcal{F} \parallel A^{k_1}, \Gamma_1 \vdash \Delta_1 \parallel \cdots \parallel A^{k_i}, B[b/x], \Gamma_i \vdash \Delta_i \parallel \cdots \parallel A^{k_n}, \Gamma_n \vdash \Delta_n} \text{Lem. 4}}{(\mathcal{G} \oplus \mathcal{F}) \parallel \Gamma, \Gamma_1 \vdash \Delta, \Delta_1 \parallel (\mathcal{H} \oplus (\Gamma_2 \vdash \Delta_2 \parallel \cdots \parallel A^{k_i}, B[b/x], \Gamma_i \vdash \Delta_i \parallel \cdots \parallel \Gamma_n \vdash \Delta_n))} \text{IH}}{(\mathcal{G} \oplus \mathcal{F}) \parallel \Gamma, \Gamma_1 \vdash \Delta, \Delta_1 \parallel (\mathcal{H} \oplus (\Gamma_2 \vdash \Delta_2 \parallel \cdots \parallel \exists x B, \Gamma_i \vdash \Delta_i \parallel \cdots \parallel \Gamma_n \vdash \Delta_n))} (\exists_l)$$

Let us consider the $(\forall_{r2})$ case

$$\cfrac{(1) \qquad \cfrac{(2) \qquad (3)}{\mathcal{F} \parallel A^{k_1}, \Gamma_1 \vdash \Delta_1 \parallel \cdots \parallel A^{k_i}, \Gamma_i \vdash \Delta_i, \forall x B \parallel A^{k_{i+1}}, \Gamma_{i+1} \vdash \Delta_{i+1} \parallel \cdots \parallel A^{k_n}, \Gamma_n \vdash \Delta_n} (\forall_{r2})}{(\mathcal{G} \oplus \mathcal{F}) \parallel \Gamma, \Gamma_1 \vdash \Delta, \Delta_1 \parallel (\mathcal{H} \oplus (\Gamma_2 \vdash \Delta_2 \parallel \cdots \parallel \Gamma_i \vdash \Delta_i, \forall x B \parallel \Gamma_{i+1} \vdash \Delta_{i+1} \parallel \cdots \parallel \Gamma_n \vdash \Delta_n))} (cut)$$

(1) $\mathcal{G} \parallel \Gamma \vdash \Delta, A \parallel \mathcal{H}_1 \parallel \Gamma_i' \vdash \Delta_i' \parallel \Gamma_{i+1}' \vdash \Delta_{i+1}' \parallel \mathcal{H}_2$

(2) $\mathcal{F} \parallel A^{k_1}, \Gamma_1 \vdash \Delta_1 \parallel \cdots \parallel A^{k_i}, \Gamma_i \vdash \Delta_i \parallel \vdash B[a/x] \parallel A^{k_{i+1}}, \Gamma_{i+1} \vdash \Delta_{i+1} \parallel \cdots \parallel A^{k_n}, \Gamma_n \vdash \Delta_n$

(3) $\mathcal{F} \parallel A^{k_1}, \Gamma_1 \vdash \Delta_1 \parallel \cdots \parallel A^{k_i}, \Gamma_i \vdash \Delta_i \parallel A^{k_{i+1}}, \Gamma_{i+1} \vdash \Delta_{i+1}, \forall x B \parallel \cdots \parallel A^{k_n}, \Gamma_n \vdash \Delta_n$

where $\mathcal{H} = \mathcal{H}_1 \parallel \Gamma_i' \vdash \Delta_i' \parallel \Gamma_{i+1}' \vdash \Delta_{i+1}' \parallel \mathcal{H}_2$. To resolve the case we invoke admissibility of (ew) (Lemma 6) on (1) to obtain a derivation of

(1)′ $\mathcal{G} \parallel \Gamma \vdash \Delta, A \parallel \mathcal{H}_1 \parallel \Gamma_i' \vdash \Delta_i' \parallel \vdash \parallel \Gamma_{i+1}' \vdash \Delta_{i+1}' \parallel \mathcal{H}_2$

Moreover, to ensure that the eigenvariable a in (2) does not occur in (1), we apply hp-admissibility of (sub) (Lemma 4) to obtain (2)′ where a has been replaced by a fresh parameter b. Applying IH between (1)′ and (2)′, and (1) and (3), followed by an application of $(\forall_{r2})$, gives the desired result. Last, note that the $(\supset_{r1})$ and $(\supset_{r2})$ cases are resolved as explained in the proof of [17, Thm. 5.16].

We assume now that one of the cut formulae A is principal in the inference yielding the right premise of (cut). The cases where A is an atomic formula $p(\vec{a})$ or is identical to $\bot$ are resolved as explained above (when $h = 0$). For the case when A is principal in an application of $(lift)$, we simply apply IH between the left premise of (cut) and the premise of the $(lift)$ rule. Also, if A is of the form $B \wedge C$, $B \vee C$, or $B \supset C$, then all such cases can be resolved as explained in the proof of [17, Thm. 5.16]. Thus, we only consider the cases where A is of the form $\exists x B$ and $\forall x B$; we begin with the former and assume our derivation ends with:

$$\cfrac{\mathcal{G} \parallel \Gamma \vdash \Delta, \exists x B \parallel \mathcal{H} \qquad \cfrac{\mathcal{F} \parallel \exists x B^{k_1}, \Gamma_1 \vdash \Delta_1 \parallel \cdots \parallel \exists x B^{k_i}, B[a/x], \Gamma_i \vdash \Delta_i \parallel \cdots \parallel \exists x B^{k_n}, \Gamma_n \vdash \Delta_n}{\mathcal{F} \parallel \exists x B^{k_1}, \Gamma_1 \vdash \Delta_1 \parallel \cdots \parallel \exists x B^{k_i+1}, \Gamma_i \vdash \Delta_i \parallel \cdots \parallel \exists x B^{k_n}, \Gamma_n \vdash \Delta_n} \; (\exists_l)}{(\mathcal{G} \oplus \mathcal{F}) \parallel \Gamma, \Gamma_1 \vdash \Delta, \Delta_1 \parallel (\mathcal{H} \oplus (\Gamma_2 \vdash \Delta_2 \parallel \cdots \parallel \Gamma_i \vdash \Delta_i \parallel \cdots \parallel \Gamma_n \vdash \Delta_n))} \; (cut)$$

Invoking IH with the left premise of (cut) and the premise of $(\exists_l)$ gives a cut-free derivation of:

$$(\mathcal{G} \oplus \mathcal{F}) \parallel \Gamma, \Gamma_1 \vdash \Delta, \Delta_1 \parallel (\mathcal{H} \oplus (\Gamma_2 \vdash \Delta_2 \parallel \cdots \parallel B[a/x], \Gamma_i \vdash \Delta_i \parallel \cdots \parallel \Gamma_n \vdash \Delta_n))$$

By invertibility of $(\exists_r)$ (Lemma 9), there exists a cut-free derivation of $\mathcal{G} \parallel \Gamma \vdash \Delta, B[a/x] \parallel \mathcal{H}$. Since $|B[a/x]| < |\exists x B|$, we can apply IH to this sequent as well as the sequent above to obtain a cut-free derivation of:

$$(\mathcal{G} \oplus \mathcal{G} \oplus \mathcal{F}) \parallel \Gamma, \Gamma, \Gamma_1 \vdash \Delta, \Delta, \Delta_1 \parallel (\mathcal{H} \oplus \mathcal{H} \oplus (\Gamma_2 \vdash \Delta_2 \parallel \cdots \parallel \Gamma_i \vdash \Delta_i \parallel \cdots \parallel \Gamma_n \vdash \Delta_n))$$

Applying admissibility of (ic_l) and (ic_r) (Lemmas 14 and 16), we obtain the desired conclusion.

Last, let us consider the case where A is of the form $\forall x B$:

$$\cfrac{\mathcal{G} \parallel \Gamma \vdash \Delta, \forall x B \parallel \mathcal{H} \qquad \cfrac{\mathcal{F} \parallel \forall x B^{k_1}, \Gamma_1 \vdash \Delta_1 \parallel \cdots \parallel \forall x B^{k_i}, B[a/x], \Gamma_i \vdash \Delta_i \parallel \cdots \parallel \forall x B^{k_n}, \Gamma_n \vdash \Delta_n}{\mathcal{F} \parallel \forall x B^{k_1}, \Gamma_1 \vdash \Delta_1 \parallel \cdots \parallel \forall x B^{k_i}, \Gamma_i \vdash \Delta_i \parallel \cdots \parallel \forall x B^{k_n}, \Gamma_n \vdash \Delta_n} \; (\forall_l)}{(\mathcal{G} \oplus \mathcal{F}) \parallel \Gamma, \Gamma_1 \vdash \Delta, \Delta_1 \parallel (\mathcal{H} \oplus (\Gamma_2 \vdash \Delta_2 \parallel \cdots \parallel \Gamma_i \vdash \Delta_i \parallel \cdots \parallel \Gamma_n \vdash \Delta_n))} \; (cut)$$

Applying IH between the left premise of (cut) and the premise of the $(\forall_l)$ rule, we obtain

$$(\mathcal{G} \oplus \mathcal{F}) \parallel \Gamma, \Gamma_1 \vdash \Delta, \Delta_1 \parallel (\mathcal{H} \oplus (\Gamma_2 \vdash \Delta_2 \parallel \cdots \parallel B[a/x], \Gamma_i \vdash \Delta_i \parallel \cdots \parallel \Gamma_n \vdash \Delta_n))$$

Depending on if $\mathcal{H}$ is empty or not, we invoke the invertibility of $(\forall_{r1})$ or $(\forall_{r2})$ (Lemmas 13 and 11), admissibility of (mrg) (Lemma 15), and hp-admissibility of (sub) (Lemma 4) to obtain a derivation of the sequent $\mathcal{G} \parallel \Gamma \vdash \Delta, B[a/x] \parallel \mathcal{H}$. Since $|B[a/x]| < |\forall x B|$ we can apply IH between this sequent and the one above to obtain a cut-free derivation of:

$$(\mathcal{G} \oplus \mathcal{G} \oplus \mathcal{F}) \parallel \Gamma, \Gamma, \Gamma_1 \vdash \Delta, \Delta, \Delta_1 \parallel (\mathcal{H} \oplus \mathcal{H} \oplus (\Gamma_2 \vdash \Delta_2 \parallel \cdots \parallel \Gamma_i \vdash \Delta_i \parallel \cdots \parallel \Gamma_n \vdash \Delta_n))$$

Admissibility of (ic_l) and (ic_r) (Lemmas 14 and 16) give the desired conclusion. $\square$

5   Conclusion

This paper presented the cut-free calculus LNIF for intuitionistic fuzzy logic within the relatively new paradigm of *linear nested sequents*. The calculus possesses highly fundamental proof-theoretic properties such as (m-)invertibility of all logical rules, admissibility of structural rules, and syntactic cut-elimination.

In future work the author aims to investigate corollaries of the cut-elimination theorem, such as a *midsequent theorem* [4]. In our context, such a theorem states that every derivable sequent containing only prenex formulae is derivable with a proof containing quantifier-free sequents, called *midsequents*, which have only propositional inferences (and potentially $(lift)$) above them in the derivation, and only quantifier inferences (and potentially $(lift)$) below them. Moreover, the present formalism could offer insight regarding which fragments interpolate (or if all of IF interpolates) by applying the so-called *proof-theoretic method* of interpolation [17,21]. Additionally, it could be fruitful to adapt linear nested sequents to other first-order Gödel logics and to investigate decidable fragments [2] by providing proof-search algorithms with implementations (e.g. [16] provides an implementation of proof-search in PROLOG for a class of modal logics within the linear nested sequent framework).

Last, [8] introduced both a nested calculus for first-order intuitionistic logic with *constant domains*, and a nested calculus for first-order intuitionistic logic with *non-constant domains*. The fundamental difference between the two calculi involves the imposition of a side condition on the left $\forall$ and right $\exists$ rules. The author aims to investigate whether such a condition can be imposed on quantifier rules in LNIF in order to readily convert the calculus into a sound and cut-free complete calculus for first-order Gödel logic with *non-constant domains*. This would be a further strength of LNIF since switching between the calculi for the constant domain and non-constant domain versions of first-order Gödel logic would result by simply imposing a side condition on a subset of the quantifier rules.

Acknowledgments. The author would like to thank his supervisor A. Ciabattoni for her continued support, B. Lellmann for his thought-provoking discussions on linear nested sequents, and K. van Berkel for his helpful comments. Work funded by FWF projects I2982, Y544-N2, and W1255-N23.

References

1. Avron, A.: Hypersequents, logical consequence and intermediate logics for concurrency. Ann. Math. Artif. Intell. **4**(3), 225–248 (1991). https://doi.org/10.1007/BF01531058
2. Baaz, M., Ciabattoni, A., Preining, N.: SAT in monadic gödel logics: a borderline between decidability and undecidability. In: Ono, H., Kanazawa, M., de Queiroz, R. (eds.) WoLLIC 2009. LNCS (LNAI), vol. 5514, pp. 113–123. Springer, Heidelberg (2009). https://doi.org/10.1007/978-3-642-02261-6_10
3. Baaz, M., Preining, N., Zach, R.: First-order gödel logics. Ann. Pure Appl. Logic **147**(1), 23–47 (2007). https://doi.org/10.1016/j.apal.2007.03.001. URL http://www.sciencedirect.com/science/article/pii/S016800720700019X
4. Baaz, M., Zach, R.: Hypersequents and the proof theory of intuitionistic fuzzy logic. In: Clote, P.G., Schwichtenberg, H. (eds.) CSL 2000. LNCS, vol. 1862, pp. 187–201. Springer, Heidelberg (2000). https://doi.org/10.1007/3-540-44622-2_12
5. Belnap Jr., N.D.: Display logic. J. Philos. Log. **11**(4), 375–417 (1982)

6. Borgwardt, S., Distel, F., Peñaloza, R.: Decidable Gödel description logics without the finitely-valued model property. In: Baral, C., De Giacomo, G., Eiter, T. (eds.) Proceedings of the 14th International Conference on Principles of Knowledge Representation and Reasoning (KR 2014), pp. 228–237. AAAI Press (2014)
7. Dummett, M.: A propositional calculus with denumerable matrix. J. Symb. Log. **24**(2), 97–106 (1959). http://www.jstor.org/stable/2964753
8. Fitting, M.: Nested sequents for intuitionistic logics. Notre Dame J. Form. Log. **55**(1), 41–61 (2014). https://doi.org/10.1215/00294527-2377869
9. Gabbay, D., Shehtman, V., Skvortsov, D.: Quantification in Non-classical Logics. Studies in Logic and Foundations of Mathematics. Elsevier, Amsterdam (2009)
10. Gentzen, G.: Untersuchungen uber das logische schliessen. Math. Z. **39**(3), 405–431 (1935)
11. Gödel, K.: Zum intuitionistischen aussagenkalkül. Anzeiger der Akademie der Wissenschaften in Wien **69**, 65–66 (1932)
12. Goré, R., Postniece, L., Tiu, A.: Cut-elimination and proof-search for bi-intuitionistic logic using nested sequents. In: Areces, C., Goldblatt, R. (eds.) Advances in Modal Logic 7, Papers from the Seventh Conference on "Advances in Modal Logic," held in Nancy, France, 9–12 September 2008, pp. 43–66. College Publications (2008). http://www.aiml.net/volumes/volume7/Gore-Postniece-Tiu.pdf
13. Hajek, P.: The Metamathematics of Fuzzy Logic. Kluwer, Dordrecht (1998)
14. Horn, A.: Logic with truth values in a linearly ordered heyting algebra. J. Symb. Log. **34**(3), 395–408 (1969). http://www.jstor.org/stable/2270905
15. Lellmann, B.: Linear nested sequents, 2-sequents and hypersequents. In: De Nivelle, H. (ed.) TABLEAUX 2015. LNCS (LNAI), vol. 9323, pp. 135–150. Springer, Cham (2015). https://doi.org/10.1007/978-3-319-24312-2_10
16. Lellmann, B.: LNSprover: modular theorem proving with linear nested sequents (2016). https://www.logic.at/staff/lellmann/lnsprover/
17. Lellmann, B., Kuznets, R.: Interpolation for intermediate logics via hyper- and linear nested sequents. In: Advances in Modal Logic, vol. 12, pp. 473–492 (2018)
18. Lifschitz, V., Pearce, D., Valverde, A.: Strongly equivalent logic programs. ACM Trans. Comput. Log. **2**(4), 526–541 (2001). https://doi.org/10.1145/383779.383783
19. Lyon, T.: Syntactic cut-elimination for intuitionistic fuzzy logic via linear nested sequents (2019). https://arxiv.org/abs/1910.06657
20. Lyon, T., van Berkel, K.: Automating agential reasoning: proof-calculi and syntactic decidability for STIT logics. In: Baldoni, M., Dastani, M., Liao, B., Sakurai, Y., Zalila Wenkstern, R. (eds.) PRIMA 2019. LNCS (LNAI), vol. 11873, pp. 202–218. Springer, Cham (2019). https://doi.org/10.1007/978-3-030-33792-6_13
21. Lyon, T., Tiu, A., Góre, R., Clouston, R.: Syntactic interpolation for tense logics and bi-intuitionistic logic via nested sequents. In: Fernández, M., Muscholl, A. (eds.) 28th EACSL Annual Conference on Computer Science Logic (CSL 2020). Leibniz International Proceedings in Informatics (LIPIcs), vol. 152. Schloss Dagstuhl-Leibniz-Zentrum fuer Informatik, Dagstuhl, Germany (2020, forthcoming)
22. Masini, A.: 2-sequent calculus: a proof theory of modalities. Ann. Pure Appl. Log. **58**(3), 229–246 (1992). https://doi.org/10.1016/0168-0072(92)90029-Y. http://www.sciencedirect.com/science/article/pii/016800729290029Y
23. Masini, A.: 2-sequent calculus: intuitionism and natural deduction. J. Log. Comput. **3**(5), 533–562 (1993). https://doi.org/10.1093/logcom/3.5.533
24. Poggiolesi, F.: A cut-free simple sequent calculus for modal logic S5. Rev. Symb. Log. **1**(1), 3–15 (2008). https://doi.org/10.1017/S1755020308080040

25. Takeuti, G., Titani, S.: Intuitionistic fuzzy logic and intuitionistic fuzzy set theory. J. Symb. Log. **49**(3), 851–866 (1984). https://doi.org/10.2307/2274139
26. Viganò, L.: Labelled Non-classical Logics. Kluwer Academic Publishers, Dordrecht (2000). With a foreword by Dov M. Gabbay
27. Visser, A.: On the completenes principle: a study of provability in heyting's arithmetic and extensions. Ann. Math. Log. **22**(3), 263–295 (1982). https://doi.org/10.1016/0003-4843(82)90024-9. http://www.sciencedirect.com/science/article/pii/0003484382900249
28. Wansing, H.: Sequent calculi for normal modal propositional logics. J. Log. Comput. **4**(2), 125–142 (1994). https://doi.org/10.1093/logcom/4.2.125

On Deriving Nested Calculi
for Intuitionistic Logics from Semantic
Systems

Tim Lyon[✉]

Institut für Logic and Computation, Technische Universität Wien,
1040 Vienna, Austria
lyon@logic.at

Abstract. This paper shows how to derive nested calculi from labelled calculi for propositional intuitionistic logic and first-order intuitionistic logic with constant domains, thus connecting the general results for labelled calculi with the more refined formalism of nested sequents. The extraction of nested calculi from labelled calculi obtains via considerations pertaining to the elimination of structural rules in labelled derivations. Each aspect of the extraction process is motivated and detailed, showing that each nested calculus inherits favorable proof-theoretic properties from its associated labelled calculus.

Keywords: Intuitionistic logic · Kripke semantics · Labelled calculi · Nested calculi · Proof theory · Structural rule elimination

1 Introduction

Numerous fruitful consequences and applications obtain through the supplementation of a logic with an *analytic* calculus. Such calculi are characterized on the basis of their inference rules, which stepwise (de)compose the formula to be proven. One of the most prominent realizations of this idea dates back to Gentzen [12], who proposed the *sequent calculus* framework for classical and intuitionistic logic. Since then, countless extensions and reformulations of Gentzen's framework have been supplied for many logics of interest. Examples of extensions include *display calculi* [2,23], *hypersequent calculi* [21], *labelled calculi* [7,24], and *nested calculi* [10,23]. Such calculi have been exploited to prove meaningful results; e.g. decidability [21,23], interpolation [18], and automated counter-model extraction [17,23]. We focus on the labelled and nested formalisms in this paper.

The labelled approach of constructing calculi may be qualified as *semantic* due to the fact that calculi are obtained through the transformation of semantic clauses and Kripke-frame properties into inference rules for a logic [7,24]. Although the approach has been criticized by some [1], it has also proven to be quite successful relative to certain criteria. For example, the labelled formalism is surprisingly modular and allows for the automated construction of

© Springer Nature Switzerland AG 2020
S. Artemov and A. Nerode (Eds.): LFCS 2020, LNCS 11972, pp. 177–194, 2020.
https://doi.org/10.1007/978-3-030-36755-8_12

analytic calculi for many intermediate and modal logics [6,7,19]. Furthermore, calculi constructed in the labelled paradigm often possess fruitful properties (e.g. contraction-admissibility, invertibility of rules, cut-elimination, etc.) that follow from general results [7,19].

In 2009, Brünnler introduced *nested sequent calculi* [3] and Poggiolesi introduced *tree hypersequent calculi* [22] for a set of modal logics. Both formalisms are essentially notational variants of one another and make use of an idea due to Bull [4] and Kashima [15] to organize sequents into treelike structures called *nested sequents*. Although nested sequents can be seen as a distinct proof-theoretic formalism, it was shown in 2012 [9] that nested sequent calculi can be viewed as 'upside-down' versions of prefixed tableaux, introduced much earlier in 1972 [8]. In contrast to labelled sequents, nested sequents are often given in a language as expressive as the language of the logic; thus, nested calculi have the advantage that they minimize the bureaucracy sufficient to prove theorems. The nested formalism continues to receive much attention, proving itself suitable for constructing analytic calculi [3], developing automated reasoning algorithms [13], and verifying interpolation [18], among other applications.

Despite the many advantages of nested calculi, constructing such calculi for logics as well as proving that they possess favorable proof-theoretic properties (admissibility of structural rules, cut-elimination, etc.) is often done on a case by case basis; i.e. the nested formalism does not—to date—offer the same generality of results that hold in the labelled paradigm (cf. [7,19]). Therefore, a significant advantage of the labelled paradigm over the nested paradigm is that labelled calculi are easily constructed on the basis of a logic's semantics and one often obtains highly favorable proof-theoretic properties of the calculi (essentially) for free via general theorems [7,19]. Nevertheless, the labelled formalism has its drawbacks: the calculi involve a complicated syntax, and labelled structural rules typically delete vital formulae from premise to conclusion, which can cause associated proof-search algorithms to be less efficient or rely on backtracking.

Since the labelled formalism is well-suited for constructing calculi and confirming properties, and the nested formalism is well-suited for applications, a general method of extracting nested calculi from labelled calculi (with properties preserved) is highly desirable. One could generate labelled calculi for a class of logics and confirm favorable proof-theoretic properties via existing general results; if such properties were preserved during an extraction procedure, then the ensuing nested calculi would possess the properties as well, yielding practical, cut-free nested calculi. Similar ideas and results have been discussed in the literature [5,14,17,20], where refined calculi (which can be viewed as nested calculi) were extracted from labelled calculi for various logics. (NB. The author has recently been made aware of [20], which mentions results strongly related to Sect. 4. Although the results presented here were discovered independently, the work of Sect. 4 can be seen as a detailed explication and expansion of the work presented in [20].). In this paper, we advance our understanding of this method, and show how to derive Fitting's nested calculi (see [10]) from labelled calculi for intuitionistic logics. The results of this paper are also worthwhile in that they clarify the connection between the intuitionistic labelled calculi and nested calculi considered, thus shedding light on the semantic roles played by certain inference rules and syntactic structures in Fitting's formalism.

This paper is organized as follows: Sect. 2 introduces the labelled and nested calculi for the intuitionistic logics considered, and Sect. 3 shows how to translate labelled sequents into nested sequents. Sections 4 and 5 show how to extract the nested calculi from the labelled calculi for propositional and first-order intuitionistic logic with constant domains, respectively. Last, Sect. 6 concludes and discusses future work.

2 Proof Calculi for Intuitionistic Logics

The language $\mathcal{L}$ for propositional intuitionistic logic (Int) is defined via the BNF grammar shown below top, and the language $\mathcal{L}_Q$ for constant domain first-order intuitionistic logic (IntQC) is defined via the BNF grammar shown below bottom:

$$A ::= p \mid \perp \mid (A \vee A) \mid (A \wedge A) \mid (A \supset A)$$

$$A ::= p(x_1, \ldots, x_n) \mid \perp \mid (A \vee A) \mid (A \wedge A) \mid (A \supset A) \mid (\forall x)A \mid (\exists x)A$$

In the language $\mathcal{L}$, p is among a denumerable set of *propositional variables* $\{p, q, r, \ldots\}$. In the language $\mathcal{L}_Q$, p is an n-ary *predicate symbol* with $x_1, \ldots, x_n, x$ *variables* ($n \in \mathbb{N}$), and when $n = 0$, p is assumed to be a propositional variable. As usual, we define $\neg A := A \supset \perp$.

We assume the reader is familiar with intuitionistic logics; for a comprehensive overview, see [11].

2.1 The Labelled Calculi **G3Int** and **G3IntQC**

We define *propositional (first-order) labelled sequents* to be syntactic objects of the form $L_1 \Rightarrow L_2$ ($L_1' \Rightarrow L_2'$, resp.), where L_1 and L_2 (L_1' and L_2', resp.) are formulae defined via the BNF grammar below top (below bottom, resp.).

$$L_1 ::= w : A \mid w \leq v \mid L_1, L_1 \qquad L_2 ::= w : A \mid L_2, L_2$$

$$L_1' ::= w : A \mid a \in D_w \mid w \leq v \mid L_1', L_1' \qquad L_2' ::= w : A \mid L_2', L_2'$$

In the propositional case, A is in the language $\mathcal{L}$ and w is among a denumerable set of labels $\{w, v, u, \ldots\}$. In the first-order case, A is in the language $\mathcal{L}_Q$, a is among a denumerable set of *parameters* $\{a, b, c, \ldots\}$, and w is among a denumerable set of labels $\{w, v, u, \ldots\}$. We refer to formulae of the forms $w \leq u$ and $a \in D_w$ as *relational atoms* (with formulae of the form $a \in D_w$ sometimes referred to as *domain atoms*, more specifically) and refer to formulae of the form $w : A$ as *labelled formulae*. Due to the two types of formulae occurring in a labelled sequent, we often use $\mathcal{R}$ to denote relational atoms, and Γ and Δ to denote labelled formulae, thus distinguishing between the two. Labelled sequents are therefore written in a general form as $\mathcal{R}, \Gamma \Rightarrow \Delta$. Moreover, we take the comma operator to be commutative and associative; for example, we identify the formula $w : A, w \leq u, u : B$ with $w \leq u, u : B, w : A$. This interpretation of comma lets us view $\mathcal{R}, \Gamma$ and Δ as multisets. Also, we allow for empty antecedents and succedents in both our labelled and nested sequents.

In the first-order setting, we syntactically distinguish between *bound variables* $\{x, y, z, \dots\}$ and *free variables*, which are replaced with *parameters* $\{a, b, c, \dots\}$, to avoid clashes between the two categories (cf. [10, Sect. 8]). Therefore, instead of using formulae directly from the first-order language, we use formulae from the first-order language where each freely occurring variable x has been replaced by a distinct parameter a. For example, we would make use of the labelled formula $w : (\forall x)p(a, x) \vee q(a, b)$ instead of $w : (\forall x)p(y, x) \vee q(y, z)$ in a first-order sequent of G3IntQC. For a formula $A \in \mathcal{L}_{\mathcal{Q}}$, we write $A[a/x]$ to mean the formula that results from substituting the parameter a for all occurrences of

$$\frac{}{\mathcal{R}, w \leq v, w : p, \Gamma \Rightarrow \Delta, v : p} \ (id) \qquad \frac{\mathcal{R}, w \leq v, v : A, \Gamma \Rightarrow \Delta, v : B}{\mathcal{R}, \Gamma \Rightarrow \Delta, w : A \supset B} \ (\supset_r)^{\dagger_1}$$

$$\frac{\mathcal{R}, w : A, w : B, \Gamma \Rightarrow \Delta}{\mathcal{R}, w : A \wedge B, \Gamma \Rightarrow \Delta} \ (\wedge_l) \qquad \frac{\mathcal{R}, \Gamma \Rightarrow \Delta, w : A \quad \mathcal{R}, \Gamma \Rightarrow \Delta, w : B}{\mathcal{R}, \Gamma \Rightarrow \Delta, w : A \wedge B} \ (\wedge_r)$$

$$\frac{\mathcal{R}, w : A, \Gamma \Rightarrow \Delta \quad \mathcal{R}, w : B, \Gamma \Rightarrow \Delta}{\mathcal{R}, w : A \vee B, \Gamma \Rightarrow \Delta} \ (\vee_l) \qquad \frac{\mathcal{R}, \Gamma \Rightarrow \Delta, w : A, w : B}{\mathcal{R}, \Gamma \Rightarrow \Delta, w : A \vee B} \ (\vee_r)$$

$$\frac{\mathcal{R}, w \leq v, w : A \supset B, \Gamma \Rightarrow \Delta, v : A \quad \mathcal{R}, w \leq v, w : A \supset B, v : B, \Gamma \Rightarrow \Delta}{\mathcal{R}, w \leq v, w : A \supset B, \Gamma \Rightarrow \Delta} \ (\supset_l)$$

$$\frac{\mathcal{R}, w \leq w, \Gamma \Rightarrow \Delta}{\mathcal{R}, \Gamma \Rightarrow \Delta} \ (ref) \qquad \frac{\mathcal{R}, w \leq v, v \leq u, w \leq u, \Gamma \Rightarrow \Delta}{\mathcal{R}, w \leq v, v \leq u, \Gamma \Rightarrow \Delta} \ (tra)$$

$$\frac{}{\mathcal{R}, w : \bot, \Gamma \Rightarrow \Delta} \ (\bot_l) \qquad \frac{}{\mathcal{R}, w \leq v, \vec{a} \in D_w, w : p(\vec{a}), \Gamma \Rightarrow \Delta, v : p(\vec{a})} \ (id_q)$$

$$\frac{\mathcal{R}, w \leq v, a \in D_v, \Gamma \Rightarrow \Delta, v : A[a/x]}{\mathcal{R}, \Gamma \Rightarrow \Delta, w : \forall x A} \ (\forall_r)^{\dagger_2} \qquad \frac{\mathcal{R}, a \in D_w, \Gamma \Rightarrow \Delta, w : A[a/x], w : \exists x A}{\mathcal{R}, a \in D_w, \Gamma \Rightarrow \Delta, w : \exists x A} \ (\exists_r)$$

$$\frac{\mathcal{R}, a \in D_w, w : A[a/x], \Gamma \Rightarrow \Delta}{\mathcal{R}, w : \exists x A, \Gamma \Rightarrow \Delta} \ (\exists_l)^{\dagger_3} \qquad \frac{\mathcal{R}, w \leq v, a \in D_v, v : A[a/x], w : \forall x A, \Gamma \Rightarrow \Delta}{\mathcal{R}, w \leq v, a \in D_v, w : \forall x A, \Gamma \Rightarrow \Delta} \ (\forall_l)$$

$$\frac{\mathcal{R}, w \leq v, a \in D_w, a \in D_v, \Gamma \Rightarrow \Delta}{\mathcal{R}, w \leq v, a \in D_w, \Gamma \Rightarrow \Delta} \ (nd) \qquad \frac{\mathcal{R}, w \leq v, a \in D_v, a \in D_w, \Gamma \Rightarrow \Delta}{\mathcal{R}, w \leq v, a \in D_v, \Gamma \Rightarrow \Delta} \ (cd)$$

Fig. 1. The labelled calculus G3Int for propositional intuitionistic logic consists of (id), $(\supset_r)$, $(\wedge_l)$, $(\wedge_r)$, $(\vee_l)$, $(\vee_r)$, $(\supset_l)$, (ref), (tra), and $(\bot_l)$ (see [7]), and all rules give the calculus G3IntQC. The side condition $\dagger_1$ states that the variable v does not occur in the conclusion, $\dagger_2$ states that neither a nor v occur in the conclusion, and $\dagger_3$ states that a does not occur in the conclusion. Labels and parameters restricted from occurring in the conclusion of an inference are called *eigenvariables*. (Note that (id) is an instance of (id_q); the same holds in the nested setting).

the free variable x in A. Last, we use the notation $A(a_0, \ldots, a_n)$, with $n \in \mathbb{N}$, to denote that the parameters $a_0, \ldots, a_n$ are all parameters occurring in the formula A. We write $A(\vec{a})$ as shorthand for $A(a_0, \ldots, a_n)$ and $\vec{a} \in D_w$ as shorthand for $a_0 \in D_w, \ldots, a_n \in D_w$. The labelled calculi are given in Fig. 1.

We define a *label substitution* $[w/v]$ on a labelled sequent in the usual way as the replacement of all labels v occurring in the sequent with the label w. Similarly, we define a *parameter substitution* $[a/b]$ on a labelled sequent as the replacement of all parameters b occurring in the sequent with the parameter a.

Theorem 1. *The calculi* G3Int *and* G3IntQC *have the following properties:*

(i) (a) *For all* $A \in \mathcal{L}$, $\vdash_{\text{G3Int}} \mathcal{R}, w \leq v, w : A, \Gamma \Rightarrow v : A, \Delta$;
   (b) *For all* $A \in \mathcal{L}$, $\vdash_{\text{G3Int}} \mathcal{R}, w : A, \Gamma \Rightarrow \Delta, w : A$;
   (c) *For all* $A \in \mathcal{L}_Q$, $\vdash_{\text{G3IntQC}} \mathcal{R}, w \leq v, \vec{a} \in D_w, w : A(\vec{a}), \Gamma \Rightarrow v : A(\vec{a}), \Delta$;
   (d) *For all* $A \in \mathcal{L}_Q$, $\vdash_{\text{G3IntQC}} \mathcal{R}, \vec{a} \in D_w, w : A(\vec{a}), \Gamma \Rightarrow \Delta, w : A(\vec{a})$;
(ii) *The* (lsub) *and* (psub) *rules are height-preserving (i.e. 'hp-') admissible;*

$$\frac{\mathcal{R}, \Gamma \Rightarrow \Delta}{\mathcal{R}[w/v], \Gamma[w/v] \Rightarrow \Delta[w/v]} \ (lsub) \qquad \frac{\mathcal{R}, \Gamma \Rightarrow \Delta}{\mathcal{R}[a/b], \Gamma[a/b] \Rightarrow \Delta[a/b]} \ (psub)$$

(iii) *All rules are hp-invertible;*
(iv) *The* (wk) *and* $\{(ctr_R), (ctr_{F_l}), (ctr_{F_r})\}$ *rules (below) are hp-admissible;*

$$\frac{\mathcal{R}, \Gamma \Rightarrow \Delta}{\mathcal{R}', \mathcal{R}, \Gamma', \Gamma \Rightarrow \Delta', \Delta} \ (wk) \qquad \frac{\mathcal{R}, \mathcal{R}', \mathcal{R}', \Gamma \Rightarrow \Delta}{\mathcal{R}, \mathcal{R}', \Gamma \Rightarrow \Delta} \ (ctr_R)$$

$$\frac{\mathcal{R}, \Gamma', \Gamma', \Gamma \Rightarrow \Delta}{\mathcal{R}, \Gamma', \Gamma \Rightarrow \Delta} \ (ctr_{F_l}) \qquad \frac{\mathcal{R}, \Gamma \Rightarrow \Delta, \Delta', \Delta'}{\mathcal{R}, \Gamma \Rightarrow \Delta, \Delta'} \ (ctr_{F_r})$$

(v) *The* (cut) *rule (below) is admissible;*

$$\frac{\mathcal{R}, \Gamma \Rightarrow \Delta, w : A \qquad \mathcal{R}, w : A, \Gamma \Rightarrow \Delta}{\mathcal{R}, \Gamma \Rightarrow \Delta} \ (cut)$$

(vi) G3Int (G3IntQC) *is sound and complete for* Int (IntQC, *resp.*).

Proof. We refer the reader to [7] for proofs of properties (i)–(vi) for G3Int; note that hp-admissibility of (psub) is trivial in the propositional setting since formulae do not contain parameters. The proofs of properties (i)–(vi) can be found in the online appended version [16] for G3IntQC. $\qquad \square$

2.2   The Nested Calculi NInt and NIntQC

We define a propositional (or, first-order) nested sequent Σ to be a syntactic object defined via the following BNF grammars:

$$X ::= A \mid X, X \qquad \Sigma ::= X \to X \mid X \to X, [\Sigma], \ldots, [\Sigma]$$

where A is in the propositional language $\mathcal{L}$ (first-order language $\mathcal{L}_Q$, resp.). As in the previous section, we take the comma operator to be commutative and associative, allowing us to view (for example) syntactic entities X as multisets.

In the first-order setting, we syntactically distinguish between bound variables and free variables in first-order formulae, using $\{x, y, z, \ldots\}$ for bound variables and replacing the occurrence of free variables in formulae with parameters $\{a, b, c, \ldots\}$. For example, we would use $p(a) \to p(b), [\bot \to \forall x q(x, b)]$ instead of the sequent $p(x) \to p(y), [\bot \to \forall x q(x, y)]$ in a nested derivation (where the free variable x has been replaced by the parameter a and y has been replaced by b).

Nested sequents are often written as $\Sigma\{X \to Y, [\Sigma_0], \ldots, [\Sigma_n]\}$, which indicates that $X \to Y, [\Sigma_0], \ldots, [\Sigma_n]$ occurs at some depth in the nestings of the sequent Σ; e.g. if Σ is taken to be $p(a) \to [\bot \to \forall x q(x, b), [\to \top]]$, then both $\Sigma\{\bot \to \forall x q(x, b)\}$ and $\Sigma\{\to \top\}$ are correct representations of Σ in our notation. The nested calculi are given in Fig. 2.

$$\frac{}{\Sigma\{X, p \to p, Y\}} \, (id) \qquad \frac{\Sigma\{X, A, B \to Y\}}{\Sigma\{X, A \wedge B \to Y\}} \, (\wedge_l) \qquad \frac{\Sigma\{X \to A, B, Y\}}{\Sigma\{X \to A \vee B, Y\}} \, (\vee_r)$$

$$\frac{\Sigma\{X, A \to Y\} \qquad \Sigma\{X, B \to Y\}}{\Sigma\{X, A \vee B \to Y\}} \, (\vee_l) \qquad \frac{\Sigma\{X \to A, Y\} \qquad \Sigma\{X \to B, Y\}}{\Sigma\{X \to A \wedge B, Y\}} \, (\wedge_r)$$

$$\frac{\Sigma\{X \to Y, [A \to]\}}{\Sigma\{X \to Y, \neg A\}} \, (\neg_r) \qquad \frac{\Sigma\{X \to A, Y\}}{\Sigma\{X, \neg A \to Y\}} \, (\neg_l) \qquad \frac{\Sigma\{X \to Y, [X', A \to Y']\}}{\Sigma\{X, A \to Y, [X' \to Y']\}} \, (lift)$$

$$\frac{\Sigma\{X \to Y, [A \to B]\}}{\Sigma\{X \to A \supset B, Y\}} \, (\supset_r) \qquad \frac{\Sigma\{X \to A, Y\} \qquad \Sigma\{X, B \to Y\}}{\Sigma\{X, A \supset B \to Y\}} \, (\supset_l)$$

$$\frac{}{\Sigma\{X, p(\vec{a}) \to p(\vec{a}), Y\}} \, (id_q) \qquad \frac{\Sigma\{X \to Y, A[a/x]\}}{\Sigma\{X \to Y, \exists x A\}} \, (\exists_r)$$

$$\frac{\Sigma\{X \to Y, A[a/x]\}}{\Sigma\{X \to Y, \forall x A\}} \, (\forall_r)^\dagger \qquad \frac{\Sigma\{X, A[a/x] \to Y\}}{\Sigma\{X, \forall x A \to Y\}} \, (\forall_l) \qquad \frac{\Sigma\{X, A[a/x] \to Y\}}{\Sigma\{X, \exists x A \to Y\}} \, (\exists_l)^\dagger$$

Fig. 2. Fitting's nested calculus NInt for propositional intuitionistic logic consists of (id), $(\wedge_l)$, $(\vee_r)$, $(\vee_l)$, $(\wedge_r)$, $(\neg_r)$, $(\neg_l)$, $(lift)$, $(\supset_r)$, and $(\supset_l)$. All rules taken together give the nested calculus NIntQC [10]. The side condition $\dagger$ states that a does not occur in the conclusion.

Theorem 2 (Soundness and Completeness [10]). *The calculus* NInt *(*NIntQC*) is sound and complete for* Int *(*IntQC*, resp.).*

3   Translating Notation: Labelled and Nested

It is instructive to observe that both nested and labelled sequents can be viewed as graphs (with the former restricted to trees and the latter more general).

Graphs of sequents are significant for two reasons: the first (technical) reason is that graphs can be leveraged to switch from labelled to nested notation; thus, graphs will play a role in deriving our nested calculi from our labelled calculi. The second reason is that graphs offer insight into *why* structural rule elimination yields nested systems, which will be discussed in the next section.

It is straightforward to define the graph of each type of sequent. To do this, we first introduce a bit of notation and define the multiset $\Gamma \restriction w := \{A \mid w : A \in \Gamma\}$. For a labelled sequent $\Lambda = \mathcal{R}, \Gamma \Rightarrow \Delta$, the *graph* $G(\Lambda)$ is the tuple (V, E, λ), where (i) $V = \{w \mid w \text{ is a label in } \Lambda.\}$, (ii) $(w, v) \in E$ iff $w \leq v \in \mathcal{R}$, and

$$(iii) \quad \lambda = \{(w, \Gamma' \Rightarrow \Delta') \mid \Gamma' = \Gamma \restriction w, \Delta' = \Delta \restriction w, \text{ and } w \in V\}.$$

For a nested sequent, the graph is defined inductively on the structure of the nestings; we use strings σ of natural numbers to denote vertices in the graph, similar to the prefixes used in prefixed tableaux [8–10].

Base Case. Let our nested sequent be of the form $X \to Y$ with X and Y multisets of formulae. Then, $G_\sigma(X \to Y) := (V_\sigma, E_\sigma, \lambda_\sigma)$, where (i) $V_\sigma := \{\sigma\}$, (ii) $E_\sigma := \emptyset$, and (iii) $\lambda_\sigma := \{(\sigma, X \to Y)\}$.

Inductive Step. Suppose our nested sequent is of the form $X \to Y, [\Sigma_0], \ldots, [\Sigma_n]$. We assume that each $G_{\sigma i}(\Sigma_i) = (V_{\sigma i}, E_{\sigma i}, \lambda_{\sigma i})$ (with $i \in \{0, \ldots, n\}$) is already defined, and define $G_\sigma(X \to Y, [\Sigma_0], \ldots, [\Sigma_n]) := (V_\sigma, E_\sigma, \lambda_\sigma)$ as follows:

$$(i) \quad V_\sigma := \{\sigma\} \cup \bigcup_{0 \leq i \leq n} V_{\sigma i} \qquad (ii) \quad E_\sigma := \{(\sigma, \sigma i) \mid 0 \leq i \leq n\} \cup \bigcup_{0 \leq i \leq n} E_{\sigma i}$$

$$(iii) \quad \lambda_\sigma := \{(\sigma, X \to Y)\} \cup \bigcup_{0 \leq i \leq n} \lambda_{\sigma i}$$

Definition 1. *Let $G_0 = (V_0, E_0, \lambda_0)$ and $G_1 = (V_1, E_1, \lambda_1)$ be two graphs. We define an* isomorphism $f : V_0 \mapsto V_1$ *between G_0 and G_1 to be a function such that: (i) f is bijective, (ii) $(x, y) \in E_0$ iff $(fx, fy) \in E_1$, (iii) $\lambda_0(x) = \lambda_1(fx)$. We say G_0 and G_1 are* isomorphic *iff there exists an isomorphism between them.*

Although the formal definitions above may appear somewhat cumbersome, the example below shows that transforming a sequent into its graph—or conversely, obtaining the sequent from its graph—is relatively straightforward.

Example 1. The nested sequent Σ is given below with its corresponding graph $G_0(\Sigma)$ shown on the left, and the labelled sequent Λ is given below with its corresponding graph $G(\Lambda)$ on the right. Regarding the labelled sequent, we assume that Γ_i and Δ_i consist solely of formulae labelled with w_i (for $i \in \{0, 1, 2, 3\}$).

$$\Sigma = X_0 \to Y_0, [X_1 \to Y_1, [X_2 \to Y_2]], [X_3 \to Y_3]$$

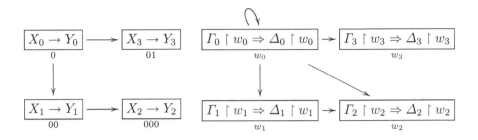

$$\Lambda = w_0 \leq w_0, w_0 \leq w_1, w_1 \leq w_2, w_0 \leq w_2, w_0 \leq w_3, \Gamma_0, \Gamma_1, \Gamma_2, \Gamma_3 \Rightarrow \Delta_0, \Delta_1, \Delta_2, \Delta_3$$

In the above example there is a loop from w_0 to itself in the graph of the labelled sequent; furthermore, there is an undirected cycle occurring between w_0, w_1, and w_2. As will be explained in the next section (specifically, Theorem 4), the (ref) and (tra) rules allow for such structures to appear in labelled derivations of theorems; however, the elimination of these rules in the labelled calculus has the effect that such structures *can no longer* occur in the labelled derivation of a theorem. Consequently, it will be seen that eliminating such rules yields a labelled derivation where every sequent has a purely *treelike* structure (see Definition 2). This implies that each labelled sequent in the derivation has a graph isomorphic to the graph of a nested sequent. It is this idea which ultimately permits the extraction of our nested calculi from our labelled calculi.

Definition 2. *Let Λ be a labelled sequent and $G(\Lambda) = (V, E, \lambda)$. We say that Λ is* treelike *iff there exists a unique vertex $w \in V$, called the* root, *such that there exists a unique path from w to every other vertex $v \in V$.*[1]

If we take the graph of a treelike labelled sequent, then it can be viewed as the graph of a nested sequent, as the example below demonstrates.

Example 2. The treelike labelled sequent Λ' and its graph are given below. We assume that Γ_i and Δ_i contain only formulae labelled with w_i (for $i \in \{0, 1, 2, 3\}$).

$$\Lambda' = w_0 \leq w_1, w_1 \leq w_2, w_0 \leq w_3, \Gamma_0, \Gamma_1, \Gamma_2, \Gamma_3 \Rightarrow \Delta_0, \Delta_1, \Delta_2, \Delta_3$$

$$\boxed{\Gamma_2' \Rightarrow \Delta_2'}_{w_2} \longleftarrow \boxed{\Gamma_1' \Rightarrow \Delta_1'}_{w_1} \longleftarrow \boxed{\Gamma_0' \Rightarrow \Delta_0'}_{w_0} \longrightarrow \boxed{\Gamma_3' \Rightarrow \Delta_3'}_{w_3}$$

Also, we assume $\Gamma_i' = \Gamma_i \upharpoonright w_i = X_i$ and $\Delta_i' = \Delta_i \upharpoonright w_i = Y_i$ (for $i \in \{0, 1, 2, 3\}$). Therefore, the above graph is isomorphic to the graph of the nested sequent in Example 1, meaning that Λ can be translated as that nested sequent.

Definition 3 (The Translation $\mathfrak{N}$). *Let Λ be a treelike labelled sequent. We define $\mathfrak{N}(\Lambda)$ to be the nested sequent obtained from the graph $G(\Lambda)$.*

---

[1] Treelike sequents are equivalently characterized as sequents with graphs that are: (i) connected, (ii) acyclic, and (iii) contain no backwards branching.

4   Deriving NInt from G3Int

We begin by presenting two useful lemmata that will be referenced in the current and next section while deriving NInt from G3Int and NIntQC from G3IntQC. All rules mentioned in the lemmata can be found in Fig. 3 below. The proofs of both lemmata can be found in the online appended version [16].

Lemma 1. *The calculus* G3Int $+ \{(id^*), (\neg_l), (\neg_r), (\supset_l^*), (lift)\}$ *and the calculus* G3IntQC $+ \{(id_q^*), (\neg_l), (\neg_r), (\supset_l^*), (\forall_l^*), (\forall_r^*), (\exists_r^*), (lift)\}$ *have the following properties:*

(i) *All sequents of the form* $\mathcal{R}, w \leq v, \vec{a} \in D_w, w : A(\vec{a}), \Gamma \Rightarrow v : A(\vec{a}), \Delta$ *and* $\mathcal{R}, \vec{a} \in D_w, w : A(\vec{a}), \Gamma \Rightarrow \Delta, w : A(\vec{a})$ *are derivable;[2]*
(ii) *The rules* $\{(lsub), (psub), (wk), (ctr_R), (ctr_{F_r})\}$ *are hp-admissible;*
(iii) *With the exception of* $\{(\wedge_l), (\exists_l)\}$, *all rules are hp-invertible;*
(iv) *The rules* $\{(\wedge_l), (\exists_l)\}$ *are invertible;*
(v) *The rule* (ctr_{F_l}) *is admissible.*

Lemma 2. *(i)* (ref) *and* (tra) *can be permuted above each rule in the set* $\{(\perp_l), (\wedge_l), (\wedge_r), (\vee_l), (\vee_r), (\supset_r), (\neg_l), (\neg_r), (\exists_l), (\exists_r), (\forall_r)\}$. *(ii)* (nd) *and* (cd) *can be permuted above* $\{(\perp_l), (\wedge_l), (\wedge_r), (\vee_l), (\vee_r), (\supset_l), (\supset_r), (\neg_l), (\neg_r), (\exists_l), (\forall_r)\}$.

Proof. Claim (i) follows from the fact that none of the rules mentioned have active relational atoms of the form $w \leq u$ in the conclusion, and so, (ref) and (tra) may be freely permuted above each rule. Claim (ii) follows from the fact that none of the rules mentioned contain active domain atoms in the conclusion, allowing for (nd) and (cd) to be permuted above each rule.   □

Deriving the calculus NInt from G3Int depends on a crucial observation made in [5] concerning labelled derivations: *rules such as* (ref) *and* (tra) *allow for theorems to be derived in proofs containing non-treelike labelled sequents.* To demonstrate this fact, observe the following derivation in G3Int:

$$\frac{\dfrac{w \leq v, v \leq v, v : p \Rightarrow v : p}{w \leq v, v : p \Rightarrow v : p} \, (ref)}{\Rightarrow w : p \supset p} \, (\supset_r)$$

The initial sequent is non-treelike due to the presence of the $v \leq v$ relational atom; however, the application of (ref) deletes this structure from the initial sequent and produces a treelike sequent as the conclusion.

In fact, it is true in general that every labelled derivation of a theorem (i.e., a derivation whose end sequent is of the form $\Rightarrow w : A$) can be partitioned into a top derivation consisting of non-treelike sequents, and a bottom derivation consisting of treelike sequents. Note that if a derivation ends with a sequent of the form $\Rightarrow w : A$, then the derivation must necessarily contain a bottom treelike

---

[2] In the propositional setting, these sequents become $\mathcal{R}, w \leq v, w : A, \Gamma \Rightarrow v : A, \Delta$ and $\mathcal{R}, w : A, \Gamma \Rightarrow \Delta, w : A$, respectively.

$$\frac{}{\mathcal{R}, a_0 \in D_{v_0}, \ldots, a_n \in D_{v_n}, w : p(\vec{a}), \Gamma \Rightarrow w : p(\vec{a}), \Delta} \ (id_q^*)^{\dagger_1} \qquad \frac{}{\mathcal{R}, w : p, \Gamma \Rightarrow \Delta, w : p} \ (id^*)$$

$$\frac{\mathcal{R}, w \leq v, v : A, \Gamma \Rightarrow \Delta}{\mathcal{R}, \Gamma \Rightarrow \Delta, w : \neg A} \ (\neg_r)^{\dagger_2} \qquad \frac{\mathcal{R}, w : \neg A, \Gamma \Rightarrow \Delta, w : A}{\mathcal{R}, w : \neg A, \Gamma \Rightarrow \Delta} \ (\neg_l)$$

$$\frac{\mathcal{R}, a \in D_v, w : A[a/x], w : \forall x A, \Gamma \Rightarrow \Delta}{\mathcal{R}, a \in D_v, w : \forall x A, \Gamma \Rightarrow \Delta} \ (\forall_l^*)^{\dagger_3} \qquad \frac{\mathcal{R}, a \in D_w, \Gamma \Rightarrow w : A[a/x], \Delta}{\mathcal{R}, \Gamma \Rightarrow w : \forall x A, \Delta} \ (\forall_r^*)^{\dagger_4}$$

$$\frac{\mathcal{R}, w \leq u, w : A, u : A, \Gamma \Rightarrow \Delta}{\mathcal{R}, w \leq u, w : A, \Gamma \Rightarrow \Delta} \ (lift) \qquad \frac{\mathcal{R}, a \in D_v, \Gamma \Rightarrow \Delta, w : A[a/x], w : \exists x A}{\mathcal{R}, a \in D_v, \Gamma \Rightarrow \Delta, w : \exists x A} \ (\exists_r^*)^{\dagger_3}$$

$$\frac{\mathcal{R}, w : A \supset B, \Gamma \Rightarrow \Delta, w : A \qquad \mathcal{R}, w : A \supset B, w : B, \Gamma \Rightarrow \Delta}{\mathcal{R}, w : A \supset B, \Gamma \Rightarrow \Delta} \ (\supset_l^*)$$

Fig. 3. Rules used to derive NInt and NIntQC from G3Int and G3IntQC, respectively. The side condition $\dagger_1$ states that there exists a path of relational atoms (not necessarily directed) from v_i to w for each $i \in \{0, \ldots, n\}$ in $\mathcal{R}$; $\dagger_2$ states that v does not occur in the conclusion; $\dagger_3$ stipulates that there exists a path of relational atoms (not necessarily directed) from v to w occurring in $\mathcal{R}$; $\dagger_4$ states that a does not occur in the conclusion. (Let $u \sim v \in \{u \leq v, v \leq u\}$. A path of relational atoms (not necessarily directed) from a label w to v occurs in a sequent Λ if and only if $w = v$, $w \sim v$, or there exist labels z_i ($i \in \{0, \ldots, n\}$) such that $w \sim z_0, \ldots, z_n \sim v$ occurs in Λ).

fragment since $G(\Rightarrow w : A)$ is a tree. By contrast, the top non-treelike fragment of the derivation may be empty (e.g. the derivation of $\Rightarrow w : \bot \supset A$).

To demonstrate why the aforementioned partition always exists, suppose you are given a labelled derivation of a theorem $w : A$ and consider the derivation in a bottom-up manner. The graph of the end sequent $\Rightarrow w : A$ is evidently treelike by Definition 2, and observe the each bottom-up application of a rule in G3Int—with the exception of (ref) and (tra)—will produce a treelike sequent (see Theorem 4 for auxiliary details). If, however, at some point in the derivation (ref) or (tra) is applied, then all sequents above the inference will inherit the (un)directed cycle produced by the rule, thus producing the non-treelike fragment of the proof.

One can therefore imagine that permuting instances of the (ref) and (tra) rules upwards in a given derivation would potentially increase the bottom treelike fragment of the derivation and decrease the top non-treelike fragment. As it so happens, this intuition is correct so long as we choose *adequate* rules—that bottom-up preserve the treelike structure of sequents—to replace certain instances of the (ref) and (tra) rules in a derivation, when necessary. We will first examine permuting instances of the (ref) rule, and motivate which adequate rules we ought to add to our calculus in order to achieve the complete elimination of (ref). After, we will turn our attention towards eliminating the (tra) rule, and conclude the section by leveraging our results to extract NInt.

Let us first observe an application of (ref) to an initial sequent obtained via the (id) rule. There are two possible cases to consider: either the relational atom principal in the initial sequent is active in the (ref) inference (shown below left), or it is not (shown below right):

$$\cfrac{\cfrac{}{\mathcal{R}, w \leq w, w : p, \Gamma \Rightarrow \Delta, w : p} \; (id)}{\mathcal{R}, w : p, \Gamma \Rightarrow \Delta, w : p} \; (ref) \qquad \cfrac{\cfrac{}{\mathcal{R}, u \leq u, w \leq v, w : p, \Gamma \Rightarrow \Delta, v : p} \; (id)}{\mathcal{R}, w \leq v, w : p, \Gamma \Rightarrow \Delta, v : p} \; (ref)$$

In the case shown above right, the end sequent is an instance of the (id) rule, regardless of if $u = w$, $u = v$, or u is distinct from w and v. The case shown above left, however, indicates that we ought to add the (id^*) rule (see Fig. 3) to our calculus if we aim to eliminate (ref) from any given derivation.

Additionally, notice that an application of (ref) to (id^*) or $(\bot_l)$ produces another instance of (id^*) or $(\bot_l)$, regardless of if $v = w$ or $v \neq w$.

$$\cfrac{\cfrac{}{\mathcal{R}, v \leq v, w : p, \Gamma \Rightarrow \Delta, w : p} \; (id^*)}{\mathcal{R}, w : p, \Gamma \Rightarrow \Delta, w : p} \; (ref) \qquad \cfrac{\cfrac{}{\mathcal{R}, v \leq v, w : \bot, \Gamma \Rightarrow \Delta} \; (\bot_l)}{\mathcal{R}, w : \bot, \Gamma \Rightarrow \Delta} \; (ref)$$

The above facts, coupled with Lemma 2, imply that any application of (ref) to an initial sequent, produces an initial sequent.

Concerning the remaining rules of G3Int, we need only investigate the permutation of (ref) above the $(\supset_l)$ rule, if we rely on Lemma 2. There are two cases: either the relational atom principal in the $(\supset_l)$ inference is active in the (ref) inference, or it is not. The latter case is easily resolved, so we observe the former:

$$\cfrac{\cfrac{\mathcal{R}, w \leq w, w : A \supset B, \Gamma \Rightarrow \Delta, w : A \qquad \mathcal{R}, w \leq w, w : A \supset B, w : B, \Gamma \Rightarrow \Delta}{\mathcal{R}, w \leq w, w : A \supset B, \Gamma \Rightarrow \Delta} \; (\supset_l)}{\mathcal{R}, w : A \supset B, \Gamma \Rightarrow \Delta} \; (ref)$$

Applying (ref) to each premise of the $(\supset_l)$ inference yields the following:

$$\cfrac{\mathcal{R}, w \leq w, w : A \supset B, \Gamma \Rightarrow \Delta, w : A}{\mathcal{R}, w : A \supset B, \Gamma \Rightarrow \Delta, w : A} \; (ref) \qquad \cfrac{\mathcal{R}, w \leq w, w : A \supset B, w : B, \Gamma \Rightarrow \Delta}{\mathcal{R}, w : A \supset B, w : B, \Gamma \Rightarrow \Delta} \; (ref)$$

The above observation suggests that we ought to add the $(\supset_l^*)$ rule (see Fig. 3) to our calculus if we wish to permute (ref) above the $(\supset_l)$ rule; a single application of the $(\supset_l^*)$ rule to the end sequents above gives the desired conclusion.

With the $(\supset_l^*)$ rule added to our calculus, we may freely permute the (ref) rule above any $(\supset_l)$ inference. Still, we must confirm that the (ref) rule is permutable with the newly introduced $(\supset_l^*)$ rule, but this is easily verifiable.

On the basis of our investigation, we may conclude the following lemma:

Lemma 3. *The (ref) rule is eliminable in* G3Int $+ \{(id^*), (\supset_l^*)\} - (tra)$.

Let us turn our attention towards eliminating the (tra) rule from a labelled derivation. Since our aim is to eliminate *both* (ref) and (tra) from any derivation, we assume that the rules $\{(id^*), (\supset_l^*)\}$ have been added to our calculus.

It is rather simple to verify that (tra) permutes with $(\perp_l)$ and (id^*), so we only consider the (id) case. As with the (ref) rule, there are two cases to consider when permuting (tra) above an (id) inference: either the active formula of (tra) is principal in (id), or it is not. In the latter case, the result of the (tra) rule is an initial sequent, implying that the (tra) rule may be eliminated from the derivation. The former case proves trickier and is explicitly given below:

$$\frac{\overline{\mathcal{R}, w \leq u, u \leq v, w \leq v, w : p, \Gamma \Rightarrow \Delta, v : p} \ (id)}{\mathcal{R}, w \leq u, u \leq v, w : p, \Gamma \Rightarrow \Delta, v : p} \ (tra)$$

Observe that the end sequent is not an initial sequent as it is not obtainable from an (id), (id^*), or $(\perp_l)$ rule. The issue is solved by considering the $(lift)$ rule (see Fig. 3), which allows us to obtain the desired end sequent without the use of (tra), as the following derivation demonstrates:

$$\frac{\dfrac{\overline{\mathcal{R}, w \leq u, u \leq v, w : p, u : p, v : p, \Gamma \Rightarrow v : p, \Delta} \ (id^*)}{\mathcal{R}, w \leq u, u \leq v, w : p, u : p, \Gamma \Rightarrow v : p, \Delta} \ (lift)}{\mathcal{R}, w \leq u, u \leq v, w : p, \Gamma \Rightarrow v : p, \Delta} \ (lift)$$

Thus, the addition of $(lift)$ to our calculus resolves the issue of permuting (tra) above any initial sequent. Nevertheless, by Lemma 2, we still need to consider the permutation of (tra) above the $(\supset_l)$, $(\supset_l^*)$, and $(lift)$ rules. The (tra) rule and $(\supset_l^*)$ are freely permutable due to the fact that (tra) solely affects relational atoms, and $(\supset_l^*)$ solely affects labelled formulae. Also, the following lemma entails that we may omit analyzing the permutation of (tra) above the $(\supset_l)$ rule.

Lemma 4. *The rule* $(\supset_l)$ *is admissible in* G3Int $+ \{(id^*), (\supset_l^*), (lift)\}$.

Proof. We derive the rule as shown below:

$$\frac{\dfrac{\dfrac{\mathcal{R}, x \leq y, x : A \supset B, \Gamma \Rightarrow \Delta, y : A}{\mathcal{R}, x \leq y, x : A \supset B, y : A \supset B, \Gamma \Rightarrow \Delta, y : A} \ \text{Lem. 1} \quad \dfrac{\mathcal{R}, x \leq y, x : A \supset B, y : B, \Gamma \Rightarrow \Delta}{\mathcal{R}, x \leq y, x : A \supset B, y : A \supset B, y : B, \Gamma \Rightarrow \Delta} \ \text{Lem. 1}}{\mathcal{R}, x \leq y, x : A \supset B, y : A \supset B, \Gamma \Rightarrow \Delta} \ (\supset_l^*)}{\mathcal{R}, x \leq y, x : A \supset B, \Gamma \Rightarrow \Delta} \ (lift)$$

$\square$

Last, the (tra) rule is permutable with the $(lift)$ rule. In the case where the principal relational atom of $(lift)$ is *not* active in the ensuing (tra) application, the two rules freely permute. The alternative case is resolved as shown below:

$$\frac{\dfrac{\mathcal{R}, w \leq u, u \leq v, w \leq v, w : A, v : A, \Gamma \Rightarrow \Delta}{\mathcal{R}, w \leq u, u \leq v, w \leq v, w : A, \Gamma \Rightarrow \Delta} \ (lift)}{\mathcal{R}, w \leq u, u \leq v, w : A, \Gamma \Rightarrow \Delta} \ (tra)$$

$$\frac{\dfrac{\dfrac{\mathcal{R}, w \leq u, u \leq v, w \leq v, w : A, v : A, \Gamma \Rightarrow \Delta}{\mathcal{R}, w \leq u, u \leq v, w : A, v : A, \Gamma \Rightarrow \Delta} \ (tra)}{\mathcal{R}, w \leq u, u \leq v, w : A, u : A, v : A, \Gamma \Rightarrow \Delta} \ \text{Lem. 1}}{\dfrac{\mathcal{R}, w \leq u, u \leq v, w : A, u : A, \Gamma \Rightarrow \Delta}{\mathcal{R}, w \leq u, u \leq v, w : A, \Gamma \Rightarrow \Delta} \ (lift)} \ (lift)$$

Hence, we obtain the following:

Lemma 5. *The* (tra) *rule is eliminable in* G3Int $+ \{(id^*), (\supset_l^*), (lift)\} - (ref)$.

Enough groundwork has been laid to state our main lemma:

Lemma 6. *The* (ref) *and* (tra) *rules are admissible in the calculus* G3Int $+$ $\{(id^*), (\supset_l^*), (lift)\}$.

Proof. Suppose we are given a proof Π in G3Int $+$ $\{(id^*), (\supset_l^*), (lift)\}$, and consider the topmost occurrence of either (ref) or (tra). If we can show that the (ref) rule permutes above the $(lift)$ rule, then we can invoke Lemmas 3 and 5 to conclude that each topmost occurrence of (ref) and (tra) can be eliminated from Π in succession. This yields a (ref) and (tra) free proof of the end sequent and establishes the claim. Thus, we prove that the (ref) rule permutes above the $(lift)$ rule.

In the case where the relational atom active in (ref) is not principal in the $(lift)$ inference, the two rules may be permuted; the alternative case is resolved as shown below:

$$\frac{\dfrac{\mathcal{R}, w \leq w, w : A, w : A, \Gamma \Rightarrow \Delta}{\mathcal{R}, w \leq w, w : A, \Gamma \Rightarrow \Delta} \ (lift)}{\mathcal{R}, w : A, \Gamma \Rightarrow \Delta} \ (ref) \qquad \frac{\mathcal{R}, w : A, w : A, \Gamma \Rightarrow \Delta}{\mathcal{R}, w : A, \Gamma \Rightarrow \Delta} \ \begin{array}{l} \text{IH} \\ \text{Lem. 1-(v)} \end{array}$$

$\square$

The addition of the rules $\{(id^*), (\supset_l^*), (lift)\}$ to our calculus and the above admissibility results demonstrate that we are readily advancing towards our goal of deriving NInt. Howbeit, our labelled calculus is still distinct since it makes use of the logical signature $\{\bot, \wedge, \vee, \supset\}$, whereas NInt uses the signature $\{\neg, \wedge, \vee, \supset\}$. Therefore, we need to show that $(\bot_l)$ (we define $\bot := p \wedge \neg p$) is admissible in the presence of (labelled versions of) the $(\neg_r)$ and $(\neg_l)$ rules (see Fig. 3). This admissibility result is explained in the main theorem below.

Theorem 3. *The rules* $\{(id), (\bot_l), (\supset_l), (ref), (tra)\}$ *are admissible in the calculus* G3Int $+$ $\{(id^*), (\neg_l), (\neg_r), (\supset_l^*), (lift)\}$.

Proof. Follows from Lemmas 2, and 6, the fact that (id) is derivable using (id^*) and $(lift)$, the fact that $(\bot_l)$ is derivable from $(\neg_l)$ and $(\wedge_l)$, and admissibility of $(\supset_l)$ is shown as in Lemma 4. $\square$

Theorem 4. *Every derivation* Π *in* G3Int $+$ $\{(id^*), (\neg_l), (\neg_r), (\supset_l^*), (lift)\}$ $-$ $\{(id), (\bot_l), (\supset_l), (ref), (tra)\}$ *of a labelled formula* $w : A$ *contains solely treelike sequents with* w *the root of each sequent in the derivation.*

Proof. To prove the claim we have to show that the graph of every sequent is (i) connected, (ii) free of directed cycles, and (iii) free of backwards branching. (NB. properties (i)–(iii) are equivalent to being treelike.) We assume that we are given a derivation Π with end sequent $\Rightarrow w : A$ and argue that every sequent in Π has properties (i)–(iii):

(i) Assume there exists a sequent Λ in Π whose graph $G(\Lambda)$ is disconnected. Then, there exist at least two distinct regions in $G(\Lambda)$ such that there does not exist an edge from a node v in one region to a node u in the other region. In other words, Λ does not contain a relational atom $v \leq u$ for some

v in one region and some u in the other region. If one observes the rules of our calculus, they will find that all rules either preserve the relational atoms $\mathcal{R}$ of a sequent or decrease it by one relation atom (as in the case of the $(\neg_r)$ and $(\supset_r)$ rules). Hence, the end sequent $\Rightarrow w : A$ will have a disconnected graph since the property will be preserved downwards, but this is a contradiction.

(ii) Assume that a sequent occurs in Π containing a relational cycle $u \leq v_1, \ldots, v_n \leq u$ (for $n \in \mathbb{N}$). Observe that the $(\neg_r)$ and $(\supset_r)$ rules are never applicable to any of the relational atoms in the cycle, since no label occurring in a relational cycle is an eigenvariable. This implies that the relational cycle will be preserved downwards into the end sequent $\Rightarrow w : A$ due to the fact that the $(\neg_r)$ and $(\supset_r)$ rules are the only rules that delete relational atoms, giving a contradiction.

(iii) Assume that a sequent occurs in Π with backwards branching it is graph, i.e. it contains relational atoms of the form $v \leq z, u \leq z$. By reasoning similar to case (ii), we obtain a contradiction.

To see that w is the root of each treelike sequent in Π, observe that applying inference rules from the calculus bottom-up to $\Rightarrow w : A$ either preserve relational structure, or add forward relational structure (e.g. $(\supset_r)$ and $(\neg_r)$), thus constructing a tree emanating from w. $\quad\square$

We refer to the labelled calculus G3Int $+ \{(id^*), (\neg_l), (\neg_r), (\supset_l^*), (lift)\} - \{(id), (\perp_l), (\supset_l), (ref), (tra)\}$ (restricted to the use of treelike sequents) under the $\mathfrak{N}$ translation as NInt*. Up to copies of principal formulae in the premise(s) of some rules, NInt* is identical to the calculus NInt (cf. [16]).

5   Deriving **NIntQC** from **G3IntQC**

To simplify notation, G3IntQC $+ \{(id_q^*), (\neg_l), (\neg_r), (\supset_l^*), (\forall_r^*), (\forall_l^*), (\exists_r^*), (lift)\}$ will be referred to as IntQCL. We begin the section by showing two lemmata that confirm the admissibility of structural rules in IntQCL, and permit the extraction of NIntQC from G3IntQC. After, we list a number of significant proof-theoretic properties inherited by our nested calculi through the extraction process.

Lemma 7. *The (ref) and (tra) rules are admissible in the calculus* IntQCL.

Proof. We consider the topmost occurrence of a (ref) or (tra) inference and eliminate each topmost occurrence in succession until we obtain a proof free of (ref) and (tra) inferences. By Lemma 2, we need only show that (ref) and (tra) permute above rules $\{(id_q), (id_q^*), (\supset_l), (\supset_l^*), (\forall_l), (\forall_l^*), (\forall_r^*), (\exists_r^*), (lift), (nd), (cd)\}$. The cases of permuting (ref) and (tra) above $(\supset_l^*)$ and $(lift)$ are similar to Lemmas 3, 5, and Theorem 3. Also, $(\supset_l)$ is admissible in the presence of $(\supset_l^*)$ (similar to Lemma 4), so the case may be omitted. Permuting (ref) and (tra) above (id_q^*) and $(\forall_r^*)$ is straightforward, so we exclude the cases. Hence, we focus only on the nontrivial cases involving the (id_q), $(\forall_l)$, $(\forall_l^*)$, $(\exists_r^*)$, (nd), and (cd) rules. We prove the elimination of (ref) and refer to reader to the online appended version [16] for the proof of (tra) elimination.

$$\dfrac{\overline{\mathcal{R}, w \leq w, \vec{a} \in D_w, w : p(\vec{a}), \Gamma \Rightarrow w : p(\vec{a}), \Delta}\ (id_q)}{\mathcal{R}, \vec{a} \in D_w, w : p(\vec{a}), \Gamma \Rightarrow w : p(\vec{a}), \Delta}\ (ref) \qquad \overline{\mathcal{R}, \vec{a} \in D_w, w : p(\vec{a}), \Gamma \Rightarrow w : p(\vec{a}), \Delta}\ (id_q^*)$$

$$\dfrac{\dfrac{\mathcal{R}, w \leq w, a \in D_w, w : A[a/x], w : \forall x A, \Gamma \Rightarrow \Delta}{\mathcal{R}, w \leq w, a \in D_w, w : \forall x A, \Gamma \Rightarrow \Delta}\ (\forall_l)}{\mathcal{R}, a \in D_w, w : \forall x A, \Gamma \Rightarrow \Delta}\ (ref) \qquad \dfrac{\dfrac{\mathcal{R}, a \in D_w, w : A[a/x], w : \forall x A, \Gamma \Rightarrow \Delta}{}\ \text{IH}}{\mathcal{R}, a \in D_w, w : \forall x A, \Gamma \Rightarrow \Delta}\ (\forall_l^*)$$

$$\dfrac{\dfrac{\mathcal{R}, u \leq u, a \in D_v, w : A[a/x], w : \forall x A, \Gamma \Rightarrow \Delta}{\mathcal{R}, u \leq u, a \in D_v, w : \forall x A, \Gamma \Rightarrow \Delta}\ (\forall_l^*)}{\mathcal{R}, a \in D_v, w : \forall x A, \Gamma \Rightarrow \Delta}\ (ref) \qquad \dfrac{\dfrac{\mathcal{R}, a \in D_v, w : A[a/x], w : \forall x A, \Gamma \Rightarrow \Delta}{}\ \text{IH}}{\mathcal{R}, a \in D_v, w : \forall x A, \Gamma \Rightarrow \Delta}\ (\forall_l^*)$$

$$\dfrac{\dfrac{\mathcal{R}, u \leq u, a \in D_v, \Gamma \Rightarrow w : A[a/x], w : \exists x A, \Delta}{\mathcal{R}, u \leq u, a \in D_v, \Gamma \Rightarrow w : \exists x A, \Delta}\ (\exists_r^*)}{\mathcal{R}, a \in D_v, \Gamma \Rightarrow w : \exists x A, \Delta}\ (ref) \qquad \dfrac{\dfrac{\mathcal{R}, a \in D_v, \Gamma \Rightarrow w : A[a/x], w : \exists x A, \Delta}{}\ \text{IH}}{\mathcal{R}, a \in D_v, \Gamma \Rightarrow w : \exists x A, \Delta}\ (\exists_l^*)$$

$$\dfrac{\dfrac{\mathcal{R}, w \leq w, a \in D_w, a \in D_w, \Gamma \Rightarrow \Delta}{\mathcal{R}, w \leq w, a \in D_w, \Gamma \Rightarrow \Delta}\ (nd)}{\mathcal{R}, a \in D_w, \Gamma \Rightarrow \Delta}\ (ref) \qquad \dfrac{\dfrac{\mathcal{R}, w \leq w, a \in D_w, a \in D_w, \Gamma \Rightarrow \Delta}{\mathcal{R}, w \leq w, a \in D_w, \Gamma \Rightarrow \Delta}\ \text{Lem. 1-(iv)}}{\mathcal{R}, a \in D_w, \Gamma \Rightarrow \Delta}\ \text{IH}$$

$$\dfrac{\dfrac{\mathcal{R}, w \leq w, a \in D_w, a \in D_w, \Gamma \Rightarrow \Delta}{\mathcal{R}, w \leq w, a \in D_w, \Gamma \Rightarrow \Delta}\ (cd)}{\mathcal{R}, a \in D_w, \Gamma \Rightarrow \Delta}\ (ref) \qquad \dfrac{\dfrac{\mathcal{R}, w \leq w, a \in D_w, a \in D_w, \Gamma \Rightarrow \Delta}{\mathcal{R}, w \leq w, a \in D_w, \Gamma \Rightarrow \Delta}\ \text{Lem. 1-(iv)}}{\mathcal{R}, a \in D_w, \Gamma \Rightarrow \Delta}\ \text{IH}$$

In the $(\forall_l^*)$ and $(\exists_r^*)$ cases, observe that the side condition continues to hold after IH is applied. If the path from w to v does not go through u, then the side condition trivially holds, and if it does go through u, then there must exist relational atoms in $\mathcal{R}$ occurring along the path from w to v, which continue to be present after the evocation of IH. □

Lemma 8. *The rules* (nd) *and* (cd) *are admissible in the calculus* IntQCL − $\{(ref), (tra)\}$.

Proof. The result is shown by induction on the height of the given derivation by permuting all instances of (nd) and (cd) upwards until all such instances are removed from the derivation. See the online appended version [16] for details. □

Theorem 5. *The rules* $\{(id_q), (\bot_l), (\supset_l), (\forall_l), (\forall_r), (\exists_r), (ref), (tra), (nd), (cd)\}$ *are admissible in* IntQCL.

Proof. Admissibility of (id_q), $(\supset_l)$, and $(\bot_l)$ is shown similarly to Lemma 4 and Theorem 3. Also, the rule $(\exists_r)$ is an instance of $(\exists_r^*)$, and the admissibility of $(\forall_r)$ and $(\forall_l)$ are witnessed by the derivations below:

$$\dfrac{\dfrac{\dfrac{\mathcal{R}, w \leq v, a \in D_v, \Gamma \Rightarrow v : A[a/x], \Delta}{\mathcal{R}, w \leq w, a \in D_w, \Gamma \Rightarrow w : A[a/x], \Delta}\ \text{Lem. 1}}{\mathcal{R}, a \in D_w, \Gamma \Rightarrow w : A[a/x], \Delta}\ \text{Lem. 7}}{\mathcal{R}, \Gamma \Rightarrow w : \forall x A, \Delta}\ (\forall_r^*) \qquad \dfrac{\dfrac{\dfrac{\mathcal{R}, w \leq v, a \in D_v, v : A[a/x], w : \forall x A, \Gamma \Rightarrow \Delta}{\mathcal{R}, w \leq w, a \in D_w, w : A[a/x], w : \forall x A, \Gamma \Rightarrow \Delta}\ \text{Lem. 1}}{\mathcal{R}, a \in D_w, w : A[a/x], w : \forall x A, \Gamma \Rightarrow \Delta}\ \text{Lem. 7}}{\mathcal{R}, a \in D_w, w : \forall x A, \Gamma \Rightarrow \Delta}\ (\forall_l^*)$$

Hence, our result follows by Lemmas 7 and 8. □

Theorem 6. *Every derivation* Π *of a labelled formula* $w : A$ *in the calculus* IntQCL − $\{(id_q), (\bot_l), (\supset_l), (\forall_l), (\forall_r), (\exists_r), (ref), (tra), (nd), (cd)\}$ *contains solely treelike sequents with* w *the root of each sequent in the derivation.*

Proof. Similar to Theorem 4.   □

We refer to $\mathsf{IntQCL} - \{(id_q), (\supset_l), (\forall_l), (\forall_r), (\exists_r), (ref), (tra), (nd), (cd)\}$ (restricted to using treelike sequents) under the $\mathfrak{N}$ translation as NIntQC^*. It is crucial to point out that by the definition of $\mathfrak{N}$ (Definition 3) and the definition of the graph of a labelled sequent, domain atoms $a \in D_w$ are omitted when translating from labelled to nested (and the $\{(id_q^*), (\forall_l^*), (\exists_r^*)\}$ side conditions become unnecessary). Hence, up to copies of principal formulae in the premises of some rules, NIntQC^* is identical to NIntQC (cf. [16]). In fact, through additional work, one can eliminate such copies of principal formulae, begetting the complete extraction of NInt and NIntQC from NInt^* and NIntQC^* (and therefore, from $\mathsf{G3Int}$ and $\mathsf{G3IntQC}$).

An interesting consequence of our work is that NInt^* and NIntQC^* inherited favorable proof-theoretic properties as a consequence of their extraction. Such properties are listed in Corollary 1 below with admissible rules found in Fig. 4.

$$\frac{\Sigma\{X \to Y, [\Sigma'], [\Sigma']\}}{\Sigma\{X \to Y, [\Sigma']\}} \ (ctr_1) \qquad \frac{\Sigma\{X, X \to Y\}}{\Sigma\{X \to Y\}} \ (ctr_2) \qquad \frac{\Sigma\{X \to Y, Y\}}{\Sigma\{X \to Y\}} \ (ctr_3)$$

$$\frac{\Sigma\{X \to Y\}}{\Sigma\{X \to Y, [\Sigma']\}} \ (wk_1) \qquad \frac{\Sigma\{X \to Y\}}{\Sigma\{X \to Y, Z\}} \ (wk_2) \qquad \frac{\Sigma\{X \to Y\}}{\Sigma\{X, Z \to Y\}} \ (wk_3)$$

Fig. 4. Examples of admissible structural rules in NInt^* and NIntQC^*.

Corollary 1. *The calculi NInt^* and NIntQC^* have inherited: (i) cut-free completeness, (ii) invertibility of all rules, and (iii) admissibility of all rules in Fig. 4.*

Proof. All properties follow from Lemma 1 and Theorems 1, 3, 4, 5, and 6. The admissibility of (ctr_1) follows from the admissibility of the rules in the set $\{(lsub), (ctr_R), (ctr_{F_l}), (ctr_{F_r})\}$, the admissibility of (ctr_2) follows from the admissibility of (ctr_{F_l}), the admissibility of (ctr_3) follows from the admissibility of (ctr_{F_r}), and the admissibility of $\{(wk_1), (wk_2), (wk_3)\}$ follows from the admissibility of (wk) in the labelled variants of NInt^* and NIntQC^*.   □

6   Conclusion

In this paper, we showed how to extract Fitting's nested calculi (up to copies of principal formulae in premises) from the labelled calculi $\mathsf{G3Int}$ and $\mathsf{G3IntQC}$. The extraction is obtained via the elimination of structural rules and through the addition of special rules to $\mathsf{G3Int}$ and $\mathsf{G3IntQC}$, necessitating the use of only treelike sequents in proofs of theorems. Consequently, the extraction of the nested calculi from the labelled calculi demonstrated that the former inherited favorable proof-theoretic properties from the latter (cut-free completeness, invertibility of rules, etc.).

Regarding future work, the author aims to investigate modal and intermediate logics that allow for the extraction of cut-free nested calculi from their labelled calculi, as well as provide new nested calculi for logics lacking one. These results could also prove beneficial in the explication of a general methodology for obtaining nested calculi well-suited for automated reasoning methods and other applications (by exploiting general results from the labelled paradigm [6,7,19]). Such results have the added benefit that they expose interesting connections between the different proof-theoretic formalisms involved.

Acknowledgments. The author would like to express his gratitude to his PhD supervisor A. Ciabattoni for her support and helpful comments. Work funded by FWF projects I2982, Y544-N2, and W1255-N23.

References

1. Avron, A.: The method of hypersequents in the proof theory of propositional non-classical logics. In: Logic: from Foundations to Applications (Staffordshire 1993), Oxford Science Publications, pp. 1–32. Oxford University Press, New York (1996)
2. Belnap Jr., N.D.: Display logic. J. Philos. Logic **11**(4), 375–417 (1982)
3. Brünnler, K.: Deep sequent systems for modal logic. Arch. Math. Logic **48**(6), 551–577 (2009)
4. Bull, R.A.: Cut elimination for propositional dynamic logic without ∗. Z. Math. Logik Grundlag. Math. **38**(2), 85–100 (1992)
5. Ciabattoni, A., Lyon, T., Ramanayake, R.: From display to labelled proofs for tense logics. In: Artemov, S., Nerode, A. (eds.) LFCS 2018. LNCS, vol. 10703, pp. 120–139. Springer, Cham (2018). https://doi.org/10.1007/978-3-319-72056-2_8
6. Ciabattoni, A., Maffezioli, P., Spendier, L.: Hypersequent and labelled calculi for intermediate logics. In: Galmiche, D., Larchey-Wendling, D. (eds.) TABLEAUX 2013. LNCS (LNAI), vol. 8123, pp. 81–96. Springer, Heidelberg (2013). https://doi.org/10.1007/978-3-642-40537-2_9
7. Dyckhoff, R., Negri, S.: Proof analysis in intermediate logics. Arch. Math. Logic **51**(1–2), 71–92 (2012)
8. Fitting, M.: Tableau methods of proof for modal logics. Notre Dame J. Formal Logic **13**(2), 237–247 (1972). https://doi.org/10.1305/ndjfl/1093894722
9. Fitting, M.: Prefixed tableaus and nested sequents. Ann. Pure Appl. Logic **163**(3), 291–313 (2012). https://doi.org/10.1016/j.apal.2011.09.004. http://www.sciencedirect.com/science/article/pii/S0168007211001266
10. Fitting, M.: Nested sequents for intuitionistic logics. Notre Dame J. Formal Logic **55**(1), 41–61 (2014)
11. Gabbay, D., Shehtman, V., Skvortsov, D.: Quantification in Non-classical Logics. Studies in Logic and Foundations of Mathematics. Elsevier, Amsterdam (2009)
12. Gentzen, G.: Untersuchungen uber das logische schliessen. Math. Z. **39**(3), 405–431 (1935)
13. Goré, R., Postniece, L., Tiu, A.: On the correspondence between display postulates and deep inference in nested sequent calculi for tense logics. Log. Methods Comput. Sci. **7**(2), 2:8, 38 (2011)

14. Goré, R., Ramanayake, R.: Labelled tree sequents, Tree hypersequents and Nested (Deep) Sequents. Advances in modal logic , Vol. 9. College Publications (2012)
15. Kashima, R.: Cut-free sequent calculi for some tense logics. Stud. Logica. **53**(1), 119–135 (1994)
16. Lyon, T.: On deriving nested calculi for intuitionistic logics from semantic systems. https://arxiv.org/abs/1910.06576 (2019)
17. Lyon, T., van Berkel, K.: Automating agential reasoning: proof-calculi and syntactic decidability for STIT logics. In: Baldoni, M., Dastani, M., Liao, B., Sakurai, Y., Zalila Wenkstern, R. (eds.) PRIMA 2019. LNCS (LNAI), vol. 11873, pp. 202–218. Springer, Cham (2019). https://doi.org/10.1007/978-3-030-33792-6_13
18. Lyon, T., Tiu, A., Góre, R., Clouston, R.: Syntactic interpolation for tense logics and bi-intuitionistic logic via nested sequents. In: Fernández, M., Muscholl, A. (eds.) 28th EACSL Annual Conference on Computer Science Logic (CSL 2020), Leibniz International Proceedings in Informatics (LIPIcs), vol. 152. Schloss Dagstuhl-Leibniz-Zentrum fuer Informatik, Dagstuhl, Germany (2020). Forthcoming
19. Negri, S.: Proof analysis beyond geometric theories: from rule systems to systems of rules. J. Logic Comput. **26**(2), 513–537 (2016). https://doi.org/10.1093/logcom/exu037
20. Pimentel, E.: A semantical view of proof systems. In: Moss, L.S., de Queiroz, R., Martinez, M. (eds.) Logic, Language, Information, and Computation, WoLLIC 2018. LNCS, vol. 10944, pp. 61–76. Springer, Heidelberg (2018). https://doi.org/10.1007/978-3-662-57669-4_3. ISBN 978-3-662-57669-4
21. Poggiolesi, F.: A cut-free simple sequent calculus for modal logic $s5$. Rev. Symb. Logic **1**(1), 3–15 (2008)
22. Poggiolesi, F.: The method of tree-hypersequents for modal propositional logic. In: Makinson, D., Malinowski, J., Wansing, H. (eds.) Towards Mathematical Philosophy. TL, vol. 28, pp. 31–51. Springer, Dordrecht (2009). https://doi.org/10.1007/978-1-4020-9084-4_3
23. Tiu, A., Ianovski, E., Goré, R.: Grammar logics in nested sequent calculus: proof theory and decision procedures. In: Bolander, T., Braüner, T., Ghilardi, S., Moss, L.S. (eds.) Advances in Modal Logic 9, Papers from the Ninth Conference on "Advances in Modal Logic," Held in Copenhagen, Denmark, 22–25 August 2012, pp. 516–537. College Publications (2012). http://www.aiml.net/volumes/volume9/Tiu-Ianovski-Gore.pdf
24. Viganò, L.: Labelled Non-classical Logics. Kluwer Academic Publishers, Dordrecht (2000). With a foreword by Dov M. Gabbay

Parameterised Complexity of Abduction in Schaefer's Framework

Yasir Mahmood[1]([✉]) [ID], Arne Meier[1] [ID], and Johannes Schmidt[2] [ID]

[1] Institut für Theoretische Informatik, Leibniz Universität Hannover,
Hanover, Germany
{mahmood,meier}@thi.uni-hannover.de

[2] Department of Computer Science and Informatics, School of Engineering,
Jönköping University, Jönköping, Sweden
johannes.schmidt@ju.se

Abstract. Abductive reasoning is a non-monotonic formalism stemming from the work of Peirce. It describes the process of deriving the most plausible explanations of known facts. Considering the positive version asking for sets of variables as explanations, we study, besides asking for existence of the set of explanations, two explanation size limited variants of this reasoning problem (less than or equal to, and equal to). In this paper, we present a thorough two-dimensional classification of these problems. The first dimension is regarding the parameterised complexity under a wealth of different parameterisations. The second dimension spans through all possible Boolean fragments of these problems in Schaefer's constraint satisfaction framework with co-clones (STOC 1978). Thereby, we almost complete the parameterised picture started by Fellows et al. (AAAI 2012), partially building on results of Nordh and Zanuttini (Artif. Intell. 2008). In this process, we outline a fine-grained analysis of the inherent parameterised intractability of these problems and pinpoint their FPT parts. As the standard algebraic approach is not applicable to our problems, we develop an alternative method that makes the algebraic tools partially available again.

Keywords: Parameterised complexity · Abduction · Schaefer's
framework · Co-clones

1 Introduction

The framework of parameterised complexity theory yields a more fine-grained complexity analysis of problems than the classical worst-case complexity may achieve. Introduced by Downey and Fellows [14,15], one associates problems with a specific *parameterisation*, that is, one studies the complexity of *parameterised problems*. Here, one aims to find parameters relevant for practice allowing to solve the problem by algorithms running in time $f(k) \cdot n^{O(1)}$, where f is a

Funded by German Research Foundation (DFG), project ME 4279/1-2.

S. Artemov and A. Nerode (Eds.): LFCS 2020, LNCS 11972, pp. 195–213, 2020.
https://doi.org/10.1007/978-3-030-36755-8_13

computable function, k is the value of the parameter and n is the input length. Problems with such a running time are called *fixed-parameter tractable* (**FPT**) and correspond to efficient computation in the parameterised setting. This is justified by the fact that parameters are usually slowly growing or even of constant value. Despite that, a different quality of runtimes is of the form $n^{f(k)}$ which are obeyed by algorithms solving problems in the class **XP**. Comparing both classes with respect to the runtimes their problems allow to be solved in, of course, both runtimes are polynomial. However, for the first type, the degree of the polynomial is independent of the parameter's value which is notable to observe. As a result, the second kind of runtimes is undesirable and usually tried to circumvented by locating different parameters. It is known that **FPT** $\subsetneq$ **XP** by diagonalisation and also that a (presumably infinite) hierarchy of parameterised intractability in between these two classes exist: the so-called **W**-hierarchy which is contained also in the class **W[P]** $\subseteq$ **XP**. These **W**-classes are regarded as a measure of intractability in the parameterised sense. Intuitively, showing **W[1]**-lower bounds corresponds to **NP**-lower bounds in the classical setting. The limit of this hierarchy, the class **W[P]** is defined via nondeterministic machines that have at most $h(k) \cdot \log n$ many nondeterministic steps, where h is a computable function, k the parameter's value, and n is the input length.

Clearly, the process of human common-sense reasoning is non-monotonic, as adding further knowledge might decrease the number of deducible facts. As a result, non-monotonic logics became a well-established approach to investigate this kind of reasoning. One of the popular formalism in this area of research is abductive reasoning which is an important concept in artificial intelligence as emphasised by Morgan [27] and Poole [33]. In particular, abduction is used in the process of medical diagnosis [21,32] and thereby relevant for practice. Intuitively, abductive reasoning describes the process of deriving the most plausible explanations of known facts and originated from the work of Peirce [31]. Formally, one uses propositional formulas to model known facts in a *knowledge base* KB together with a set of *manifestations* M and a set of *hypotheses* H. In this paper, H and M are sets of propositions as studied by Fellows et al. [18] as well as Eiter and Gottlob [17]. Formally, one tries to find a preferably small set of propositions $E \subseteq H$ such that $E \wedge KB$ is satisfiable and $E \wedge KB \models M$. E is then called an explanation for M. In this context, we distinguish three kinds of problems: the first just asks for such a very set E that fulfils these properties (ABD), the second tries to find a set of size less than or equal to a specific size (ABD$_\leq$), and the third one wants to spot a set of exactly a given size (ABD$_=$). Classically, ABD is complete for the second level of the polynomial hierarchy Σ_2^P [17] and its difficulty is very well understood [9,13,28,39]. As a result, under reasonable complexity-theoretic assumptions, the problem is highly intractable posing the question in turn for sources of this complexity. In this direction, there exists research that aims to better understand the structure and difficulty of this problem, that is, in the context of parameterised complexity. Here, Fellows et al. [18] initiated an investigation of possible parameters and classified CNF-induced fragments of the reasoning problems with respect to a multitude of parameters.

The authors study the CNF-fragments HORN, KROM, and DEFHORN. They studied the parameterisations $|M|$ (number of manifestations), $|H|$ (number of hypotheses), $|V|$ (number of variables), $|E|$ (number of explanations which is equivalent to their solution size k) directly stemming from problem components, as well as the tree-width [36], and the size of the smallest vertex cover. In their classification, besides showing several **para-NP-/W[P]**-complete/**FPT** cases, they also focus on the existence of polynomial kernels and present a complete picture regarding their CNF-classes.

Universal algebra yields a systematic way to rigorously classify fragments of a problem induced by restricting its Boolean connectives. This technique is built around Post's lattice [34] which bases on the notion of (co-)clones. Intuitively, given a set of Boolean functions B, the *clone* of B is the set of functions that are expressible by compositions of functions from B (plus introducing fictive variables). The most prominent result under this approach is the dichotomy theorem of Lewis [22] which classifies propositional satisfiability into polynomial-time solvable cases and intractable ones depending merely on the existence of specific Boolean operators. This approach has been followed many times in a wealth of different contexts [2,3,7,12,25,26,35] as well as in the context of abduction itself [10,28]. Interestingly, in the scope of constraint satisfaction problems, the investigation of co-clones (or relational clones) allows one to proceed a similar kind of classification (see, e.g., the work of Nordh and Zanuttini [28]). The reason for that lies in the concept of invariance of relations under some function f (one defines this property via *polymorphisms* where f is applied component-wisely to the columns of the relation). In view of this, Post's lattice supplies a similar lattice, now for sets of relations which are invariant under respective functions. With respect to constraint satisfaction, the most prominent classification is due to Schaefer [37] who similarly divides the constraint satisfaction problem restricted to co-clones into polynomial-time solvable and **NP**-complete cases. The algebraic approach has been successfully applied to abduction by Nord and Zanuttini [28]. For the problems that we consider, it is less obvious how to use the algebraic tools: the standard trick to obtain reductions preserves the existence of explanations, but not their size. Due to this, we develop an alternative method that makes the algebraic tools partially available again (see Sect. 2.1).

Much in the vein of Schaefer's classification, we present a thorough study directly pinpointing those restrictions of the abductive reasoning problem which yield efficiency under the parameterised approach. In a sense, we present an almost complete picture which has been initiated by Fellows et al. [18] except for some minor cases around the affine co-clones. Their classification is covered by our study now, as HORN cases correspond to the co-clones below IE_2, DEFHORN conforms IE_1, and KROM matches with ID_2. The motivation of our research is to draw a finer line than Fellow et al. did and to present complete picture with respect to all possible constraint languages now. From this classification, we draw some surprising results. Regarding the essentially negative cases for the parameter $|M|$, $\mathrm{ABD}_=$ is **para-NP**-complete whereas $\mathrm{ABD}_\leq$ is **FPT**. Also for this parameter, IE_1 and IE are hard for $\mathrm{ABD}_=$ and $\mathrm{ABD}_\leq$ (both para-**NP**-complete) but ABD is **FPT**. Regarding $|E|$ as parameterisation, the behaviour is similarly unexpected for the essentially negative cases: **FPT** for $\text{ABD}_\le$ versus **W**[1]-hardness for $\text{ABD}_=$. For the parameters $|V|$ as well as $|H|$ the classifications for all three problems are the same. Figure 1 shows our results for all problems and parameterisations in a single picture. Due to space constraints, proof details symbolised by '$\star$' can be found in the full version of the paper [23].

2   Preliminaries

We require standard notions from classical complexity theory [30]. We encounter the classical complexity classes **P**, **NP**, **DP** $= \{A \setminus B \mid A, B \in \mathbf{NP}\}$, **coNP**, $\mathbf{\Sigma_2^P} = \mathbf{NP^{NP}}$ and their respective completeness notions, employing polynomial time many-one reductions ($\le_m^P$).

Parameterised Complexity Theory. A *parameterised problem* (PP) $P \subseteq \Sigma^* \times \mathbb{N}$ is a subset of the crossproduct of an alphabet and the natural numbers. For an instance $(x, k) \in \Sigma^* \times \mathbb{N}$, k is called the (value of the) *parameter*. A *parameterisation* is a polynomial-time computable function that maps a value from $x \in \Sigma^*$ to its corresponding $k \in \mathbb{N}$. The problem P is said to be *fixed-parameter tractable* (or in the class **FPT**) if there exists a deterministic algorithm $\mathcal{A}$ and a computable function f such that for all $(x, k) \in \Sigma^* \times \mathbb{N}$, algorithm $\mathcal{A}$ correctly decides the membership of $(x, k) \in P$ and runs in time $f(k) \cdot |x|^{O(1)}$. The problem P belongs to the class **XP** if $\mathcal{A}$ runs in time $|x|^{f(k)}$. There exists a hierarchy of complexity classes in between **FPT** and **XP** which is called **W**-hierarchy (for details see the textbook of Flum and Grohe [19]). We will make use of the classes **W**[1] and **W**[2]. Complete problems characterising these classes are introduced later in Proposition 4. Also, we work with classes that can be defined via a precomputation on the parameter.

Definition 1. *Let $\mathcal{C}$ be any complexity class. Then* **para-**$\mathcal{C}$ *is the class of all PPs $P \subseteq \Sigma^* \times \mathbb{N}$ such that there exists a computable function $\pi \colon \mathbb{N} \to \Delta^*$ and a language $L \in \mathcal{C}$ with $L \subseteq \Sigma^* \times \Delta^*$ such that for all $(x, k) \in \Sigma^* \times \mathbb{N}$ we have that $(x, k) \in P \Leftrightarrow (x, \pi(k)) \in L$.*

Notice that **para-P** = **FPT**. The complexity classes $\mathcal{C} \in \{\mathbf{NP}, \mathbf{coNP}, \mathbf{DP}, \mathbf{\Sigma_2^P}\}$ are used in the **para-**$\mathcal{C}$ context by us.

Let $c \in \mathbb{N}$ and $P \subseteq \Sigma^* \times \mathbb{N}$ be a PP, then the *c-slice* of P, written as P_c is defined as $P_c := \{(x, k) \in \Sigma^* \times \mathbb{N} \mid k = c\}$. Notice that P_c is a classical problem then. Observe that, regarding our studied complexity classes, showing membership of a PP P in the complexity class **para-**$\mathcal{C}$, it suffices to show that each slice $P_c \in \mathcal{C}$.

Definition 2. *Let $P \subseteq \Sigma^* \times \mathbb{N}, Q \subseteq \Gamma^*$ be two PPs. One says that P is* fpt-reducible *to Q, $P \le^{\mathbf{FPT}} Q$, if there exists an fpt-computable function $f \colon \Sigma^* \times \mathbb{N} \to \Gamma^* \times \mathbb{N}$ such that*

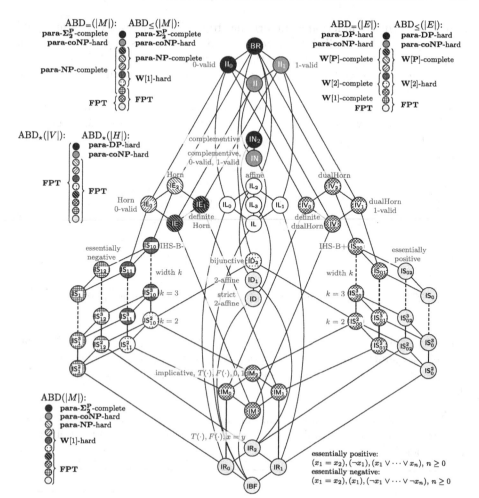

Fig. 1. Complexity landscape of abductive reasoning with respect to the studied parameters $|M|, |H|, |V|, |E|$. Notice, that due to presentation reasons, some completeness results are just mentioned with their lower bound, e.g., case $\mathrm{ABD}_\leq(\mathsf{IS}_{11}^2, |E|)$ is $\mathbf{W}[2]$-complete (Theorem 22). White colouring means unclassified. $\mathrm{ABD}_\star$ means same result for all three variants.

- *for all $(x, k) \in \Sigma^* \times \mathbb{N}$ we have that $(x, k) \in P \Leftrightarrow f(x, k) \in Q$,*
- *there exists a computable function $g \colon \mathbb{N} \to \mathbb{N}$ such that for all $(x, k) \in \Sigma^* \times \mathbb{N}$ and $f(x, k) = (x', k')$ we have that $k' \leq g(k)$.*

Propositional Logic. We assume familiarity with propositional logic. A *literal* is a variable x or its negation $\neg x$. A *clause* is a disjunction of literals and a *term* is a conjunction of literals. We denote by $\mathrm{var}(\varphi)$ the variables of a formula φ. Analogously, for a set of formulas F, $\mathrm{var}(F)$ denotes $\bigcup_{\varphi \in F} \mathrm{var}(\varphi)$. We identify finite F with the conjunction of all formulas from F, that is, $\bigwedge_{\varphi \in F} \varphi$. A mapping $\sigma : \mathrm{var}(\varphi) \mapsto \{0, 1\}$ is called an *assignment* to the variables of φ. A *model* of a formula φ is an assignment to $\mathrm{var}(\varphi)$ that satisfies φ. The *weight* of an assignment σ is the number of variables x such that $\sigma(x) = 1$. For two formulas ψ, φ we write $\psi \models \varphi$ if every model of ψ also satisfies φ. A formula is positive (resp. negative) if every literal appears positively (negatively) and a negation symbol appears only in front of a variable. The class of all propositional formulas is denoted by *PROP*. Occasionally, in this paper, we will consider special subclasses of formulas, namely

$$\Gamma_{0,d} = \{\ell_1 \wedge \ldots \wedge \ell_c \mid \ell_1, \ldots, \ell_c \text{ are literals and } c \leq d\},$$
$$\Delta_{0,d} = \{\ell_1 \vee \ldots \vee \ell_c \mid \ell_1, \ldots, \ell_c \text{ are literals and } c \leq d\},$$
$$\Gamma_{t,d} = \left\{ \bigwedge_{i \in I} \alpha_i \,\middle|\, \alpha_i \in \Delta_{t-1,d} \text{ for } i \in I \right\}, \Delta_{t,d} = \left\{ \bigvee_{i \in I} \alpha_i \,\middle|\, \alpha_i \in \Gamma_{t-1,d}, \ i \in I \right\}.$$

Finally, $\Gamma_{t,d}^+$ (resp. $\Gamma_{t,d}^-$) denote the class of all positive (negative) formulas in $\Gamma_{t,d}$.

Example 3. Let $\phi = \bigwedge_{i \leq m} (\neg x_{i,1} \vee \cdots \vee \neg x_{i,n_i})$ for $1 \leq n_i \leq d$ and $d, m \in \mathbb{N}$. That is, ϕ is a conjunction of the clauses containing negative literals. Then $\phi \in \Gamma_{1,d}$, the so-called d-CNF. Note also that ϕ is an IS_1^d-formula using only negative clauses.

We will often reduce a problem instance to (and from) a parameterised weighted satisfiability problem for propositional formulas. This problem is defined below.

Problem:	p-WSAT$(\Gamma_{t,d})$
Input:	A $\Gamma_{t,d}$-formula α over variables V with $t, d \geq 1$ and $k \in \mathbb{N}$.
Parameter:	k.
Question:	Is there a satisfying assignment for α of weight k?

Two similarly defined problems are p-WSAT$(\Gamma_{t,1}^+)$ and p-WSAT$(\Gamma_{t,1}^-)$ where an instance α comes from classes $\Gamma_{t,1}^+$ (resp. $\Gamma_{t,1}^-$). The classes of the **W**-hierarchy can be defined in terms of these problems as proved by Downey and Fellows [19].

Proposition 4 ([19]). *The following problems are* **W**[t]*-complete for each* $t \geq 1$, *under* $\leq^{\mathbf{FPT}}$*-reductions:*

- p-WSAT$(\Gamma_{t,1}^+)$ *if* t *is even,*
- p-WSAT$(\Gamma_{t,1}^-)$ *if* t *is odd,*
- p-WSAT$(\Gamma_{t,d})$ *for every* t *and* $d \geq 1$.

Table 1. Overview of bases [4] and clause descriptions [28] for co-clones, where $\text{EVEN}^4 = x_1 \oplus x_2 \oplus x_3 \oplus x_4 \oplus 1$.

co-clone	base	clause type	name/indication
$\text{BR}\ (\text{II}_2)$	1-IN-3 $= \{001,010,100\}$	all clauses	all Boolean relations
II_1	$x \vee (y \oplus z)$	at least one positive literal per clause	1-valid
II_0	DUP, $x \to y$	at least one negative literal per clause	0-valid
II	$\text{EVEN}^4,\ x \to y$	at least one negative and one positive literal per clause	1- and 0-valid
IN_2	NAE $= \{0,1\}^3 \setminus \{000,111\}$	cf. previous column	complementive
IN	DUP $= \{0,1\}^3 \setminus \{101,010\}$	cf. previous column	complementive and 1- and 0-valid
IE_2	$x \wedge y \to z, x, \neg x$	clauses with at most one positive literal	Horn
IE_1	$x \wedge y \to z, x$	clauses with exactly one positive literal	definite Horn
IE_0	$x \wedge y \to z, \neg x$	$(x_1 \vee \neg x_2 \vee \cdots \vee \neg x_n), n \geq 2, (\neg x_1 \vee \cdots \vee \neg x_n), n \geq 1$	Horn and 0-valid
IE	$x \wedge y \to z$	$(x_1 \vee \neg x_2 \vee \cdots \vee \neg x_n), n \geq 2$	Horn and 1- and 0-valid
IV_2	$x \vee y \vee \neg z, x, \neg x$	clauses with at most one negative literal	dualHorn
IV_1	$x \vee y \vee \neg z, x$	$(\neg x_1 \vee x_2 \vee \cdots \vee x_n), n \geq 2, (x_1 \vee \cdots \vee x_n), n \geq 1$	dualHorn and 1-valid
IV_0	$x \vee y \vee \neg z, \neg x$	clauses with exactly one negative literal	definite dualHorn
IV	$x \vee y \vee \neg z$	$(\neg x_1 \vee x_2 \vee \cdots \vee x_n), n \geq 2$	dualHorn and 1- and 0-valid
IL_2	$\text{EVEN}^4, x, \neg x$	all affine clauses (all linear equations)	affine
IL_1	EVEN^4, x	$(x_1 \oplus \cdots \oplus x_n = a), n \geq 0, a = n \pmod 2$	affine and 1-valid
IL_0	$\text{EVEN}^4, \neg x$	$(x_1 \oplus \cdots \oplus x_n = 0), n \geq 0$	affine and 0-valid
IL_3	$\text{EVEN}^4, x \oplus y$	$(x_1 \oplus \cdots \oplus x_n = a), n$ even, $a \in \{0,1\}$	-
IL	EVEN^4	$(x_1 \oplus \cdots \oplus x_n = 0), n$ even	affine and 1- and 0-valid
ID_2	$x \oplus y, x \to y$	clauses of size 1 or 2	bijunctive, KROM, 2CNF
ID_1	$x \oplus y, x, \neg x$	affine clauses of size 1 or 2	2-affine
ID	$x \oplus y$	affine clauses of size 2	strict 2-affine
IM_2	$x \to y, x, \neg x$	$(x_1 \to x_2), (x_1), (\neg x_1)$	implicative
IM_1	$x \to y, x$	$(x_1 \to x_2), (x_1)$	implicative and 1-valid
IM_0	$x \to y, \neg x$	$(x_1 \to x_2), (\neg x_1)$	implicative and 0-valid
IM	$x \to y$	$(x_1 \to x_2)$	implicative and 1- and 0-valid
IS_{10}	cf. next column	$(x_1), (x_1 \to x_2), (\neg x_1 \vee \cdots \vee \neg x_n), n \geq 0$	IHS-B-
IS_{10}^k	cf. next column	$(x_1), (x_1 \to x_2), (\neg x_1 \vee \cdots \vee \neg x_n), k \geq n \geq 0$	IHS-B- of width k
IS_{12}	cf. next column	$(x_1), (\neg x_1 \vee \cdots \vee \neg x_n), n \geq 0, (x_1 = x_2)$	essentially negative
IS_{12}^k	cf. next column	$(x_1), (\neg x_1 \vee \cdots \vee \neg x_n), k \geq n \geq 0, (x_1 = x_2)$	essentially negative of width k
IS_{11}	cf. next column	$(x_1 \to x_2), (\neg x_1 \vee \cdots \vee \neg x_n), n \geq 0$	-
IS_{11}^k	cf. next column	$(x_1 \to x_2), (\neg x_1 \vee \cdots \vee \neg x_n), k \geq n \geq 0$	-
IS_1	cf. next column	$(\neg x_1 \vee \cdots \vee \neg x_n), n \geq 0, (x_1 = x_2)$	negative
IS_1^k	cf. next column	$(\neg x_1 \vee \cdots \vee \neg x_n), k \geq n \geq 0, (x_1 = x_2)$	negative of width k
IS_{00}	cf. next column	$(\neg x_1), (x_1 \to x_2), (x_1 \vee \cdots \vee x_n), n \geq 0$	IHS-B+
IS_{00}^k	cf. next column	$(\neg x_1), (x_1 \to x_2), (x_1 \vee \cdots \vee x_n), k \geq n \geq 0$	IHS-B+ of width k
IS_{02}	cf. next column	$(\neg x_1), (x_1 \vee \cdots \vee x_n), n \geq 0, (x_1 = x_2)$	essentially positive
IS_{02}^k	cf. next column	$(\neg x_1), (x_1 \vee \cdots \vee x_n), k \geq n \geq 0, (x_1 = x_2)$	essentially positive of width k
IS_{01}	cf. next column	$(x_1 \to x_2), (x_1 \vee \cdots \vee x_n), n \geq 0$	-
IS_{01}^k	cf. next column	$(x_1 \to x_2), (x_1 \vee \cdots \vee x_n), k \geq n \geq 0$	-
IS_0	cf. next column	$(x_1 \vee \cdots \vee x_n), n \geq 0, (x_1 = x_2)$	positive
IS_0^k	cf. next column	$(x_1 \vee \cdots \vee x_n), k \geq n \geq 0, (x_1 = x_2)$	positive of width k
IR_2	$x_1, \neg x_2$	$(x_1), (\neg x_1), (x_1 = x_2)$	-
IR_1	x_1	$(x_1), (x_1 = x_2)$	-
IR_0	$\neg x_1$	$(\neg x_1), (x_1 = x_2)$	-
$\text{IR}\ (\text{IBF})$	$\emptyset$	$(x_1 = x_2)$	-

Constraints and S-Formulas. A *logical relation* of arity k is a relation $R \subseteq \{0,1\}^k$. A *constraint* is a formula $R(x_1, \ldots, x_k)$, where R is a logical relation of arity k and the x_i's are (not necessarily distinct) variables. An assignment σ to the x_i's satisfies the constraint if $(\sigma(x_1), \ldots, \sigma(x_k)) \in R$. A *constraint language* S is a finite set of logical relations. An *S-formula* φ is a conjunction of constraints built upon logical relations only from S, and accordingly can be seen as a quantifier-free first-order formula. An assignment σ is called a *model* of φ if σ satisfies all constraints in φ simultaneously. Whenever an S-formula or constraint is logically equivalent to a single clause or term, we treat it as such.

Definition 5. *1. The set $\langle S \rangle$ is the smallest set of relations that contains S, the equality constraint, $=$, and which is closed under primitive positive first order definitions, that is, if ϕ is an $S \cup \{=\}$-formula and $R(x_1, \ldots, x_n) \equiv$*

$\exists y_1 \ldots \exists y_l \phi(x_1, \ldots, x_n, y_1, \ldots, y_l)$, then $R \in \langle S \rangle$. In other words, $\langle S \rangle$ is the set of relations that can be expressed as an $S \cup \{=\}$-formula with existentially quantified variables.

2. The set $\langle S \rangle_{\neq}$ is the set of relations that can be expressed as an S-formula with existentially quantified variables (no equality relation is allowed).

The set $\langle S \rangle$ is called a *relational clone* or *co-clone* with *base* S [4]. Throughout the text, we refer to different types of Boolean relations and corresponding co-clones following Schaefer's terminology [37]. For an overview of co-clones and bases, see Table 1. Note that $\langle S \rangle_{\neq} \subseteq \langle S \rangle$ by definition. The other direction does not hold in general. However, if $(x = y) \in \langle S \rangle_{\neq}$, then $\langle S \rangle_{\neq} = \langle S \rangle$.

Abduction. An instance of the abduction problem for S-formulas is given by $\langle V, H, M, KB \rangle$, where V is the set of variables, H is the set of hypotheses, M is the set of manifestations, and KB is the knowledge base (or theory) built upon variables from V. A knowledge base KB is a set of S-formulas that we assimilate with the conjunction of all formulas it contains. We define the following abduction problems for S-formulas.

Problem:	$\mathrm{ABD}(S, k)$—the abductive reasoning problem for S-formulas parameterised by k
Input:	$\langle V, H, M, KB, k \rangle$, where KB is a set of S-formulas, H, M are each set of propositions, and $V = \mathrm{var}(H) \cup \mathrm{var}(M) \cup \mathrm{var}(KB)$.
Parameter:	k.
Question:	Is there a set $E \subseteq H$ such that $E \wedge KB$ is satisfiable and $E \wedge KB \models M$?

Similarly, the problem $\mathrm{ABD}(S)$ is the classical pendant of $\mathrm{ABD}(S, k)$. Additionally, we consider size restrictions for a solution and define the following problems.

Problem:	$\mathrm{ABD}_{\leq}(S, k)$
Input:	$\langle V, H, M, KB, s, k \rangle$, where KB is a set of S-formulas, H, M are each set of propositions, and $V = \mathrm{var}(H) \cup \mathrm{var}(M) \cup \mathrm{var}(KB)$, and $s \in \mathbb{N}$.
Parameter:	k.
Question:	Is there a set $E \subseteq H$ with $|E| \leq s$ such that $E \wedge KB$ is satisfiable and $E \wedge KB \models M$?

Analogously, $\mathrm{ABD}_{=}(S, k)$ requires the size of E to be exactly s and $\mathrm{ABD}_{=}(S), \mathrm{ABD}_{\leq}(S)$ are the classical counterparts. Notice that, for instance, in cases where the parameter is the size of solutions, then $s = k$.

Example 6. Sitting in a train you realise that it is still not moving even though the clock suggests it should be. You start reasoning about it. Either some door is open, the train has delayed, or that engine has failed. This form of reasoning is called abductive reasoning. Having some additional information that the operator of train usually announces in case the train is delayed or engine has failed, you deduce that some door must be opened and that train will start moving soon when all the doors are closed. Formally, one is interested in an explanation for the observed event (manifestation) {stop}. The knowledge base includes following statements:

- ¬moving ↔ stop
- ¬announcement,
- moving → time,
- engineFailed → announcement,

- trainDelayed → newTime,

- (engineFailed ∨ trainDelayed ∨ doorOpen) → stop.

Then the set of hypotheses {time, doorOpen, announcement} has an explanation, namely, {doorOpen}. On the other hand, {time} does not explain the event {stop}, whereas, {announcement} is not consistent with the knowledge base. Consequently, an explanation of size 1 exists. There also exists an explanation of size 2 since {time, doorOpen} is consistent with KB and explains M. Note that having the set of hypotheses {engineFailed, doorOpen} facilitates only one explanation of size 1, namely, {doorOpen}, even though the hypotheses set has size 2.

2.1 Base Independence

We present now a number of technical expressivity results (Lemma 7). They allow us in the sequel to prove a crucial property for the whole classification endeavour (Lemma 8). To prove the following lemma, we need to express equality by some other construction.

Lemma 7 ($\star$). *Let S be a constraint language. If S is not essentially negative and not essentially positive, then $(x = y) \in \langle S \rangle_{\neq}$, and $\langle S \rangle = \langle S \rangle_{\neq}$.*

The following property is crucial for presented results in the course of this paper. It supplies generalised upper as well as lower bounds (independence of the base of a co-clone), as long as the constraint language is not essentially negative and not essentially positive. The proof idea is to implement the previous lemma.

Lemma 8 ($\star$). *Let S, S' be two constraint languages such that S' is neither essentially positive nor essentially negative. Let $\mathrm{ABD}_* \in \{\mathrm{ABD}, \mathrm{ABD}_=, \mathrm{ABD}_\leq\}$. If $S \subseteq \langle S' \rangle$, then $\mathrm{ABD}_*(S) \leq_m^{\mathbf{P}} \mathrm{ABD}_*(S')$.*

The last lemma in this section takes care of the essentially positive cases. The proof idea is to remove the equality clauses maintaining the size counts and the satisfiability property.

Lemma 9 ($\star$). *Let S, S' be two constraint languages such that S' is essentially positive. Let $\mathrm{ABD}_* \in \{\mathrm{ABD}, \mathrm{ABD}_=, \mathrm{ABD}_\leq\}$. If $S \subseteq \langle S' \rangle$, then $\mathrm{ABD}_*(S) \leq_m^{\mathbf{P}} \mathrm{ABD}_*(S')$.*

Remark 10. Notice that Lemmas 8 and 9 are stated with respect to the classical and unparameterised decision problems. However, these reductions can be generalised to $\leq^{\mathbf{FPT}}$-reductions whenever the parameters are bound as required by Definition 2. That is, in our case, for any parameterisation $k \in \{|H|, |E|, |M|\}$ the reductions are valid. Even more, the values of the parameters stay the same as in the reduction the sizes of H, E, and M remain unchanged.

Remark 11. It is rather cumbersome to mention the base independence results in almost every single proof. As a result, we omit this reference and show the results only for concrete bases, thereby, implicitly using the above lemmas. In cases where we deal with essentially negative constraint languages, we do not have a general base independence result, but direct constructions showing hardness in our cases for all bases (e.g., [23, Lem. 32]).

Let SAT and IMP denote the classical satisfiability and implication problems. Given a constraint language S then an instance of SAT(S) is an S-formula φ and the question is whether there exists a satisfying assignment for φ. On the other hand, an instance of IMP(S) is (ϕ, ψ) such that ϕ, ψ are two S-formulas and the question is whether $\phi \models \psi$. We have the following observation regarding the classical SAT and IMP problems.

Proposition 12 ([37,38]). *Let S be a constraint language such that $\langle S \rangle \subseteq \mathsf{C}$ where $\mathsf{C} \in \{\mathsf{ID}_2, \mathsf{IV}_2, \mathsf{IE}_2, \mathsf{IL}_2\}$. Then SAT($S$) and IMP($S$) are both in $\mathbf{P}$.*

3   Complexity Results for Abductive Reasoning

In this section, we start with general observations and reductions between the defined problems. Then we prove some immediate (parameterised) complexity results. We provide two results which help us to consider fewer cases to solve.

Lemma 13. *For every constraint language S we have ABD(S) $\leq^{\mathbf{P}}_m$ ABD$_\leq$(S).*

Proof. Clearly, $(V, H, M, KB) \in$ ABD(S) $\Leftrightarrow$ $(V, H, M, KB, s) \in$ ABD$_\leq$(S), where $s = |H|$. That is, there is an explanation for an abduction instance if and only if there is one with size at most that of the hypotheses set.   □

Lemma 14. *ABD$_\leq$(S) = ABD$_=$(S) for any S such that IBF $\subseteq \langle S \rangle \subseteq$ IV$_2$.*

Proof. "$\subseteq$": Every positive instance $(V, H, M, KB, s) \in$ ABD$_\leq$(S) has a solution E of size exactly s. We show that a solution of size $< s$ can always be extended to a size s solution. Given a solution of size $\leq s$ then a solution of size $= s$ can be constructed from it (in even polynomial time) w.r.t. $|H|$ by adding one element h at a time from H to E and checking that $\neg h \notin KB$.

"$\supseteq$": Every solution of size exactly s is a solution of size $\leq s$.   □

Intractable Cases. It turns out that for 0-valid, 1-valid and complementive languages, all three problems remain hard under any parametrisation except the case $|V|$.

Lemma 15. *The problems* $\mathrm{ABD}(S, k)$, $\mathrm{ABD}_\leq(S, k)$, $\mathrm{ABD}_=(S, k)$ *are*

1. **para-coNP**-*hard if* $\mathsf{IN} \subseteq \langle S \rangle \subseteq \mathsf{II}_1$ *and* $k \in \{|H|, |E|, |M|\}$,
2. **para-DP**-*hard if* $\mathsf{C} \subseteq \langle S \rangle \subseteq \mathsf{BR}$ *and* $\mathsf{C} \in \{\mathsf{IN}_2, \mathsf{II}_0\}$ *and* $k \in \{|H|, |E|\}$.
3. **para-**$\mathbf{\Sigma_2^P}$-*hard if* $\mathsf{C} \subseteq \langle S \rangle \subseteq \mathsf{BR}$ *and* $k = |M|$ *for* $\mathsf{C} \in \{\mathsf{IN}_2, \mathsf{II}_0\}$.

Proof. (1) We prove the case for IN regarding all three parameters simultaneously. Notice that $\mathrm{IMP}(\mathsf{II}_1)$ is **coNP**-hard [28, Thm. 34] even if the right side contains only a single variable. We describe in the following a modified proof from [28, Prop. 48]. Since $\langle \mathsf{IN} \cup \{T\} \rangle = \mathsf{II}_1$ (define $T(x) \equiv x$) we have that $\mathrm{IMP}(\mathsf{IN} \cup \{T\})$ is **coNP**-hard, even if the right side contains only a single variable. We reduce $\mathrm{IMP}(\mathsf{IN} \cup \{T\})$ to our abduction problems with $|H| = 1$, $|M| = 1$, and $|E| = 1$. Let (KB_T, q) be an instance of $\mathrm{IMP}(\mathsf{IN} \cup \{T\})$, where $KB_T = KB \wedge \bigwedge_{x \in V_T} T(x)$ with KB being an IN-formula. We map (KB_T, q) to $(V, \{h\}, \{q\}, KB')$, where $V = \mathrm{var}(KB) \cup \{h\}$, h is a fresh variable, and KB' is obtained from KB by replacing any variable from V_T by h. Note that $KB_T \equiv KB' \wedge h$. Since KB and KB' are 1-valid, clearly, $KB' \wedge h$ is always satisfiable and there exists an explanation iff $KB' \wedge h \models q$, iff $KB_T \models q$. Furthermore, observe that $KB_T \models q$ if and only if $(V, \{h\}, \{q\}, KB', |H|) \in \mathrm{ABD}(\mathsf{IN}, |H|)$ if and only if $(V, \{h\}, \{q\}, KB', 1, |H|) \in \mathrm{ABD}_\leq(\mathsf{IN}, |H|)$ if and only if $(V, \{h\}, \{q\}, KB', 1, |H|) \in \mathrm{ABD}_=(\mathsf{IN}, |H|)$. The latter is true also when replacing $|H|$ by $|E|$ or $|M|$. This proves the claimed **para-coNP**-hardnesses.

(2) From Fellows et al. [18, Prop. 4] we know that all three problems for BR are **DP**-complete for $|H| = 0$ even if $|M| = 1$. We argue that the hardness can be extended to IN_2. Note that $\langle \mathsf{IN}_2 \cup \{F\} \rangle = \mathsf{BR}$ where $F(x) \equiv \neg x$. Creignou & Zanuttini [13] prove that $\mathrm{ABD}(S \cup \{F\}) \leq_m^P \mathrm{ABD}(S \cup \{\mathrm{SymOR}_{2,1}\})$ where $\mathrm{SymOR}_{2,1}(x, y, z) = ((x \to y) \wedge T(z)) \vee ((y \to x) \wedge F(z))$. Moreover, they also prove that $\mathrm{SymOR}_{2,1} \in \langle S \rangle$ such that $\mathsf{IN}_2 \subseteq \langle S \rangle$ [13, Lem. 21/27]. Finally, having $|M| = 1$ allows us to use their proof and, as a consequence, $\mathrm{ABD}(\mathsf{BR}) \leq_m^P \mathrm{ABD}(S)$ such that $\mathsf{IN}_2 \subseteq \langle S \rangle$. This gives the desired lower bound for IN_2. Regarding II_0, the proof follows by similar arguments using the observations that $\langle \mathsf{II}_0 \cup \{T\} \rangle = \mathsf{BR}$ and $\mathrm{OR}_{2,1} \in \langle S \rangle$ such that $\mathsf{II}_0 \subseteq \langle S \rangle$ where $\mathrm{OR}_{2,1}(x, y) = x \to y$ [13, Lem. 19/27].

(3) Nordh and Zanuttini [28, Prop. 46/47] prove $\mathbf{\Sigma_2^P}$-hardness for both IN_2 as well as II_0 with positive literal manifestations. This implies that the 1-slice of each of $\mathrm{ABD}(\mathsf{IN}_2, |M|)$ and $\mathrm{ABD}(\mathsf{II}_0, |M|)$ is $\mathbf{\Sigma_2^P}$-hard, which gives the desired result. For $\mathrm{ABD}_\leq(S, |M|)$ and $\mathrm{ABD}_=(S, |M|)$, the results follow from Lemma 13. $\qquad\square$

Fixed-Parameter Tractable Cases. The following corollary is immediate because the classical questions corresponding to these cases are in **P** due to Nordh and Zanuttini [28].

Corollary 16. *The problem* $\mathrm{ABD}(S, k)$ *is* **FPT** *for any parameterisation k and* $\langle S \rangle \subseteq \mathsf{C}$ *with* $\mathsf{C} \in \{\mathsf{IV}_2, \mathsf{ID}_1, \mathsf{IE}_1, \mathsf{IS}_{12}\}$.

The next result is already due to Fellows et al. [18, Prop. 13].

Corollary 17. *The problems* $\mathrm{ABD}(S, |V|)$, $\mathrm{ABD}_{\leq}(S, |V|)$, $\mathrm{ABD}_{=}(S, |V|)$ *are all* **FPT** *for all Boolean constraint languages S.*

Now, we prove **P**-membership for some cases of the classical problems and start with the essentially positive cases. The proof idea is to start with unit propagation. The positive clauses do not explain anything and one just only checks whether the elements of M appear either in KB or H. Then, we need to adjust the size accordingly.

Lemma 18 ($\star$). *The classical problems* $\mathrm{ABD}_{=}(S)$ *and* $\mathrm{ABD}_{\leq}(S)$ *are in* **P** *for* $\langle S \rangle \subseteq \mathsf{IS}_{02}$.

The following lemma proves that essentially negative languages for $\mathrm{ABD}_{\leq}$ also remain tractable.

Lemma 19. *The classical problem* $\mathrm{ABD}_{\leq}(S)$ *is in* **P** *if* $\langle S \rangle \subseteq \mathsf{IS}_{12}$.

Proof. First, we prove the result with respect to $\langle S \rangle_{\neq} \subseteq \mathsf{IS}_{12}$. Let P denote the set of positive unit clauses from KB and denote $E_{MP} = M \setminus P$. Now, we have the following two observations.

Observation 1. There exists an explanation iff $E_{MP} \subseteq H$ and M is consistent with KB. That is, what is not yet explained by P must be explainable directly by H because negative clauses can not contribute to explaining anything, they can only contribute to 'rule out' certain subsets of H as possible explanations.

Observation 2. If there exists an explanation, then any explanation contains E_{MP}.

As a result, E_{MP} represents a cardinality-minimal and a subset-minimal explanation. We conclude that there exists an explanation E with $|E| \leq s$ iff E_{MP} constitutes an explanation and $|E_{MP}| \leq s$. Now, we proceed with base independence for this case.

Claim. $\mathrm{ABD}_{\leq}(S \cup \{=\}) \leq_m^{\mathbf{P}} \mathrm{ABD}_{\leq}(S)$ for $\langle S \rangle \subseteq \mathsf{IS}_{12}$.

Proof of Claim. The reduction gets rid of the equality clauses by removing them and deleting the duplicating occurrences of variables. This decreases only the size of H and might also the size of an explanation E. Notice that $x = y \in KB$ does not enforce both x and y into E. ∎

This completes the proof to lemma. □

Finally, the 2-affine cases are also tractable as we prove in the following lemma. The idea is, similarly to Creignou et al. [8, Prop. 1], to change the representation of the knowledge base.

Lemma 20 ($\star$). *The classical problems* $\mathrm{ABD}_{=}(S)$ *and* $\mathrm{ABD}_{\leq}(S)$ *are in* **P** *if* $\langle S \rangle \subseteq \mathsf{ID}_1$.

3.1 Parameter 'Number of Hypotheses' $|H|$

For this parameter, it turns out that the only intractable cases are those pointed out in Lemma 15.

Theorem 21. $\mathrm{ABD}(S, |H|)$, $\mathrm{ABD}_{\leq}(S, |H|)$ and $\mathrm{ABD}_{=}(S, |H|)$ are

1. **para-DP**-*hard if* $\mathsf{C} \subseteq \langle S \rangle \subseteq \mathsf{BR}$ *and* $\mathsf{C} \in \{\mathsf{IN}_2, \mathsf{II}_0\}$,
2. **para-coNP**-*hard if* $\mathsf{IN} \subseteq \langle S \rangle \subseteq \mathsf{BR}$,
3. **FPT** *if* $\langle S \rangle \subseteq \mathsf{C} \in \{\mathsf{IE}_2, \mathsf{IV}_2, \mathsf{ID}_2, \mathsf{IL}_2\}$.

Proof. **1.+2.** We proved these cases in Lemma 15.
3. Recall that $\mathrm{SAT}(S)$ and $\mathrm{IMP}(S)$ are both in **P** for every S in the question (Proposition 12). By $|H| \geq |E|$, we have that $\binom{|H|}{|E|} = |H|^{|E|} \in O(k^k)$, where $k = |H|$. Consequently, we brute-force the candidates for E and verify them in polynomial time. This yields **FPT** membership. $\qquad\square$

3.2 Parameter 'Number of Explanations' $|E|$

In this subsection, we consider the solution size as a parameter. Notice that, because of the parameter $|E|$, the problem ABD is not meaningful anymore. As a result, we only consider the size limited variants $\mathrm{ABD}_{=}$ and $\mathrm{ABD}_{\leq}$. The following theorem provides a classification into six different complexity degrees.

Theorem 22. *The problems* $\mathrm{ABD}_{\leq}(S, |E|)$ *and* $\mathrm{ABD}_{=}(S, |E|)$ *are*

1. **para-DP**-*hard if* $\mathsf{C} \subseteq \langle S \rangle \subseteq \mathsf{BR}$ *and* $\mathsf{C} \in \{\mathsf{IN}_2, \mathsf{II}_0\}$
2. **para-coNP**-*hard if* $\mathsf{IN} \subseteq \langle S \rangle \subseteq \mathsf{II}_1$,
3. **W**[**P**]-*complete if* $\mathsf{IE} \subseteq \langle S \rangle \subseteq \mathsf{IE}_2$,
4. **W**[2]-*complete if* $\mathsf{IM} \subseteq \langle S \rangle \subseteq \mathsf{C}$ *for* $\mathsf{C} \in \{\mathsf{ID}_2, \mathsf{IV}_2\}$ *and* **W**[2]-*hard if* $\mathsf{IM} \subseteq \langle S \rangle \subseteq \mathsf{IS}_{10}$,
5. **FPT** *if* $\langle S \rangle \subseteq \mathsf{ID}_1$ *or* $\langle S \rangle \subseteq \mathsf{IS}_{02}$,

Moreover, if $\mathsf{IS}_1^2 \subseteq \langle S \rangle \subseteq \mathsf{IS}_{12}$*, then* $\mathrm{ABD}_{\leq}(S, |E|) \in$ **FPT** *and* $\mathrm{ABD}_{=}(S, |E|)$ *is* **W**[1]-*complete.*

Proof Ideas. **1.+2.** This is a corollary to Theorem 21.
3. Upper bound for IE_2 follows from the fact that $\mathrm{SAT}(\mathsf{IE}_2)$ and $\mathrm{IMP}(\mathsf{IE}_2)$ are in **P** (cf. Proposition 12). Guessing E takes $k \cdot \log n$ non-deterministic steps and verification can be done in polynomial time. For the lower bound, we argue that the proof in [18, Thm. 8] can be extended. Details are presented in [23, Lem. 29].
4. Note that the difficult part of the abduction problem for $\langle S \rangle$ such that $\mathsf{IM} \subseteq \langle S \rangle$ is the case when a solution of size larger than k is found. This solution must be reduced to one of size $\leq k$ (resp. $= k$). For **W**[2]-membership of $\mathrm{ABD}_{=}(\mathsf{IM}, |E|)$, we reduce our problem to p-WSAT$(\Gamma_{2,1})$ which is **W**[2]-complete. For hardness, we reduce from p-WSAT$(\Gamma_{2,1}^+)$ which is again **W**[2]-complete. Details of the completeness proof for $\mathrm{ABD}_{=}(\mathsf{IM}, E)$ can be found

in [23, Lem. 30]. The $\mathbf{W}[2]$-membership for IV_2 uses a little modification of the same reduction. Proof details can be found in [23, Lem. 33]. For these two cases, $\mathrm{ABD}_\leq(S,|E|)$ follow from the monotone argument from Lemma 14. For ID_2, the result follows from [18, Thm. 21]. Finally, the hardness for IS_{10} is a consequence of the $\mathbf{W}[2]$-hardness for IM. However, [23, Lem. 31] strengthens this results to $\mathbf{W}[2]$-completeness by showing membership in $\mathbf{W}[2]$ for $\mathrm{ABD}_=$. Regarding $\mathrm{ABD}_\leq(\mathsf{IS}_{10},|E|)$, we also believe in $\mathbf{W}[2]$-completeness but have not proved it yet.

5. This follows from the fact that the classical problems are in $\mathbf{P}$ (Lemmas 18/20).

Finally, $\mathbf{FPT}$ membership for $\mathrm{ABD}_\leq(\mathsf{IS}_{12},|E|)$ follows from Lemma 19. Note that this is the only case with $|E|$ when the two problems $\mathrm{ABD}_\leq$ and $\mathrm{ABD}_=$ have different complexity. We prove $\mathbf{W}[1]$-hardness for the languages S, such that $\neg x \vee \neg y \in \langle S \rangle_{\neq}$. The membership for $\mathrm{ABD}_=(S,|E|)$ with $\langle S \rangle \subseteq \mathsf{IS}_{12}$, this means also arbitrary bases, then follows as a corollary (for details see [23, Lem. 34/35]).   □

3.3   Parameter 'Number of Manifestations' $|M|$

The complexity landscape regarding the parameter $|M|$ is more diverse. The classification differs for each of the investigated problem variants. Consequently, we treat each case separately and start with the general abduction problem which provides a hexachotomy.

Theorem 23. *The problem* $\mathrm{ABD}(S,|M|)$ *is*

1. **para-$\Sigma_2^{\mathbf{P}}$**-*hard if* $\mathsf{C} \subseteq \langle S \rangle \subseteq \mathsf{BR}$ *and* $\mathsf{C} \in \{\mathsf{IN}_2, \mathsf{II}_0\}$,
2. **para-coNP**-*hard if* $\mathsf{IN} \subseteq \langle S \rangle \subseteq \mathsf{II}_1$,
3. **para-NP**-*complete if* $\langle S \rangle = \mathsf{IE}_2$,
4. $\mathbf{W}[1]$-*complete if* $\mathsf{IS}_{11}^2 \subseteq \langle S \rangle \subseteq \mathsf{ID}_2$,
5. $\mathbf{W}[1]$-*hard if* $\mathsf{IS}_{11}^3 \subseteq \langle S \rangle$,
6. **FPT** *if* $\langle S \rangle \subseteq \mathsf{C} \in \{\mathsf{ID}_1, \mathsf{IS}_{12}, \mathsf{IE}_1, \mathsf{IV}_2\}$.

Proof. **1.+2.** We proved this in Lemma 15 using the fact that 1-slice of each problem is hard for respective classes.

3. Membership is easy to see since the classical problem is $\mathbf{NP}$-complete. For hardness, notice that the 1-slice of the problem is $\mathbf{NP}$-complete [17].

4.+5. The first result follows from Fellows et al. [18, Thm. 26]. Notice that they prove this for ID_2, but using the fact that the formulas (or clauses) in their reduction are IS_{11}^2-formulas, we derive the hardness for IS_{11}^2. The second statement is then a consequence.

6. Follows from classical problems being in $\mathbf{P}$ (Corollary 16).

□

For $\mathrm{ABD}_\leq$, definite Horn cases surprisingly behave different and are much harder than for the general case.

Theorem 24. *The problem* $\mathrm{ABD}_{\leq}(S, |M|)$ *is*

1. **para-$\Sigma_2^{\mathbf{P}}$-***hard if* $\mathsf{C} \subseteq \langle S \rangle \subseteq \mathsf{BR}$ *and* $\mathsf{C} \in \{\mathsf{IN}_2, \mathsf{II}_0\}$,

2. **para-coNP-***hard if* $\mathsf{IN} \subseteq \langle S \rangle \subseteq$ II_1,

3. **para-NP-***complete if* $\mathsf{IE} \subseteq \langle S \rangle \subseteq$ IE_2,

4. $\mathbf{W}[1]$-*complete if* $\mathsf{IS}_{11}^2 \subseteq \langle S \rangle \subseteq \mathsf{ID}_2$,

5. $\mathbf{W}[1]$-*hard if* $\mathsf{IS}_{11}^3 \subseteq \langle S \rangle$,

6. **FPT** *if* $\langle S \rangle \subseteq \mathsf{C} \in \{\mathsf{ID}_1, \mathsf{IS}_{12}, \mathsf{IV}_2\}$.

Proof Ideas. **1.+2.** Follows from Theorem 23 in conjunction with Lemma 13.

3. We reduce VERTEXCOVER to our problem similar to the approach of Fellows et al. [18, Thm. 5]. The problem can be translated into an abduction instance with IE knowledge base, consequently giving the desired hardness result.

4.+5. The first result follows from [18, Thm. 25]. Notice that they prove this for ID_2, but using the fact that the formulas (or clauses) in their reduction are IS_{11}^2-formulas, we derive a hardness result for IS_{11}^2. The second statement is then a consequence.

6. We prove this for IM by reducing our problem to the MAXSATs problem which asks, given m clauses, is it possible to set at most s variables to true so that at least k clauses are satisfied (details are presented in [23, Lem. 36]). This problem when parametrised by k, the number of clauses to be satisfied, is **FPT**. Moreover, this reduction can be extended to the languages in IV_2. The problematic part is the presence of positive and unit negative clauses which need to be taken care of (for details, see [23, Lem. 37]). Accordingly, the result for IV_2 follows. The remaining cases are due to Lemmas 19 and 20. □

Now, we end by stating results for $\mathrm{ABD}_{=}$. Interestingly to observe, the majority of the intractable cases is already much harder with large parts being **para-NP**-complete. Even the case of the essentially negative co-clones which are **FPT** for $\mathrm{ABD}_{\leq}$ yield **para-NP**-completeness in this situation. Merely the 2-affine and dualHorn cases are **FPT**.

Theorem 25. *The problem* $\mathrm{ABD}_{=}(S, |M|)$ *is*

1. **para-$\Sigma_2^{\mathbf{P}}$-***hard if* $\mathsf{C} \subseteq \langle S \rangle \subseteq \mathsf{BR}$ *and* $\mathsf{C} \in \{\mathsf{IN}_2, \mathsf{II}_0\}$,

2. **para-coNP-***hard if* $\mathsf{IN} \subseteq \langle S \rangle \subseteq \mathsf{II}_1$,

3. **para-NP-***complete if* $\mathsf{IS}_1^2 \subseteq \langle S \rangle$ *and* $\langle S \rangle \subseteq \mathsf{C} \in \{\mathsf{IE}_2, \mathsf{ID}_2\}$,

4. **FPT** *if* $\langle S \rangle \subseteq \mathsf{C} \in \{\mathsf{ID}_1, \mathsf{IV}_2\}$.

Proof Ideas. **1.+2.** Follows from Theorem 23 in conjunction with Lemma 13.

3. In [23, Lem. 39], we prove that the problem $\mathrm{ABD}_{=}(S, |M|)$ is **para-NP**-hard as long as $\neg x \vee \neg y \in \langle S \rangle_{\neq}$. The case for $\mathrm{ABD}_{=}(\mathsf{IS}_1^2, |M|)$, so also arbitrary bases, then follows as a corollary. The hardness for $\mathsf{IE} \subseteq \langle S \rangle$ follows from arguments used in the proof of Theorem 24 for the IE case. The upper bounds for IE_2 and ID_2 follow trivially since the classical problems are in **NP**.

4. The proof for IV_2 is due to the monotone argument of Lemma 14 and Theorem 24. For ID_1, we proved in Lemma 20 that the classical problem is in **P**. □

4   Conclusion

In this paper, we presented a two-dimensional classification of three central abductive reasoning problems (unrestricted explanation size, $=$, and $\leq$). In one dimension, we consider the different parameterisations $|H|, |M|, |V|, |E|$, and in the other dimension we consider all possible constraint languages defined by corresponding co-clones except the affine co-clones. Often in the past, problems regarding the affine co-clones (resp., clones) resisted a complete classification [1,2,9,11,20,24,35,40]. Also the result of Durand and Hermann [16] underlines how restive problems around affine functions are. It is difficult to explain why exactly these cases are so problematic but the notion of the Fourier expansion [29] of Boolean functions gives a nice and fitting view on that. Informally, the Fourier expansion of a Boolean function is a probability measure mimicking how likely a flip of a variable changes the function value. For instance, disjunctions have a very low Fourier expansion value whereas the exclusive-or function has the maximum. Affine functions can though be seen as rather counterintuitive as every variable influences the function value dramatically.

For all three studied problems, we exhibit the same trichotomy for the parameter $|H|$ (IN is **para-coNP**-hard, IN_2 is **para-DP**-hard, and the remaining are **FPT**). The parameter $|V|$ always allows for **FPT** algorithms independent of the co-clone. Regarding $|E|$, only the two size restricted variants are meaningful. For '$\leq$' we achieve a pentachotomy between **FPT**, **W**[2]-hard, **W**[P]-complete, **para-coNP**-, and **para-DP**-hard. Whereas, for '$=$', we achieve a hexachotomy additionally having **W**[1]-hardness for the essentially negative cases. These **W**[1]-hard cases are also surprising in the sense that for '$\leq$' they are easy and **FPT**. Similarly, the same easy/hard-difference has been observed as well for $|M|$ as the studied parameter. However, here, we distinguish between **para-NP**-complete for '$=$' and **FPT** for '$\leq$'. The complete picture for '$=$' and $|M|$ is a tetrachotomy ranging through **FPT**, **para-NP**-complete, **para-coNP**-hard, and **para-Σ_2^P**-complete. With respect to '$\leq$' and the unrestrictied cases, we also have some **W**[1]-hard cases which lack a precise classification.

Additionally, we already started a bit to study the parameterised enumeration complexity [6] of these problems yielding **FPT-enum** algorithms for $|V|$ and BR as well as for $|H|$ and IE_2, IV_2, ID_2, and IL_2. Furthermore, IL_1 even allows **FPT** algorithms for any parameterisation (so it extends Corollary 16 in that way).

Notice that in this paper, we did not require $H \cap M$ to be empty. However, one can require this (as, for instance, Fellows et al. [18] did). All our proofs (e.g., Lemma 18) can easily be adapted in that direction. Furthermore, we believe that the **para-DP**-hardness for $|H|$ and IN_2 should be extendable to **para-Σ_2^P**-hardness but do not have a full proof yet.

Finally, we want to attack the affine co-clones as well as present matching upper and lower bounds for all cases. Also, parameterised enumeration complexity [5,6] is the next object of our investigations.

References

1. Bauland, M., Mundhenk, M., Schneider, T., Schnoor, H., Schnoor, I., Vollmer, H.: The tractability of model checking for LTL: the good, the bad, and the ugly fragments. ACM Trans. Comput. Logic (TOCL) **12**(2), 13:1–13:28 (2011)
2. Bauland, M., Schneider, T., Schnoor, H., Schnoor, I., Vollmer, H.: The complexity of generalized satisfiability for linear temporal logic. Log. Methods Comput. Sci. **5**(1) (2009). http://arxiv.org/abs/0812.4848
3. Beyersdorff, O., Meier, A., Thomas, M., Vollmer, H.: The complexity of reasoning for fragments of default logic. J. Log. Comput. **22**(3), 587–604 (2012). https://doi.org/10.1093/logcom/exq061
4. Böhler, E., Reith, S., Schnoor, H., Vollmer, H.: Bases for Boolean co-clones. Inf. Process. Lett. **96**(2), 59–66 (2005). https://doi.org/10.1016/j.ipl.2005.06.003
5. Creignou, N., Ktari, R., Müller, J.S., Olive, F., Vollmer, H.: Parameterised enumeration for modification problems. Algorithms **12**(9) (2019). https://doi.org/10.3390/a12090189
6. Creignou, N., Meier, A., Müller, J.S., Schmidt, J., Vollmer, H.: Paradigms for parameterized enumeration. Theory Comput. Syst. **60**(4), 737–758 (2017). https://doi.org/10.1007/s00224-016-9702-4
7. Creignou, N., Meier, A., Thomas, M., Vollmer, H.: The complexity of reasoning for fragments of autoepistemic logic. ACM Trans. Comput. Logic **13**(2), 17:1–17:22 (2012). https://doi.org/10.1145/2159531.2159539
8. Creignou, N., Olive, F., Schmidt, J.: Enumerating all solutions of a Boolean CSP by non-decreasing weight. In: Sakallah, K.A., Simon, L. (eds.) SAT 2011. LNCS, vol. 6695, pp. 120–133. Springer, Heidelberg (2011). https://doi.org/10.1007/978-3-642-21581-0_11
9. Creignou, N., Schmidt, J., Thomas, M.: Complexity of propositional abduction for restricted sets of Boolean functions. In: Lin, F., Sattler, U., Truszczynski, M. (eds.), Principles of Knowledge Representation and Reasoning: Proceedings of the Twelfth International Conference, KR 2010, Toronto, Ontario, Canada, 9–13 May 2010. AAAI Press (2010). http://aaai.org/ocs/index.php/KR/KR2010/paper/view/1201
10. Creignou, N., Schmidt, J., Thomas, M.: Complexity classifications for propositional abduction in post's framework. J. Log. Comput. **22**(5), 1145–1170 (2012). https://doi.org/10.1093/logcom/exr012
11. Creignou, N., Schmidt, J., Thomas, M., Woltran, S.: Sets of Boolean connectives that make argumentation easier. In: Janhunen, T., Niemelä, I. (eds.) JELIA 2010. LNCS (LNAI), vol. 6341, pp. 117–129. Springer, Heidelberg (2010). https://doi.org/10.1007/978-3-642-15675-5_12
12. Creignou, N., Schmidt, J., Thomas, M., Woltran, S.: Complexity of logic-based argumentation in post's framework. Argum. Comput. **2**(2–3), 107–129 (2011). https://doi.org/10.1080/19462166.2011.629736
13. Creignou, N., Zanuttini, B.: A complete classification of the complexity of propositional abduction. SIAM J. Comput. **36**(1), 207–229 (2006). https://doi.org/10.1137/S0097539704446311
14. Downey, R.G., Fellows, M.R.: Parameterized Complexity. Monographs in Computer Science. Springer, Heidelberg (1999). https://doi.org/10.1007/978-1-4612-0515-9
15. Downey, R.G., Fellows, M.R.: Fundamentals of Parameterized Complexity. Texts in Computer Science. Springer, Heidelberg (1999). https://doi.org/10.1007/978-1-4471-5559-1

16. Durand, A., Hermann, M.: The inference problem for propositional circumscription of afine formulas Is coNP-complete. In: Alt, H., Habib, M. (eds.) STACS 2003. LNCS, vol. 2607, pp. 451–462. Springer, Heidelberg (2003). https://doi.org/10.1007/3-540-36494-3_40

17. Eiter, T., Gottlob, G.: The complexity of logic-based abduction. J. ACM **42**(1), 3–42 (1995). https://doi.org/10.1145/200836.200838

18. Fellows, M.R., Pfandler, A., Rosamond, F.A., Rümmele, S.: The parameterized complexity of abduction. In: Hoffmann, J., Selman, B. (eds.), Proceedings of the Twenty-Sixth AAAI Conference on Artificial Intelligence, 2 July 2012, Toronto, Ontario, Canada. AAAI Press (2012). http://www.aaai.org/ocs/index.php/AAAI/AAAI12/paper/view/5048

19. Flum, J., Grohe, M.: Parameterized Complexity Theory. Texts in Theoretical Computer Science. An EATCS Series. Springer, Heidelberg (2006). https://doi.org/10.1007/3-540-29953-X

20. Hemaspaandra, E., Schnoor, H., Schnoor, I.: Generalized modal satisfiability. CoRR, abs/0804.2729:1–32 (2008). http://arxiv.org/abs/0804.2729

21. Josephson, J.R., Chandrasekaran, B., Smith, J.W., Tanner, M.C.: A mechanism for forming composite explanatory hypotheses. IEEE Trans. Syst. Man Cybern. **17**(3), 445–454 (1987). https://doi.org/10.1109/TSMC.1987.4309060

22. Lewis, H.R.: Satisfiability problems for propositional calculi. Math. Sys. Theory **13**, 45–53 (1979)

23. Mahmood, Y., Meier, A., Schmidt, J.: Parameterised complexity for abduction. CoRR, abs/1906.00703 (2019)

24. Meier, A., Mundhenk, M., Schneider, T., Thomas, M., Weber, V., Weiss, F.: The complexity of satisfiability for fragments of hybrid logic—Part I. In: Královič, R., Niwiński, D. (eds.) MFCS 2009. LNCS, vol. 5734, pp. 587–599. Springer, Heidelberg (2009). https://doi.org/10.1007/978-3-642-03816-7_50

25. Meier, A., Schneider, T.: Generalized satisfiability for the description logic ALC. Theor. Comput. Sci. **505**, 55–73 (2013). https://doi.org/10.1016/j.tcs.2013.02.009

26. Meier, A., Thomas, M., Vollmer, H., Mundhenk, M.: The complexity of satisfiability for fragments of CTL and ctl*. Int. J. Found. Comput. Sci. **20**(5), 901–918 (2009). https://doi.org/10.1142/S0129054109006954

27. Morgan, C.G.: Hypothesis generation by machine. Artif. Intell. **2**(2), 179–187 (1971). https://doi.org/10.1016/0004-3702(71)90009-9

28. Nordh, G., Zanuttini, B.: What makes propositional abduction tractable. Artif. Intell. **172**(10), 1245–1284 (2008). https://doi.org/10.1016/j.artint.2008.02.001

29. O'Donnell, R.: Analysis of Boolean Functions. Cambridge University Press, Cambridge (2014). http://www.cambridge.org/de/academic/subjects/computer-science/algorithmics-complexity-computer-algebra-and-computational-g/analysis-boolean-functions

30. Papadimitriou, C.H.: Computational Complexity. Addison-Wesley, Boston (1994)

31. Peirce, C.S.: Collected Papers of Charles Sanders Peirce. Oxford University Press, London (1958)

32. Peng, Y., Reggia, J.A.: Abductive Inference Models for Diagnostic Problem-Solving. Artificial Intelligence. Springer. New York (1990). https://doi.org/10.1007/978-1-4419-8682-5

33. Poole, D.: Normality and faults in logic-based diagnosis. In: Sridharan, N.S. (ed.), Proceedings of the 11th International Joint Conference on Artificial Intelligence. Detroit, MI, USA, August 1989, pp. 1304–1310. Morgan Kaufmann (1989). http://ijcai.org/Proceedings/89-2/Papers/073.pdf

34. Post, E.L.: The two-valued iterative systems of mathematical logic. Ann. Math. Stud. **5**, 1–122 (1941)
35. Reith, S.: Generalized satisfiability problems. Ph.D. thesis, Julius Maximilians University Würzburg, Germany (2001). http://opus.bibliothek.uni-wuerzburg.de/opus/volltexte/2002/7/index.html
36. Robertson, N., Seymour, P.D.: Graph minors. II. Algorithmic aspects of tree-width. J. Algorithms **7**(3), 309–322 (1986). https://doi.org/10.1016/0196-6774(86)90023-4
37. Schaefer, T.J.: The complexity of satisfiability problems. In: Lipton, R.J., Burkhard, W.A., Savitch, W.J., Friedman, E.P., Aho, A.V. (eds.), Proceedings of the 10th Annual ACM Symposium on Theory of Computing, San Diego, California, USA, 1–3 May 1978, pp. 216–226. ACM (1978). https://doi.org/10.1145/800133.804350
38. Schnoor, H., Schnoor, I.: Partial polymorphisms and constraint satisfaction problems. In: Creignou, N., Kolaitis, P.G., Vollmer, H. (eds.) Complexity of Constraints. LNCS, vol. 5250, pp. 229–254. Springer, Heidelberg (2008). https://doi.org/10.1007/978-3-540-92800-3_9
39. Selman, B., Levesque, H.J.: Abductive and default reasoning: a computational core. In: Shrobe, H.E., Dietterich, T.G., Swartout, W.R. (eds.), Proceedings of the 8th National Conference on Artificial Intelligence, Boston, Massachusetts, USA, 29 July–3 August 1990, vol. 2, pp. 343–348. AAAI Press/The MIT Press (1990). http://www.aaai.org/Library/AAAI/1990/aaai90-053.php
40. Thomas, M.: The complexity of circumscriptive inference in post's lattice. In: Erdem, E., Lin, F., Schaub, T. (eds.) LPNMR 2009. LNCS (LNAI), vol. 5753, pp. 290–302. Springer, Heidelberg (2009). https://doi.org/10.1007/978-3-642-04238-6_25

Tracking Computability
of GPAC-Generable Functions

Diogo Poças[1]([✉])[iD] and Jeffery Zucker[2]

[1] Operations Research, Technical University of Munich, Munich, Germany
diogo.pocas@tum.de
[2] Department of Computing and Software, McMaster University, Hamilton, Canada
zucker@cas.mcmaster.ca

Abstract. Analog computation attempts to capture any type of computation, that can be realized by any type of physical system or physical process, including but not limited to computation over continuous measurable quantities. A pioneering model is the General Purpose Analog Computer (GPAC), initially presented by Shannon in 1941. The GPAC is capable of manipulating real-valued data streams; however, it has been shown to be strictly less powerful than other models of computation on the reals, such as computable analysis.

In previous work, we proposed an extension of the Shannon GPAC, denoted LGPAC, designed to overcome its limitations. Not only is the LGPAC model capable of expressing computation over general data spaces $\mathcal{X}$, it also directly incorporates approximating computations by means of a limit module. In this paper, we compare the LGPAC with a digital model of computation based on effective representations (tracking computability). We establish general conditions under which LGPAC-generable functions are tracking computable.

Keywords: Generalized computability · Generalized recursion theory · Computation on the reals · Analog computation · Shannon GPAC · Tracking computability

1 Introduction

A central goal in computability theory is to establish equivalences between disparate notions of computation; such equivalence results serve as strong indications of the validity of the theory as a whole, as they suggest robustness (or perhaps, indifference) against the choice of a particular model of computation.

In the framework of digital computation, such considerations have led to the celebrated Church-Turing thesis that asserts that any realizable method of computation has the same computational power as the Turing machine. However, the picture is not so clear in the case of computation over more general data spaces, or analog computation. Analog computation, as conceived by Kelvin [18], Bush [2], and Hartree [6], is a form of experimental computation with physical systems called analog devices or analog computers. Historically, data are

© Springer Nature Switzerland AG 2020
S. Artemov and A. Nerode (Eds.): LFCS 2020, LNCS 11972, pp. 214–235, 2020.
https://doi.org/10.1007/978-3-030-36755-8_14

represented by measurable physical quantities, including lengths, shaft rotation, voltage, current, resistance, etc.

The General Purpose Analog Computer (GPAC) was introduced by Shannon [15] as a model of Bush's Differential Analyzer [2]. Shannon discovered that a function can be generated by a GPAC if, and only if, it is differentially algebraic. In particular, this implies that non-differentially algebraic functions, such as the gamma function, cannot be generated by the Shannon GPAC.

In previous work [11,12], we proposed different extensions of the Shannon GPAC, attempting to overcome its limitations. In particular, our models express computation over general data spaces $\mathcal{X}$ beyond real numbers, and directly incorporate approximating computations by means of a *limit module*. The goal of this paper is to connect the LGPAC (GPAC + limits) with other such models of computation. Specifically, we shall consider the notion of *tracking computability* [16,19], which we take as a paradigm for digital computation. The idea of tracking computability comes from Mal'cev [9], and it has been found to be equivalent (under reasonable conditions) to a number of other well-known digital computation models [16,17,19,21]. In this work, we find suitable conditions that guarantee that a function generated by an LGPAC is also tracking computable.

We begin by introducing both notions of computability: for the GPAC model, we describe the channels, modules, input-output operator and fixed point semantics; for tracking computability, we study computable structures and effective representations. In the most technical part of the paper, we prove tracking computability of the functions associated with the LGPAC modules and of the input-output operator of an LGPAC. Finally, we attempt to prove tracking computability of LGPAC-generable functions; in order to achieve this, we assume an additional condition, which we call *effective well-posedness*.

This research is part of a project to compare the strengths of various models of analog and digital computation. In the present case (LGPAC and tracking computability) we have been successful in one direction, while the other direction remains an open problem.

In regard to the original content of this paper, we remark that the paper [1] already shows an equivalence between a GPAC-like model for *real* computation (which includes approximability) and computable analysis (which is closely related to tracking computability; papers [19] and [20] have some equivalence results). However, our model differs from the model in that paper in two critical ways: computation on general data spaces is allowed, and approximability is directly incorporated by means of limit modules. Hence, the two models are not obviously comparable. The technical notion of effective well-posedness (Definition 8) is also an original idea.

2   Preliminaries

Our model of computation is built over a data space $\mathcal{X}$ which represents the space of possible data points. Typical spaces of interest are $\mathbb{R}$ (the real numbers), $C(\mathbb{R})$ (continuous real functions of one real variable), $C^1(\mathbb{R})$ (continuously differentiable real functions) and so on. The results in this paper will be stated for *separable Fréchet spaces*, which satisfy the following assumptions.

1. $\mathcal{X}$ is a Fréchet space with respect to a family of pseudonorms $\|\cdot\|_n$ (indexed by $n \in \mathbb{N}$); in particular, it is equipped with the vector space operations of addition and scalar multiplication, as well as a zero element $0 \in \mathcal{X}$. We recall the pseudonorm axioms:

1a. each pseudonorm $\|\cdot\|_n : \mathcal{X} \to \mathbb{R}_{\geq 0}$ is positive semidefinite ($\|x\|_n \geq 0$), scalable ($\|rx\|_n = |r|\|x\|_n$ for $r \in \mathbb{R}$) and subadditive ($\|x + y\|_n \leq \|x\|_n + \|y\|_n$);

1b. the family of pseudonorms separates points: if $\|x - y\|_n = 0$ for all n, then $x = y$;

1c. $\mathcal{X}$ is complete, that is, Cauchy sequences are convergent: for any sequence $(x_m) \in \mathcal{X}^{\mathbb{N}}$, if $\lim_{s,t\to\infty} \|x_s - x_t\|_n = 0$ for all n, then there exists $x \in \mathcal{X}$ such that $\lim_{s\to\infty} \|x_s - x\|_n = 0$;

1d. for convenience, we additionally assume without loss of generality that the pseudonorms are nondecreasing: if $n \leq m$ then $\|x\|_n \leq \|x\|_m$.

2. Finally, $\mathcal{X}$ is separable, i.e. it has a countable dense subset. We fix an enumeration $\alpha_{\mathcal{X}} : \mathbb{N} \to \mathcal{X}_c$ of a countable dense subset $\mathcal{X}_c = \alpha_{\mathcal{X}}(\mathbb{N}) \subseteq \mathcal{X}$. We also assume for convenience that $\alpha_{\mathcal{X}}(0) = 0$.

The main reason for considering Fréchet spaces instead of Banach or Hilbert spaces is that the data spaces we wish to consider, such as $C(\mathbb{R})$ or $C(\mathbb{T}, C(\mathbb{R}))$, are generally not equipped with norms. We refer the reader to [13, Ch. V], where a detailed exposition of Fréchet spaces can be found. Finally, we note that the family of pseudonorms induces a metric on the Fréchet space as follows.

Definition 1 (Metric from pseudonorms). *Let $\mathcal{X}$ be a Fréchet space with pseudonorms $\|\cdot\|_n$, $n \in \mathbb{N}$. We define the metric*

$$d(x, y) = \sup_{n \in \mathbb{N}} 2^{-n} \min\left(\|x - y\|_n, 1\right), \quad x, y \in \mathcal{X}. \tag{1}$$

Proposition 1 (Bounds on the pseudonorms and bounds on the metric). *Let $\mathcal{X}$ be a Fréchet space with pseudonorms $\|\cdot\|_n$, $n \in \mathbb{N}$. Then for the metric defined by (1), the following hold for any $x, y \in \mathcal{X}$ and $n, m \in \mathbb{N}$:*

$$if\ d(x, y) < 2^{-n-m},\ then\ \|x - y\|_n < 2^{-m};$$
$$if\ \|x - y\|_n < 2^{-m},\ then\ d(x, y) < 2^{-\min(n,m)}.$$

3   The LGPAC

We give a formal definition of the LGPAC model[1]. The main objects of our study are analog networks or analog systems, whose main components can be viewed as follows:

---

[1] More details can be found in [11, 12].

Analog network = data + time + channels + modules.

As already mentioned, we model data as elements of a separable Fréchet space $\mathcal{X}$. We will use the nonnegative real numbers as a continuous model of *time* $\mathbb{T} = [0, \infty)$. We consider two types of *channels*: *scalar channels* carry constant values in $\mathcal{X}$, whereas *stream channels* carry continuously differentiable streams in $C^1(\mathbb{T}, \mathcal{X})$.

We also remark that if $\mathcal{X}$ is a Fréchet space, so is $C^1(\mathbb{T}, \mathcal{X})$; this however does not hold in general for Banach spaces, which again explains our choice for Fréchet spaces. In particular, we can define the following family of nondecreasing pseudonorms on $C^1(\mathbb{T}, \mathcal{X})$,[2]

$$\|u\|_n = \|u(0)\|_n + \sup_{0 \le t \le n} \|u'(t)\|_n. \tag{2}$$

Note that many useful properties of integration can be extended to $C^1(\mathbb{T}, \mathcal{X})$; in particular, for $u \in C^1(\mathbb{T}, \mathcal{X})$, $\nu \in \mathbb{N}$, and $t \in [0, \nu]$, we have the bound (which we will use later)

$$\|u(t)\|_\nu \le \|u(0)\|_\nu + \int_0^t \|u'(s)\|_\nu ds \le \|u(0)\|_\nu + t \sup_{t \le \nu} \|u'(t)\|_\nu \le \nu \|u\|_\nu. \tag{3}$$

Each *module* M has zero, one or more input channels, and must have a single output channel; thus it can be specified by a (possibly partially defined) *stream function*

$$F_M : A_1 \times \ldots \times A_k \rightharpoonup A_{k+1} \quad (k \ge 0),$$

where each of A_i, $i = 1 \ldots k+1$ is either $\mathcal{X}_i$ or $C^1(\mathbb{T}, \mathcal{X}_i)$ for some data space $\mathcal{X}_i$; and we use the symbol $\rightharpoonup$ to mean that F_M may be partial-valued. The Shannon GPAC is obtained if all $\mathcal{X}_i = \mathbb{R}$, and the following four types of modules are considered.

Definition 2 (Shannon modules). *The* Shannon modules *are defined as follows:*

- *for each $c \in \mathbb{R}$, there is a* constant *module with zero inputs and one output $v(t) = c$;*
- *the* adder *module has two inputs u, v and one output w, given by $w(t) = u(t) + v(t)$;*
- *the* multiplier *module has two inputs u, v and one output w, given by $w(t) = u(t)v(t)$;*
- *the* integrator *module has a scalar input c (also called the initial setting), two stream inputs u, v and one output w, given by the Riemann-Stieltjes integral $w(t) = c + \int_0^t u(s)v'(s)ds$;*

---

[2] Here, assumption (1d), that the original family of pseudonorms on $\mathcal{X}$ is nondecreasing, is required; alternatively, one could introduce a double-indexing family such as $\|u\|_{n,m} = \|u(0)\|_n + \sup_{0 \le t \le m} \|u'(t)\|_n$.

We have previously extended the Shannon GPAC in two different ways.

1. General data spaces. In [11] we defined the $\mathcal{X}$-GPAC, allowing the study of functions of more than one variable. The main idea present in that paper is to extend the *output space*, that is, replacing $C^1(\mathbb{T}, \mathbb{R})$ with $C^1(\mathbb{T}, \mathcal{X})$, where $\mathcal{X}$ is a metric vector space. For example, we can think of $\mathcal{X}$ as the space of continuous real-valued functions on $\mathbb{R}^n$, that is, $\mathcal{X} = C(\mathbb{R}^n, \mathbb{R})$. In this way, our channels will now carry $\mathcal{X}$-valued streams of data $u : \mathbb{T} \to \mathcal{X}$, which correspond to functions of $n + 1$ real variables, under the "uncurrying" $\mathbb{T} \to (\mathbb{R}^n \to \mathbb{R}) \simeq \mathbb{T} \times \mathbb{R}^n \to \mathbb{R}$. Evidently, one of the independent variables, namely "time", plays a different role from the others - it can be used as a variable for integration and taking limits.

This leads us to consider a multityped GPAC, which means that different channels may carry values over different data spaces. In particular, we shall fix one separable Fréchet space $\mathcal{X}$ and allow four channel types: $\mathbb{R}$-*variables*, $\mathcal{X}$-*variables*, $\mathbb{R}$-*streams* and $\mathcal{X}$-*streams*, which carry values in $\mathbb{R}$, $\mathcal{X}$, $C^1(\mathbb{T}, \mathbb{R})$ and $C^1(\mathbb{T}, \mathcal{X})$, respectively.

We generalize the Shannon modules to $C^1(\mathbb{T}, \mathcal{X})$, obtaining the four basic modules depicted in Fig. 1 as box diagrams.[3] We also introduce the symbol '$\int$' to denote the operator associated with the integrator module; we can then write $\int (c, u, r) = c + \int u dr$.

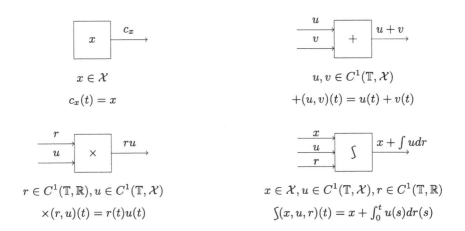

$x \in \mathcal{X}$

$c_x(t) = x$

$u, v \in C^1(\mathbb{T}, \mathcal{X})$

$+(u, v)(t) = u(t) + v(t)$

$r \in C^1(\mathbb{T}, \mathbb{R}), u \in C^1(\mathbb{T}, \mathcal{X})$

$\times(r, u)(t) = r(t)u(t)$

$x \in \mathcal{X}, u \in C^1(\mathbb{T}, \mathcal{X}), r \in C^1(\mathbb{T}, \mathbb{R})$

$\int(x, u, r)(t) = x + \int_0^t u(s)dr(s)$

Fig. 1. The four basic modules.

Observe that we are using *scalar* multiplication (of type $\mathbb{R} \times \mathcal{X} \to \mathcal{X}$) as the basis for the multiplication and integrator modules. At this level of generality, we cannot use multiplication (of type $\mathcal{X} \times \mathcal{X} \to \mathcal{X}$) because such an operation does not arise from the Fréchet space axioms. One can extend our model, assuming $\mathcal{X}$ is equipped with a multiplication operator, under suitable additional assumptions: of course, multiplication should be bilinear (i.e. distributive with respect

---

[3] By assumption, addition and scalar multiplication are defined on $\mathcal{X}$. The integral can be generalized to $C^1(\mathbb{T}, \mathcal{X})$ via Riemann sums: see, for example, [14, p. 89].

to addition, and compatible with scalar multiplication), but more importantly, it should be *bounded* (for example, one could assume $\|u \times v\|_\nu \leq \|u\|_\nu \|v\|_\nu$ for each pseudonorm ν). In Sect. 7, we will consider the case $\mathcal{X} = C(\mathbb{R})$ and extend the GPAC with a function multiplication module.

2. Limit modules. We introduced in [12] a *limit module* in order to incorporate approximating computations by means of *effective convergence*. If $M : \mathbb{N} \to \mathbb{N}$ is nondecreasing and $(g_n) \in \mathcal{X}^{\mathbb{N}}$, we say that (g_n) is an *M-convergent Cauchy sequence* if for all $\nu \in \mathbb{N}$ and $m, n \geq M(\nu)$ one has $d(g_m, g_n) < 2^{-\nu}$. Similarly, if $T \in C^1(\mathbb{T}, \mathbb{R})$ is nondecreasing and $u \in C^1(\mathbb{T}, \mathcal{X})$, we say that u is a *T-convergent Cauchy stream* if for all $\tau \in \mathbb{T}$ and $s, t \geq T(\tau)$ one has $d(u(s), u(t)) < 2^{-\tau}$.

We call such a non-decreasing function M (resp. T) a *discrete* (resp. *continuous*) *modulus of convergence*. A typical example is the identity function, either discrete (id $: \mathbb{N} \to \mathbb{N}$) or continuous (id $\in C^1(\mathbb{T}, \mathbb{R})$). We note that any M-convergent Cauchy sequence may be replaced by an id-convergent Cauchy sequence via a composition with its modulus of convergence. Similarly, a T-convergent Cauchy stream may be replaced by an id-convergent Cauchy stream. This brings us to the notion of a *limit operator*.

Definition 3 (Limit modules). *For the data type $\mathcal{X}$, there is a continuous limit module with one input of type $C^1(\mathbb{T}, \mathcal{X})$ and one output of type $\mathcal{X}$. For input u, it outputs the id-convergent limit $\lim_{t \to \infty} u(t)$ (if it exists).*

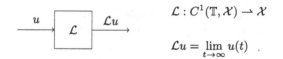

$$\mathcal{L} : C^1(\mathbb{T}, \mathcal{X}) \rightharpoonup \mathcal{X}$$

$$\mathcal{L}u = \lim_{t \to \infty} u(t) \ .$$

Fig. 2. The limit module.

The limit module is depicted in Fig. 2. Observe that it defines a partial-valued operator; it is only defined for those functions in $C^1(\mathbb{T}, \mathcal{X})$ that have an id-convergent limit.

Definition 4 (LGPAC). *Let $\mathcal{X}$ be a separable Fréchet space. A limit general purpose analog computer (LGPAC) is a network built with $\mathbb{R}$-channels, $\mathcal{X}$-channels (carrying either constants or streams), the basic modules (constants, adders, multipliers, integrators) and the continuous limit module. Moreover, the channels connect the inputs and outputs of the modules, with the following restrictions: the only connections allowed are between an output and an input; each input may be connected to either zero or one output.*

Thus, a GPAC channel may appear as an unconnected input (*proper input*), unconnected output (*proper output*), or connect an input with an output (*mixed input/output*).

Definition 5 (LGPAC semantics [11]). *Any LGPAC $\mathcal{G}$ induces an* input-output operator $\Phi : \mathcal{I} \times \mathcal{M} \rightharpoonup \mathcal{M} \times \mathcal{O}$, where $\mathcal{I}$, $\mathcal{M}$, $\mathcal{O}$ *denote the spaces of proper input, mixed input/output, and proper output channels respectively;*

1. *for variables* $\mathbf{u}^I \in \mathcal{I}$, $\mathbf{u}^M \in \mathcal{M}$, $\mathbf{u}^O \in \mathcal{O}$, *the* fixed point equation *is given by*

$$\Phi(\mathbf{u}^I, \mathbf{u}^M) = (\mathbf{u}^M, \mathbf{u}^O); \tag{4}$$

2. $\mathcal{G}$ *is* well-posed *on an open subset* $U \subseteq \mathcal{I}$ *if for all* $\mathbf{u}^I \in U$ *there is a unique* $(\mathbf{u}^M, \mathbf{u}^O)$ *such that* (4) *holds; and moreover, the solution map* $\mathbf{u}^I \mapsto (\mathbf{u}^M, \mathbf{u}^O)$ *describes a continuous function* $F : U \to \mathcal{M} \times \mathcal{O}$ *with domain* U*; we further say that* $\mathcal{G}$ generates F*, or that* F *is* LGPAC-generable.

Although Definitions 4 and 5 refer to networks built with the LGPAC modules, it is not hard to see how they generalize to any choice of arbitrary modules, which would define a more abstract notion of *multityped GPAC*. Some of our results (namely Lemma 1) can be stated in this more general form.

4   Tracking Computability

The procedure for defining tracking computability in general spaces has been extensively documented by many authors (see, e.g. [16,17,20,21]). The basic construction consists of taking an enumeration of a countable dense subset, defining *computable* elements as those given by effective Cauchy sequences, and considering *tracking functions*. We assume that we have fixed an enumeration $\alpha_{\mathcal{X}} : \mathbb{N} \to \mathcal{X}_c$ of a (countable) dense subset $\mathcal{X}_c \subseteq \mathcal{X}$.

Let us also fix a family of computable bijections $\langle \cdot, \cdots, \cdot \rangle : \mathbb{N}^k \to \mathbb{N}$, for $k \in \mathbb{N}^+$ (say, the Cantor pairing function $\langle \cdot, \cdot \rangle$ for $k = 2$ and its generalizations to higher dimensions), as well as an enumeration $\{\cdot\} : \mathbb{N} \to (\mathbb{N} \rightharpoonup \mathbb{N})$ of the recursive functions (say, for $T \in \mathbb{N}$ the encoding of a one-input, one-output Turing machine, $\{T\}$ is the corresponding recursive function).

Definition 6 (Computability structure). *Let* $\mathcal{X}$ *be a complete metric space and* $(\mathcal{X}_c, \alpha)$ *an enumerated countable dense subset. A* computability structure $(\Omega_{\bar{\alpha}}, C_{\bar{\alpha}}, \bar{\alpha})$ *is defined as follows.*

1. *The set of* valid codes*,* $\Omega_{\bar{\alpha}}$*, is the subset of* $\mathbb{N}$ *given by encodings of pairs of numbers* $c = \langle T, M \rangle$ *such that* T *is the index for a total recursive function* $\{T\}$*,* M *is the index for a total recursive discrete modulus of convergence* $\{M\}$ *and* $(\alpha\{T\}(n))$ *is an* $\{M\}$*-convergent Cauchy sequence.[4]*
2. *The* partial enumeration $\bar{\alpha} : \mathbb{N} \rightharpoonup \mathcal{X}$ *is the function with domain* $\Omega_{\bar{\alpha}}$ *such that for any* $c = \langle T, M \rangle \in \Omega_{\bar{\alpha}}$*,* $\bar{\alpha}(c) = \lim_{n \to \infty} \alpha\{T\}(n)$*.*
3. *The set of* computable elements $C_{\bar{\alpha}} \subseteq \mathcal{X}$ *is the range of* $\bar{\alpha}$*, i.e.* $C_{\bar{\alpha}} = \bar{\alpha}(\mathbb{N})$*.*

*Example 1 (**Computability on** $\mathbb{R}$).* To construct a computability structure on the space $\mathcal{X} = \mathbb{R}$, we can take $\mathcal{X}_c = \mathbb{Q}$, and $\alpha = \alpha_{\mathbb{R}}$ as any standard enumeration of the rationals. This gives the set $C_{\bar{\alpha}}$ of computable reals.

---

[4] For ease of notation we write $\alpha\{T\}(n)$ instead of $\alpha(\{T\}(n))$.

*Example 2 (**Computability on** $C(\mathbb{R})$).* We define a computability structure on $\mathcal{X} = C(\mathbb{R})$, which is a Fréchet space with pseudonorms $\|f\|_n = \sup_{-n \leq x \leq n} |f(x)|$. We take $\mathcal{X}_c$ to be a countable subset of piecewise linear rational functions, defined as follows. For each $N \in \mathbb{N}$ and each tuple $(p_{-N^2}, \ldots, p_{-1}, p_0, p_1, \ldots, p_{N^2})$ of $2N^2 + 1$ rational numbers, we can consider a function $f : \mathbb{R} \to \mathbb{R}$ such that: $f(x) = p_{-N^2}$ for $x \leq -N$; $f(x) = p_{N^2}$ for $x \geq N$; $f(j/N) = p_j$ for $j \in \{-N^2, \ldots, 0, \ldots, N^2\}$; and f is piecewise linear on each interval $[j/N, (j+1)/N]$ for $j \in \{-N^2, \ldots, 0, \ldots, N^2 - 1\}$.

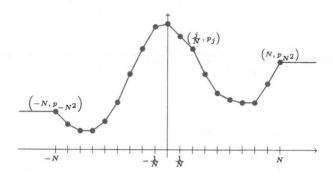

Fig. 3. A piecewise linear rational function.

In this way, the role of N is both to increase the 'window size' and decrease the 'step size' of our approximation (see Fig. 3). By using the bijections of type $\mathbb{N}^2 \to \mathbb{N}$ and $\mathbb{N}^{2N^2+1} \to \mathbb{N}$, and the enumeration $\alpha_\mathbb{R}$ from the previous example, we can define an enumeration $\alpha_\mathcal{X} : \mathbb{N} \to \mathcal{X}_c$. Specifically, the enumeration is as follows: for $e = \langle N, \langle m_{-N^2}, \ldots, m_{N^2} \rangle \rangle$, we define $\alpha_\mathcal{X}(e)$ to be the stream u built from N and the tuple $(p_{-N^2}, \ldots, p_{N^2})$ where $p_j = \alpha_\mathbb{R}(m_j)$ for each $j \in \{-N^2, \ldots, 0, \ldots, N^2\}$. Finally, we can apply the construction of Definition 6 and consider the set of computable elements $C_{\bar\alpha}$. In this case, this set coincides with the familiar set of computable real functions, as seen in [10,22], among others.

*Example 3 (**Computability on** $C^1(\mathbb{T}, \mathcal{X})$).* Given a computability structure on a separable Fréchet space $\mathcal{X}$, say with an enumeration $(\alpha_\mathcal{X}, \mathcal{X}_c)$, we shall construct a computability structure on the space of $\mathcal{X}$-streams $\mathcal{Z} = C^1(\mathbb{T}, \mathcal{X})$. We apply the same idea as in Example 2, but now we need to account for continuous differentiability. The idea is to construct an interpolant from a finite amount of 'data points'. If $u \in C^1(\mathbb{T}, \mathcal{X})$ then its derivative $u' \in C(\mathbb{T}, \mathcal{X})$. Therefore, we can approximate u' by a piecewise linear function and then integrate the approximation with respect to the time variable.

Formally, for each $N \in \mathbb{N}$ and each tuple $(x_0, y_0, \ldots, y_{N^2})$ of $N^2 + 2$ elements in $\mathcal{X}_c$, we consider the functions $u, v : \mathbb{T} \to \mathcal{X}$ such that: $v(t) = y_{N^2}$ for $t \geq N$; $v(j/N) = y_j$ for $j \in \{0, \ldots, N^2\}$; v is piecewise linear (as a function of t) and given by $v(t) = y_j + (y_{j+1} - y_j)(Nt - j)$ on each interval $[j/N, (j+1)/N]$, for $j \in \{0, \ldots, N^2 - 1\}$; finally, $u(t) = x_0 + \int_0^t v(s)ds$.

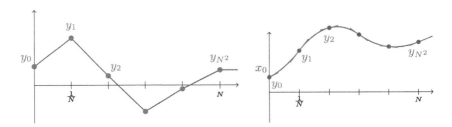

Fig. 4. A continuous piecewise linear function v (left) and its integral, a C^1 piecewise quadratic function u (right). The data consist of an initial value x_0 and derivative values $y_0, \ldots, y_{N^2}$ at equispaced points.

By construction, each u is continuously differentiable and piecewise quadratic (Fig. 4). Now let $\mathcal{Z}_c \subseteq \mathcal{Z}$ be the space of functions u considered above. Using the bijections of type $\mathbb{N}^2 \to \mathbb{N}$ and $\mathbb{N}^{N^2+2} \to \mathbb{N}$, and the enumeration $\alpha_{\mathcal{X}}$, we can define an enumeration $\alpha_{\mathcal{Z}} : \mathbb{N} \to \mathcal{Z}_c$. Specifically: for $e = \langle N, \langle m_0, m_0', \ldots, m_{N^2}' \rangle \rangle$, we define $\alpha_{\mathcal{Z}}(e)$ to be the stream u built from N and the tuple $(x_0, y_0, \ldots, y_{N^2})$ where $x_0 = \alpha_{\mathcal{X}}(m_0)$ and $y_j = \alpha_{\mathcal{X}}(m_j')$ for each $j = 0, \ldots N^2$.

$\mathcal{Z}_c$ is easily seen to be countable and dense in $\mathcal{Z}$. Thus, we can apply the construction of Definition 6 and obtain the computability structure $(\Omega_{\bar{\alpha}_{\mathcal{Z}}}, C_{\bar{\alpha}_{\mathcal{Z}}}, \bar{\alpha}_{\mathcal{Z}})$.

*Example 4 (**Computability on** $\mathcal{X} \times \mathcal{Y}$).* Given computability structures on spaces $\mathcal{X}, \mathcal{Y}$, one can define a computability structure on the product $\mathcal{X} \times \mathcal{Y}$ using the enumeration $\alpha_{\mathcal{X} \times \mathcal{Y}}(\langle \ell, r \rangle) = (\alpha_{\mathcal{X}}(\ell), \alpha_{\mathcal{Y}}(r))$. Note that pseudonorms (and a metric) on $\mathcal{X} \times \mathcal{Y}$ can be easily induced from $\mathcal{X}$ and $\mathcal{Y}$ as, for example, $\|(x_1, y_1) - (x_2, y_2)\|_n = \|x_1 - x_2\|_n + \|y_1 - y_2\|_n$. It is also not hard to see that $C_{\bar{\alpha}_{\mathcal{X} \times \mathcal{Y}}} = C_{\bar{\alpha}_{\mathcal{X}}} \times C_{\bar{\alpha}_{\mathcal{Y}}}$, and that this construction can be generalized to finite products of the form $\mathcal{X}_1 \times \ldots \times \mathcal{X}_N$ and to X^N.

Definition 7 (Tracking computability). *Let $\mathcal{X}$ and $\mathcal{Y}$ be complete metric spaces with enumerated countable dense subsets $(\mathcal{X}_c, \alpha_{\mathcal{X}})$, $(\mathcal{Y}_c, \alpha_{\mathcal{Y}})$ and computability structures $(\Omega_{\bar{\alpha}_{\mathcal{X}}}, C_{\bar{\alpha}_{\mathcal{X}}}, \bar{\alpha}_{\mathcal{X}})$, $(\Omega_{\bar{\alpha}_{\mathcal{Y}}}, C_{\bar{\alpha}_{\mathcal{Y}}}, \bar{\alpha}_{\mathcal{Y}})$. Let $f : \mathcal{X} \rightharpoonup \mathcal{Y}$, and consider a function $\varphi : \mathbb{N} \rightharpoonup \mathbb{N}$ (see Fig. 5 for a graphical depiction).*

We say that φ is a tracking function with respect to $(\alpha_{\mathcal{X}}, \alpha_{\mathcal{Y}})$, or an $(\alpha_{\mathcal{X}}, \alpha_{\mathcal{Y}})$-tracking function, for f, if for all $c \in \Omega_{\bar{\alpha}_{\mathcal{X}}}$ with $\bar{\alpha}_{\mathcal{X}}(c) \in \mathrm{dom}\, f$, we have that $c \in \mathrm{dom}\, \varphi$ and $\varphi(c) \in \Omega_{\bar{\alpha}_{\mathcal{Y}}}$ and $\bar{\alpha}_{\mathcal{Y}}(\varphi(c)) = f(\bar{\alpha}_{\mathcal{X}}(c))$. When a function f has a recursive tracking function φ, we say that f is tracking computable with respect to $(\alpha_{\mathcal{X}}, \alpha_{\mathcal{Y}})$, or $(\alpha_{\mathcal{X}}, \alpha_{\mathcal{Y}})$-computable.

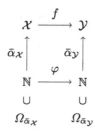

Fig. 5. Tracking function.

5 Computability of the Input-Output Operator

The goal of this section is to demonstrate that the input-output operator of an LGPAC is tracking computable. We first show that this follows directly from the tracking computability of the basic modules.

Lemma 1 (Tracking computability of the input-output operator). *Let $\mathcal{G}$ be a multityped GPAC with input-output operator $\Phi : \mathcal{I} \times \mathcal{M} \rightharpoonup \mathcal{M} \times \mathcal{O}$. Suppose that each of the modules occuring in $\mathcal{G}$ defines a tracking computable function. Then Φ is tracking computable.*

Proof. Let $M_1, \ldots, M_\ell$ be the modules appearing in $\mathcal{G}$, each defining a corresponding function $F_1, \ldots, F_\ell$, and having a corresponding tracking function $\varphi_1, \ldots, \varphi_\ell$. Note that Φ can be obtained from $F_1, \ldots, F_\ell$ via composition, projection and pairing. More formally, we can write $\Phi(x_1, \ldots, x_k) = (y_1, \ldots, y_\ell)$ where each $y_j = F_j(\mathbf{x}_j)$, and $\mathbf{x}_j$ is a subset of the inputs $x_1, \ldots, x_k$. Since composition, projection and pairing preserve tracking computability,[5] we obtain a recursive tracking function φ for Φ from $\varphi_1, \ldots, \varphi_\ell$. Thus, Φ is tracking computable.   □

Hence, we only need to prove that each of the basic modules considered in Sect. 3 is tracking computable, which can be done under suitable assumptions.

Lemma 2 (Tracking computability of the LGPAC modules). *Let $\mathcal{X}$ be a separable Fréchet space. Suppose that addition, scalar multiplication and pseudonorm evaluation are all tracking computable on $\mathcal{X}$. Let $\mathcal{Z} = C^1(\mathbb{T}, \mathcal{X})$ be the space of $\mathcal{X}$-streams with the computable structure induced by $\alpha_{\mathcal{X}}$, as in Example 3. Then (1) for each computable element $x \in \mathcal{X}$, the constant stream $u(t) = x$ is a computable element in $C^1(\mathbb{T}, \mathcal{X})$; and (2) each of the nonconstant modules from Sect. 3 (adder, multiplier, integrator and continuous limit) defines a tracking computable function.*

---

[5] The computability of basic algebraic operations is usually one of the first results to be proved for a model of computation. For example, in the framework of computable analysis, this is proved in [10, Sect. 0.4]; and in the framework of type-2 theory of effectivity, this is proved in [22, Sect. 2.1]. The techniques carry over to the tracking computability framework in this paper.

Proof. We sketch the proof outline; additional technical details are given in the Appendix. Recall that an element $u \in \mathcal{Z}_c$ is described by a data tuple $(x_0, y_0, \ldots, y_{N^2})$ and can be encoded by $e = \langle N, \langle m_0, m_0', \ldots, m_{N^2}' \rangle \rangle$, where m_0, m_0', ..., m_{N^2}' encode elements in $\mathcal{X}_c$.

Constants: given an element $x \in \mathcal{X}_c$, the constant stream $u(t) = x$ is in $\mathcal{Z}_c$; in particular, it is encoded by $N = 1$ and the tuple $(x, 0, 0)$. Recall that, by assumption, we have $\alpha_{\mathcal{X}}(0) = 0$. Thus, given a code $c = \langle T, M \rangle$ for a computable $x \in \mathcal{X}$, we can consider the code $c' = \langle T', M \rangle$ in which $\{T'\}(j) = \langle 1, \langle \{T\}(j), 0, 0 \rangle \rangle$. To verify that the same modulus of convergence works, let $x_j = \alpha_{\mathcal{X}}\{T\}(j)$ and $u_j = \alpha_{\mathcal{Z}}\{T'\}(j)$. Then $u_j(t) \equiv x_j$, so that $u_j'(t) \equiv 0$ and

$$\|u_i - u_j\|_n = \|u_i(0) - u_j(0)\|_n + \sup_{0 \leq t \leq n} \|u_i'(t) - u_j'(t)\|_n = \|x_i - x_j\|_n.$$

Hence c' is a code for the desired constant stream, so that u is a computable element in $\mathcal{Z}$.

Addition: essentially, we need to approximately compute addition at "two levels". At the "first level", we create a procedure that receives codes e_1 and e_2 for computable $\mathcal{X}$-streams $u_1 = \alpha(e_1)$ and $u_2 = \alpha(e_2)$, as well as a natural number ℓ; it produces a code e_+ of some element $u_+ = \alpha(e_+)$ that approximates $u_1 + u_2$ to precision $2^{-\ell}$. This is done by building a large *common refinement*, i.e. codes for approximations $\tilde{u}_1, \tilde{u}_2$ of u_1, u_2 on a finer common grid. Each value in the new discretization can be seen as a convex combination of two consecutive values in the old discretization. Since addition and scalar multiplication are tracking computable in $\mathcal{X}$, these convex combinations can be approximated to arbitrarily high precision. Then, in order to compute the addition on the common refinement, we can simply compute the pointwise addition with sufficiently high precision.

At the "second level", assume we have codes $c_1 = \langle T_1, M_1 \rangle, c_2 = \langle T_2, M_2 \rangle$ for computable elements u and v respectively; we wish to find a code $c_+ = \langle T_+, M_+ \rangle$ for their sum $w = u + v$. If we write $u_i = \alpha\{T_1\}(i)$, $v_i = \alpha\{T_2\}(i)$, $w_i = \alpha\{T_+\}(i)$, then the main idea is to define w_j as a (sufficiently good) approximation of $u_{k_1(j)} + v_{k_2(j)}$, for some choice of k_1 and k_2 (depending on the moduli of convergence $\{M_1\}$ and $\{M_2\}$) that ensures w_j is id-convergent to $u + v$.

Scalar multiplication: compared to addition, there are two additional sources of error that we have to control. At the "first level", we recall the product rule for derivatives, $(ru)'(t) = r(t)u'(t) + r'(t)u(t)$. After finding approximate $\tilde{r}$, $\tilde{u}$ on a large common refinement, we can approximately evaluate the above expression at equispaced points, as long as we are able to compute $\tilde{r}(j/N)$ and $\tilde{u}(j/N)$. Since $\tilde{r}$ and $\tilde{u}$ are piecewise quadratic, these can be retrieved by integration using the trapezoid rule $\tilde{x}_{j+1} \approx \tilde{x}_j + \frac{1}{2N}(\tilde{y}_j + \tilde{y}_{j+1})$, and hence computed to arbitrary precision. Yet another source of error appears in the analysis, since any approximation of ru is piecewise quadratic whereas ru itself is piecewise quartic (as functions of t). This additional error can be controlled by first finding an upper bound K_ℓ on $\|r\|_\ell$ and $\|u\|_\ell$ and then choosing a suitable large discretization $\bar{N}$.

At the "second level", assume we have codes $c_1 = \langle T_1, M_1 \rangle, c_2 = \langle T_2, M_2 \rangle$ for sequences $r_i = \alpha\{T_1\}(i)$, $u_i = \alpha\{T_2\}(i)$ converging to computable elements r and u respectively; we wish to find a code $c_\times = \langle T_\times, M_\times \rangle$ for a sequence $v_i = \alpha\{T_\times\}(i)$ converging to their product $v = ru$. Again, we define v_j to be an approximation of the product $r_{k_1(j)} u_{k_2(j)}$ to sufficiently high precision, computed at the "first level". By choosing a suitable $k_1(j)$, $k_2(j)$ we can ensure (v_j) is id-convergent to ru.

Integration: the case of integration is quite similar to multiplication: if $x \in \mathcal{X}$ and $r \in C^1(\mathbb{T}, \mathbb{R})$, $u \in C^1(\mathbb{T}, \mathcal{X})$ are represented by the tuples of data $(p_0^1, q_0^1, \ldots, q_{N^2}^1)$ and $(x_0^2, y_0^2, \ldots, y_{N^2}^2)$, respectively, then the integral $w(t) = x + \int_0^t u(s) dr(s)$ is a function with $w(0) = x$ and $w'(t) = u(t) r'(t)$. Since the values of $u(t)$ at equispaced points can be approximated by the trapezoid rule, this again yields a natural way to approximately compute a data tuple representation for w.

Continuous limit: if $u_n \in C^1(\mathbb{T}, \mathcal{X})$ is an effective Cauchy sequence converging to a stream $u \in C^1(\mathbb{T}, \mathcal{X})$ which in turn has an id-convergent limit $x \in \mathcal{X}$, then x equals $\lim_{t \to \infty} \lim_{n \to \infty} u_n(t)$. Thus, a candidate for an approximation of x is $u_{k_n}(t_n)$, where t_n and k_n are large enough integers. By effectivizing this line of thought, we produce a tracking function for the continuous limit module as well. □

6   Computability of LGPAC-Generable Functions

In this section we prove the main result of this paper. The goal is to find out under which conditions the function generated by an LGPAC is tracking computable. We recall that, in our terminology, an LGPAC induces an operator and fixed point problem

$$\Phi : \mathcal{I} \times \mathcal{M} \rightharpoonup \mathcal{M} \times \mathcal{O}, \quad \Phi(\mathbf{u}^I, \mathbf{u}^M) = (\mathbf{u}^M, \mathbf{u}^O); \tag{5}$$

for the LGPAC to generate a valid function, we require the fixed point problem to be *well-posed*, that is, (5) has a unique, continuous, solution map $F : \mathbf{u}^I \mapsto (\mathbf{u}^M, \mathbf{u}^O)$.

Our goal is to find conditions on Φ that imply that F is tracking computable. The idea is to find F by solving an *approximate fixed point problem*

Given $\mathbf{u}^I$ and $\epsilon > 0$, find $(\mathbf{u}^M, \mathbf{u}^O)$ such that $d(\Phi(\mathbf{u}^I, \mathbf{u}^M), (\mathbf{u}^M, \mathbf{u}^O)) < \epsilon$.

Moreover, from the point of view of tracking computability, we look for desired $\mathbf{u}^M, \mathbf{u}^O$ in the enumerated, countable dense subset. Then, by using a sequence of ϵ converging to 0, and under additional assumptions on F (namely, we will require a notion of *effective well-posedness*), this yields a sequence of $\mathbf{u}^M, \mathbf{u}^O$ converging to the desired $F(\mathbf{u}^I)$.

Let us now focus on the first step of this construction. Namely, we prove that it is possible to construct *approximate fixed points*.

Lemma 3. *Let $\mathcal{G}$ be an LGPAC with input-output operator $\Phi : \mathcal{I} \times \mathcal{M} \rightharpoonup \mathcal{M} \times \mathcal{O}$. Assume that Φ is tracking computable, and that $\mathcal{G}$ is well-posed on an open subset $U \subseteq \mathcal{I}$. Then there exists a computable procedure $\mathbf{FixPt} : (n, \ell) \mapsto m$ such that, if n is the code for an element $\mathbf{u}^I = \bar{\alpha}_\mathcal{I}(n) \in U$ and $\ell \in \mathbb{N}$, then m is the code for an enumerated element $(\mathbf{u}^M, \mathbf{u}^O) = \alpha_{\mathcal{M} \times \mathcal{O}}(m) \in \mathcal{M} \times \mathcal{O}$; and also $d(\Phi(\mathbf{u}^I, \mathbf{u}^M), (\mathbf{u}^M, \mathbf{u}^O)) < 2^{-\ell}$.*

Proof. The procedure works as follows. For a given input n, ℓ, let us write $\mathbf{u}^I = \bar{\alpha}_\mathcal{I}(n)$. We perform the following dovetailing loop. First, guess an index $m \in \mathbb{N}$ for an element in $(\mathcal{M} \times \mathcal{O})_c$. Second, find m_1, m_2 such that $\alpha_{\mathcal{M} \times \mathcal{O}}(m) = (\alpha_\mathcal{M}(m_1), \alpha_\mathcal{O}(m_2))$, via the pairing bijections. For clarity, let us write $\mathbf{u}^M = \alpha_\mathcal{M}(m_1)$, $\mathbf{u}^O = \alpha_\mathcal{O}(m_2)$. Third, find n' such that $\bar{\alpha}_{\mathcal{I} \times \mathcal{M}}(n') = (\mathbf{u}^I, \mathbf{u}^M)$, using the code n for $\mathbf{u}^I$ and a code for the constant function $\{T_1\}(n) = m_1$. Fourth, find $m' = \varphi(n') = \langle T', M' \rangle$, where φ is a tracking function for Φ. Notice that

$$\bar{\alpha}_{\mathcal{M} \times \mathcal{O}}(m') = \bar{\alpha}_{\mathcal{M} \times \mathcal{O}}(\varphi(n')) = \Phi(\bar{\alpha}_{\mathcal{I} \times \mathcal{M}}(n')) = \Phi(\mathbf{u}^I, \mathbf{u}^M).$$

Fifth, find $m'' = \{T'\}(\{M'\}(\ell + 2))$. Since $\{M'\}$ is a module of convergence, it follows that for $k \geq \{M'\}(\ell + 2)$, one has

$$d(\alpha_{\mathcal{M} \times \mathcal{O}}(m''), \alpha_{\mathcal{M} \times \mathcal{O}}(\{T'\}(k))) < 2^{-\ell-2}.$$

In particular, since $\bar{\alpha}_{\mathcal{M} \times \mathcal{O}}(m')$ is the limit of $\alpha_{\mathcal{M} \times \mathcal{O}}(\{T'\}(n))$, then

$$d(\bar{\alpha}_{\mathcal{M} \times \mathcal{O}}(m'), \alpha_{\mathcal{M} \times \mathcal{O}}(m'')) \leq 2^{-\ell-2}.$$

For clarity, let us write $(\tilde{\mathbf{u}}^M, \tilde{\mathbf{u}}^O) = \alpha_{\mathcal{M} \times \mathcal{O}}(m'')$.

Finally, check if $d(\alpha_{\mathcal{M} \times \mathcal{O}}(m''), \alpha_{\mathcal{M} \times \mathcal{O}}(m)) < 2^{-\ell-1}$; if yes, then break the loop and return m. Observe that the distance function is tracking computable (to get a close enough approximation, it is enough to evaluate sufficiently but finitely many pseudonorms).

Observe that, for some values of m, the corresponding execution of the loop may not terminate. This may happen if $\bar{\alpha}_{\mathcal{I} \times \mathcal{M}}(n')$ is not an element in the domain of Φ, so that $\varphi(n')$ may be a divergent computation, or if the value of $d(\alpha_{\mathcal{M} \times \mathcal{O}}(m''), \alpha_{\mathcal{M} \times \mathcal{O}}(m))$ is exactly $2^{-\ell-1}$ (equality may not be a computable predicate). However, if a certain value of m happens to pass our test, then that value satisfies the desired property: indeed,

$$d(\Phi(\mathbf{u}^I, \mathbf{u}^M), (\mathbf{u}^M, \mathbf{u}^O)) \leq d(\Phi(\mathbf{u}^I, \mathbf{u}^M), (\tilde{\mathbf{u}}^M, \tilde{\mathbf{u}}^O)) + d((\tilde{\mathbf{u}}^M, \tilde{\mathbf{u}}^O), (\mathbf{u}^M, \mathbf{u}^O))$$
$$= d(\bar{\alpha}_{\mathcal{M} \times \mathcal{O}}(m'), \alpha_{\mathcal{M} \times \mathcal{O}}(m'')) + d(\alpha_{\mathcal{M} \times \mathcal{O}}(m''), \alpha_{\mathcal{M} \times \mathcal{O}}(m))$$
$$< 2^{-\ell-2} + 2^{-\ell-1} = 2^{-\ell}.$$

Moreover, such a value of m can always be found by our algorithm, due to our assumption that $\mathcal{G}$ is well-posed. To see this, let $\mathbf{u}^I = \bar{\alpha}_\mathcal{I}(n) \in U$. By well-posedness, there exists (a unique) $(\mathbf{u}^M_*, \mathbf{u}^O_*) \in \mathcal{M} \times \mathcal{O}$ with $\Phi(\mathbf{u}^I, \mathbf{u}^M_*) = (\mathbf{u}^M_*, \mathbf{u}^O_*)$, and thus $d(\Phi(\mathbf{u}^I, \mathbf{u}^M_*), (\mathbf{u}^M_*, \mathbf{u}^O_*)) = 0$. Now the left hand side of this equality is a continuous expression in $\mathbf{u}^M_*, \mathbf{u}^O_*$ (the continuity of Φ follows from the continuity the module functions, and every metric d is continuous over its topology); thus there exists $\delta > 0$ such that for any $\mathbf{u}^M, \mathbf{u}^O \in \mathcal{M} \times \mathcal{O}$ one has

if $d((\mathbf{u}_*^M, \mathbf{u}_*^O), (\mathbf{u}^M, \mathbf{u}^O)) < \delta$ then $d(\Phi(\mathbf{u}^I, \mathbf{u}^M), (\mathbf{u}^M, \mathbf{u}^O)) < 2^{-\ell-2}$.

By density of the enumerated subset, there exists $m \in \mathbb{N}$ such that $\alpha_{\mathcal{M} \times \mathcal{O}}$ $(m) = (\mathbf{u}^M, \mathbf{u}^O)$ with $d((\mathbf{u}_*^M, \mathbf{u}_*^O), (\mathbf{u}^M, \mathbf{u}^O)) < \delta$, and thus $d(\Phi(\mathbf{u}^I, \mathbf{u}^M), (\mathbf{u}^M, \mathbf{u}^O)) < 2^{-\ell-2}$, or in other words,

$$d(\bar{\alpha}_{\mathcal{M} \times \mathcal{O}}(m'), \alpha_{\mathcal{M} \times \mathcal{O}}(m)) < 2^{-\ell-2}.$$

Moreover, the value of m'', computed on step 4, will be such that

$$d(\bar{\alpha}_{\mathcal{M} \times \mathcal{O}}(m'), \alpha_{\mathcal{M} \times \mathcal{O}}(m'')) \leq 2^{-\ell-2},$$

and a simple application of the triangle inequality yields that

$$d(\alpha_{\mathcal{M} \times \mathcal{O}}(m''), \alpha_{\mathcal{M} \times \mathcal{O}}(m)) < 2^{-\ell-1},$$

so the condition on step 5 is met. Thus the dovetailing loop will effectively succeed in finding a valid m. □

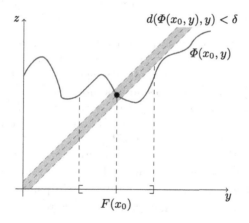

Fig. 6. Approximate fixed points vs approximations of the fixed point. Intuitively, assume that the fixed point equation $\Phi(x, y) = y$ has a continuous solution operator $y = F(x)$. Then, in a neighborhood of x, approximate fixed points are 'near' the exact fixed point.

We have shown that it is possible to find approximate fixed points of the input-output operator. In the next step, we would like to argue that *approximate fixed points* are in fact 'near' *exact fixed points*, which is by no means a trivial statement (rather, there is extensive research on this problem; see for example [7,8]). Intuitively, we want to establish conditions on the input-output operator Φ (and the corresponding solution functional F) that effectively ensure the following: for each ϵ there is δ such that if $d(\Phi(\mathbf{u}^I, \mathbf{u}^M), (\mathbf{u}^M, \mathbf{u}^O)) < \delta$, then $d(F(\mathbf{u}^I), (\mathbf{u}^M, \mathbf{u}^O)) < \epsilon$ (see Fig. 6 for an intuition). This is captured in the following notion, that we use as an assumption towards proving our main theorem.

Definition 8 (Effective well-posedness). *Let* $\mathcal{G}$, Φ, U *be as in Definition 5, with* $\mathcal{G}$ *well-posed on* U *and generating some function* F. *We say that* $\mathcal{G}$ *is* effectively well-posed *if there is a computable modulus of convergence* $M : \mathbb{N} \to \mathbb{N}$ *such that for all* $\nu \in \mathbb{N}$ *and all* $\mathbf{u}^I \in U$, $\mathbf{u}^M \in \mathcal{M}$, $\mathbf{u}^O \in \mathcal{O}$,

$$d(\Phi(\mathbf{u}^I, \mathbf{u}^M), (\mathbf{u}^M, \mathbf{u}^O)) < 2^{-M(\nu)} \Rightarrow d(F(\mathbf{u}^I), (\mathbf{u}^M, \mathbf{u}^O)) < 2^{-\nu}. \qquad (6)$$

*Remark 1 (**Well-posedness and effective well-posedness**).* The well-posedness of $\mathcal{G}$ implies that, for any $\mathbf{u}^I \in U$, $\mathbf{u}^M \in \mathcal{M}$, $\mathbf{u}^O \in \mathcal{O}$,

$$d(\Phi(\mathbf{u}^I, \mathbf{u}^M), (\mathbf{u}^M, \mathbf{u}^O)) = 0 \quad \text{iff} \quad d(F(\mathbf{u}^I), (\mathbf{u}^M, \mathbf{u}^O)) = 0.$$

Thus effective well-posedness can be understood as an *effective strengthening* of this equivalence.

Theorem 1 (Tracking computability of LGPAC generable functions). *Let* $\mathcal{G}$ *be an effectively well-posed LGPAC generating some function* F *on domain* U. *Suppose also that each of the modules in* $\mathcal{G}$ *are tracking computable. Then* F *is tracking computable.*

Proof. By Lemma 1, the input-output operator $\Phi : \mathcal{I} \times \mathcal{M} \rightharpoonup \mathcal{M} \times \mathcal{O}$ of $\mathcal{G}$ is tracking computable. By Lemma 3, there exists a procedure $\mathbf{FixPt} : (e_I, \ell) \mapsto e_{\mathcal{M} \times \mathcal{O}}$ that maps codes of computable elements $\mathbf{u}^I \in U$ to $2^{-\ell}$-approximate fixed points. Let M_W be a computable modulus of convergence witnessing the effective well-posedness of $\mathcal{G}$. Then, given a code $c = \langle T, M \rangle$ of an element $\mathbf{u}^I \in U$, we can construct a code $\varphi(c) = \langle T', M' \rangle$ for $F(\mathbf{u}^I)$ by letting T' be a code for the function $\{T'\}(j) = \mathbf{FixPt}(c, M_W(j+1))$ and M' be a code for the identity function. Indeed, the above procedure is effective and, letting $(\mathbf{u}_j^M, \mathbf{u}_j^O) = \alpha\{T'\}(j)$, we have by construction that $d(\Phi(\mathbf{u}^I, \mathbf{u}_j^M), (\mathbf{u}_j^M, \mathbf{u}_j^O)) < 2^{-M_W(j+1)}$, so that $d(F(\mathbf{u}^I), (\mathbf{u}_j^M, \mathbf{u}_j^O)) < 2^{-j-1}$. In particular, $(\mathbf{u}_j^M, \mathbf{u}_j^O)$ is an id-convergent Cauchy sequence that converges to $F(\mathbf{u}^I)$. $\qquad \square$

7   Some Applications of Theorem 1

We proceed to give two applications of our main result.

1. Computability over continuous real functions. Let us consider the data space $\mathcal{X} = C(\mathbb{R})$ of continuous real functions. This can be considered a basic example of how the Shannon GPAC can be generalized beyond real-valued computation. Moreover, $C(\mathbb{R})$ is equipped with a multiplication operation $(fg)(x) = f(x)g(x)$, which naturally induces multiplication and integration over $C(\mathbb{R})$-streams, $\times(u, v)(t) = u(t)v(t)$ and $\int (c, u, v) = c + \int u dv$. As one would expect, all these operations are tracking computable.

Lemma 4. *Let $\mathcal{X} = C(\mathbb{R})$ be the class of continuous real functions with the computability structure defined in Example 2. Then addition, scalar multiplication, multiplication, and pseudonorm evaluation are tracking computable operations on $\mathcal{X}$. Moreover, multiplication and integration over $C^1(\mathbb{T}, \mathcal{X})$, defined as $\times(u, v)(t) = u(t)v(t)$ and $\int (c, u, v) = c + \int u\,dv$ are also tracking computable operations.*

Proof. For addition, multiplication and integration, we can adapt the proof from Lemma 2. Note that, due to our choice of computability structure on $\mathbb{R}$ (Example 1), addition and multiplications by rationals can be performed *exactly* (on their codes), so a lot of the error analysis disappears. Moreover, functions in $C(\mathbb{R})$ are approximated by piecewise linear functions instead of the more complicated piecewise quadratic functions that we used in $C^1(\mathbb{T}, \mathcal{X})$. Thus, the proofs become much simpler and we omit the details.

Now consider pseudonorm evaluation. Given a function $f \in \mathcal{X}_c$ via its code $e_1 = \langle N, \langle m_{-N^2}, \ldots, m_0, \ldots, m_{N^2}\rangle\rangle$, and an integer n, note that $\|f\|_n$ can be computed exactly: it simply corresponds to the maximum of the rational numbers $\alpha_{\mathbb{R}}(m_j)$, where j ranges either: between $-N^2$ and N^2 (if $N \leq n$); or between $-Nn$ and Nn (if $N \geq n$).

Next, let $c = \langle T, M\rangle$ be a code for a function in $\mathcal{X}$ and n be an integer. Let $f_j = \alpha_{\mathcal{X}}\{T\}(j)$. Define a code $\langle T_n, M_n\rangle$ where $\{T_n\}(j)$ is a code for the value $\|f_j\|_n$ (which can be computed exactly) and $\{M_n\}(\nu) = \{M\}(\nu + n)$. Note that for $i, j \geq \{M_n\}(\nu)$ we have that $d_{\mathcal{X}}(f_i, f_j) < 2^{-(\nu+n)}$ and hence $\|f_i - f_j\|_n < 2^{-\nu}$ (by Proposition 1). By the triangular inequality we conclude that $|\|f_i\|_n - \|f_j\|_n| < 2^{-\nu}$ as desired. $\qquad\square$

Combining Lemmas 2 and 4, we conclude that each nonconstant module on a multityped GPAC over $C(\mathbb{R})$ is tracking computable. Together with Theorem 1, we obtain:

Corollary 1. *Let $\mathcal{G}$ be a multityped GPAC with channels over $\mathbb{R}$ and $\mathcal{X} = C(\mathbb{R})$, constructed with the following types of modules: constants (over $\mathbb{R}$ and $\mathcal{X}$), adders (over $\mathbb{R}$-streams and $\mathcal{X}$-streams), multipliers $\times(u, v)$ and integrators $\int (c, u, v)$, (where each of u, v is either an $\mathbb{R}$-stream or an $\mathcal{X}$-stream), and continuous limits (over $\mathbb{R}$-streams and $\mathcal{X}$-streams). Suppose that each of the constant modules appearing in $\mathcal{G}$ is tracking computable, and that $\mathcal{G}$ is effectively well-posed on domain U, generating a function F. Then F is tracking computable.*

2. Contracting operators. We show that the condition of effective well-posedness (Definition 8) is automatically achieved for contracting operators, which form an important class in fixed point theory. Formally, an input-output operator Φ is *contracting* if there is a constant $\lambda \in [0, 1)$ such that

$$d(\Phi(\mathbf{u}^I, \mathbf{u}^M), \Phi(\mathbf{u}^I, \tilde{\mathbf{u}}^M)) \leq \lambda d(\mathbf{u}^M, \tilde{\mathbf{u}}^M).$$

Lemma 5. *Let $\mathcal{G}$ be a multityped GPAC and assume that its input-output operator Φ is contracting. Then $\mathcal{G}$ is well-posed iff it is effectively well-posed.*

Proof. Clearly, effective well-posedness implies well-posedness. For the converse direction, assume $\mathcal{G}$ is well-posed and let F be the function generated by $\mathcal{G}$. For each $\mathbf{u}^I \in \mathcal{I}$ define $\hat{\Phi} : \mathcal{M} \times \mathcal{O} \to \mathcal{M} \times \mathcal{O}$ as $\hat{\Phi}(\mathbf{u}^M, \mathbf{u}^O) = \Phi(\mathbf{u}^I, \mathbf{u}^M)$. In this way, $\hat{\Phi}$ is a contracting self-map on $\mathcal{M} \times \mathcal{O}$ with the same constant λ. As a consequence of the Banach fixed point theorem [13, Th. V.18], for any $(\mathbf{u}^M, \mathbf{u}^O) \in \mathcal{M} \times \mathcal{O}$ we have that $d(F(\mathbf{u}^I), (\mathbf{u}^M, \mathbf{u}^O)) \leq \frac{1}{1-\lambda} d(\hat{\Phi}(\mathbf{u}^M, \mathbf{u}^O), (\mathbf{u}^M, \mathbf{u}^O))$. Let us take $M(\nu) = \nu + C$ as a modulus of convergence, where C is any natural such that $2^{-C} \leq 1 - \lambda$. Then for any $\mathbf{u}^I \in \mathcal{I}, \mathbf{u}^M \in \mathcal{M}, \mathbf{u}^O \in \mathcal{O}$ such that $d(\Phi(\mathbf{u}^I, \mathbf{u}^M), (\mathbf{u}^M, \mathbf{u}^O)) < 2^{-M(\nu)}$, we get

$$d(F(\mathbf{u}^I), (\mathbf{u}^M, \mathbf{u}^O)) \leq \frac{1}{1-\lambda} d(\hat{\Phi}(\mathbf{u}^M, \mathbf{u}^O), (\mathbf{u}^M, \mathbf{u}^O)) < \frac{2^{-\nu-C}}{1-\lambda} < 2^{-\nu}.$$

$\square$

The following corollary is immediate from Theorem 1 and Lemma 5.

Corollary 2. *Let $\mathcal{G}$ be a well-posed LGPAC generating some function F on domain U. Suppose that each of the modules in $\mathcal{G}$ are tracking computable, and that the input-output operator Φ is contracting. Then F is tracking computable.*

8   Discussion

In this paper we presented partial results towards a comparison between the GPAC model of computation and tracking computability on separable Fréchet spaces $\mathcal{X}$. Two important questions are left for further research.

1. Effective well-posedness. Our main result hinges on this extra assumption, allowing us to use approximate fixed points to obtain approximations of the exact fixed point. The question of whether this condition can be relaxed remains an open problem. Our difficulty stems from the usage of arbitrary data spaces $\mathcal{X}$, which in particular can be infinite-dimensional. In the case of $\mathcal{X} = \mathbb{R}^k$, i.e. finite-dimensional spaces, standard results in analysis (e.g. the Picard-Lindelöf Theorem [4]) allow us to consider iterative methods to obtain such fixed points. Related to this observation, we have argued that effective well-posedness comes 'for free' when the input-output operator is contracting. It would be interesting to extend this argument to a larger class of 'typical' operators appearing in Analysis.

2. Converse of Theorem 1. Investigating under which conditions tracking computable functions are LGPAC-generable remains a major open problem. The most likely approach to answer this question is to first simulate the behavior of a Turing machine (or any other discrete model of computation) in an analog network. As relevant literature, papers [1,3,5] provide a way to embed states, transitions, and the discrete evolution of a Turing machine into real numbers, continuous real functions, and the continuous evolution of a dynamical system respectively. With some care, their techniques may be adaptable to our framework.

We hope that in tackling these problems new insights can be acquired about the power of analog networks, and in particular the GPAC, as a model for analog computation.

Acknowledgements. The research of Diogo Poças was supported by the Alexander von Humboldt Foundation with funds from the German Federal Ministry of Education and Research (BMBF). The research of Jeffery Zucker was supported by the Natural Sciences and Engineering Research Council of Canada.

Appendix: Technical Details in the Proof of Lemma 2

Addition: let e_1, e_2, ℓ be natural numbers, where e_1 and e_2 encode computable $\mathcal{X}$-streams $u_1 = \alpha(e_1)$ and $u_2 = \alpha(e_2)$. We need to show how to effectively compute a code e_+ of some element $u_+ = \alpha(e_+)$ that approximates $u_1 + u_2$ to precision $2^{-\ell}$, that is, such that $\|u_+ - (u_1 + u_2)\|_\ell < 2^{-\ell-2}$.

We know that u_1 and u_2 are given by some data tuples $(x_0^1, y_0^1, \ldots, y_{N_1}^1)$ and $(x_0^2, y_0^2, \ldots, y_{N_2}^2)$ respectively. First, we build a large common refinement, that is, a large discretization parameter $\bar{N}$ which is a multiple of both N_1 and N_2, and data tuples $(\tilde{x}_0^1, \tilde{y}_0^1, \ldots, \tilde{y}_{\bar{N}}^1)$, $(\tilde{x}_0^2, \tilde{y}_0^2, \ldots, \tilde{y}_{\bar{N}}^2)$ that correspond to approximations $\tilde{u}_1, \tilde{u}_2$ of u_1, u_2 on a finer grid. For example, if $\bar{N} = k \times N_1$, $\tilde{u}_1$ can be obtained by setting $\tilde{y}_{ki+\ell}^1 \approx \frac{k-\ell}{k} y_i^1 + \frac{\ell}{k} y_{i+1}^1$; each value in the new discretization is a convex combination of two consecutive values in the old discretization. Since addition and scalar multiplication are tracking computable in $\mathcal{X}$, these convex combinations can be approximated to arbitrarily high precision.

To compute the addition on the common refinement, we can simply add the pointwise values, that is, set $x_0^+ \approx \tilde{x}_0^1 + \tilde{x}_0^2$ and $y_j^+ \approx \tilde{y}_j^1 + \tilde{y}_j^2$. By computing these sums with sufficiently high precision, we have indeed produced the desired code e_+.

By the previous discussion, we have a procedure **add** $: (e_1, e_2, \ell) \mapsto e_+$ such that, for $u_1 = \alpha(e_1), u_2 = \alpha(e_2), u_+ = \alpha(e_+)$, we have $\|u_+ - (u_1 + u_2)\|_\ell < 2^{-\ell-2}$. Next, assume we have codes $c_1 = \langle T_1, M_1 \rangle, c_2 = \langle T_2, M_2 \rangle$ for computable elements u and v respectively; we wish to find a code $c_+ = \langle T_+, M_+ \rangle$ for their sum $w = u + v$. Let us introduce the notation $u_i = \alpha\{T_1\}(i)$, $v_i = \alpha\{T_2\}(i)$, $w_i = \alpha\{T_+\}(i)$.

We shall set $\{T_+\}(j) = \mathbf{add}(\{T_1\}(k_1(j)), \{T_2\}(k_2(j)), j)$, where $k_1(j) = \{M_1\}(2j+2)$ and $k_2(j) = \{M_2\}(2j+2)$. Intuitively, w_j is a (sufficiently good) approximation of $u_{k_1(j)} + v_{k_2(j)}$. Furthermore, we set M_+ as a code for the identity function. To show that (w_j) is id-convergent, fix ν and suppose that $i, j \geq \nu$. Observe that

$$\|w_i - w_j\|_\nu \leq \|w_i - (u_{k_1(i)} + v_{k_2(i)})\|_\nu + \|u_{k_1(i)} - u_{k_1(j)}\|_\nu$$
$$+ \|v_{k_2(i)} - v_{k_2(j)}\|_\nu + \|w_j - (u_{k_1(j)} + v_{k_2(j)})\|_\nu.$$

To bound the first term above, we observe that $\|w_i - (u_{k_1(i)} + v_{k_2(i)})\|_\nu \leq \|w_i - (u_{k_1(i)} + v_{k_2(i)})\|_i < 2^{-i-2} \leq 2^{-\nu-2}$; a similar argument holds for the fourth term. For the second term, note that by our choice of $k_1(\nu)$ we have $d(u_{k_1(i)}, u_{k_1(j)}) < 2^{-2\nu-2}$, and by Proposition 1 this implies $\|u_{k_1(i)} - u_{k_1(j)}\|_\nu < 2^{-\nu-2}$; similarly for the third term. Putting all this together yields $\|w_i - w_j\|_\nu < 2^{-\nu}$, which again by Proposition 1 implies $d(w_i, w_j) < 2^{-\nu}$, as desired. A similar reasoning also proves that w_i converges to $u + v$. Hence addition is tracking computable.

Scalar multiplication: in the same way as for addition, we show how to approximately compute the scalar multiplication at "two levels". At the "first level", let e_1, e_2, ℓ be natural numbers encoding a computable $\mathbb{R}$-stream $r = \alpha(e_1)$ and $\mathcal{X}$-stream $u = \alpha(e_2)$ respectively. We need to show that we can compute the scalar multiplication ru to an arbitrary precision. In particular, we will show how to effectively compute a code $e_\times$ of some element $u_\times = \alpha(e_\times)$ such that $\|u_\times - ru\|_\ell < 2^{-\ell-2}$.

We know that r and u are given by some data tuples $(p_0, q_0, \ldots, q_{N_1^2})$ and $(x_0, y_0, \ldots, y_{N_2^2})$ respectively. First, we effectively find an upper bound K_ℓ on the pseudonorms $\|r\|_\ell$ and $\|u\|_\ell$ by (approximately) computing the maximum of $|p_0|, |q_j|, \|x_0\|_\ell, \|y_0\|_\ell$ (by assumption, pseudonorm evaluation is tracking computable on $\mathcal{X}$).

Next, we construct a large common refinement, say $(\tilde{p}_0, \tilde{q}_0, \ldots, \tilde{q}_{\bar{N}^2})$ and $(\tilde{x}_0, \tilde{y}_0, \ldots, \tilde{y}_{\bar{N}^2})$, corresponding to approximations $\tilde{r}, \tilde{u}$ of r, u on a finer grid, as we did for addition. To compute the multiplication on the common refinement, we recall the product rule for derivatives, $(ru)'(t) = r(t)u'(t) + r'(t)u(t)$. To compute this expression at equispaced values of t, we must first find the values of $\tilde{r}(j/N), \tilde{u}(j/N)$. Since $\tilde{r}, \tilde{u}$ are piecewise quadratic, these can be recursively obtained by integration using the trapezoid rule,

$$\tilde{p}_{j+1} \approx \tilde{p}_j + \frac{1}{2N}(\tilde{q}_j + \tilde{q}_{j+1}), \qquad \tilde{x}_{j+1} \approx \tilde{x}_j + \frac{1}{2N}(\tilde{y}_j + \tilde{y}_{j+1}). \qquad (7)$$

Again, $\tilde{p}_j, \tilde{x}_j$ can be approximated to arbitrarily high precision. Therefore, $\tilde{r}\tilde{u}$ can be approximated by the function $u_\times$ given by $(x_0^\times, y_0^\times, \ldots, y_{\bar{N}^2}^\times)$, where $x_0^\times$ is (the approximating computation of) $\tilde{p}_0\tilde{x}_0$; and each $y_j^\times$ is (the approximating computation of) $\tilde{p}_j\tilde{y}_j + \tilde{q}_j\tilde{x}_j$.

There is one more error term appearing in our analysis, since $u_\times$ is piecewise quadratic whereas ru is piecewise quartic (as functions of t). To describe an effective bound on the approximation error $\|u_\times - ru\|_\ell$, we need to take into account: the approximation errors for the refinement and the multiplications over $\mathcal{X}_c$, the upper bound K_ℓ on $\|r\|_\ell$ and $\|u\|_\ell$; the consecutive differences $\max \|\tilde{q}_{j+1} - \tilde{q}_j\|_n, \max \|\tilde{y}_{j+1} - \tilde{y}_j\|_n$; and the discretization $\bar{N}$. Ultimately, we can bound this error in an effective way by choosing $\bar{N}$ large enough.

By the previous discussion, we have a procedure **mult** : $(e_1, e_2, \ell) \mapsto e_\times$ such that, for $r = \alpha(e_1), u = \alpha(e_2), u_\times = \alpha(e_\times)$, we have $\|u_\times - ru\|_\ell < 2^{-\ell-2}$. At the "second level", assume we have codes $c_1 = \langle T_1, M_1 \rangle, c_2 = \langle T_2, M_2 \rangle$ for computable elements r and u respectively; we wish to find a code $c_\times = \langle T_\times, M_\times \rangle$ for their product $v = ru$. Let us introduce the notation $r_i = \alpha\{T_1\}(i), u_i = \alpha\{T_2\}(i), v_i = \alpha\{T_\times\}(i)$.

First, for any $\nu \in \mathbb{N}$, we can effectively find a *uniform bound* $K(\nu)$ such that $\|r_i\|_\nu, \|u_i\|_\nu < K(\nu)$ independently of i. This is because, letting $\mu = \{M_1\}(\nu)$, we know that for $i > \mu$ one has $d(r_i, r_\mu) < 2^{-\nu}$ and hence $\|r_i - r_\mu\|_\nu < 1$ by Proposition 1, so that $\|r_i\|_\nu < \|r_\mu\|_\nu + 1$. On the other hand, we can approximately compute $\|r_i\|_\nu$ for each of the finitely many $i \leq \mu$. A similar analysis holds for $\|u_i\|_\nu$. Taking (a sufficiently close approximation of) the maximum of these values gives the desired uniform bound.

Next, observe that for any $r, \tilde{r} \in \mathbb{R}$, $x, \tilde{x} \in \mathcal{X}$, $\nu \in \mathbb{N}$, we have $\|rx - \tilde{r}\tilde{x}\|_\nu \leq |r|\|x - \tilde{x}\|_\nu + |r - \tilde{r}|\|\tilde{x}\|_n$; together with (3), we can derive the useful bound

$$\|r_{i_1}u_{j_1} - r_{i_2}u_{j_2}\|_\nu \leq (\nu + 1)K(\nu)\left(\|r_{i_1} - r_{i_2}\|_\nu + \|u_{j_1} - u_{j_2}\|_\nu\right). \tag{8}$$

We are now in condition to describe how to compute $\{T_\times\}(\nu)$ for a given ν. First, find a uniform bound $K(\nu)$ as described above. Second, find an integer C such that $2^C > K(\nu)(\nu + 1)$. Third, compute $k_1(\nu) = \{M_1\}(2\nu + C + 2)$ and $k_2(\nu) = \{M_2\}(2\nu + C + 2)$. Finally, return

$$\{T_\times\}(\nu) = \mathbf{mult}(\{T_1\}(k_1(\nu)), \{T_2\}(k_2(\nu)), \nu).$$

Intuitively, this means that v_i is a (sufficiently good) approximation of $r_{k_1(i)}u_{k_2(i)}$. We show that the sequence v_i constructed in this way is id-convergent. Fix ν and suppose that $i, j \geq \nu$. Observe that

$$\|v_i - v_j\|_\nu \leq \|v_i - r_{k_1(i)}u_{k_2(i)}\|_\nu + \|r_{k_1(i)}u_{k_2(i)} - r_{k_1(j)}u_{k_2(j)}\|_\nu + \|r_{k_1(j)}u_{k_2(j)} - v_j\|_\nu.$$

The first term above, by construction, can be bounded as $\|v_i - r_{k_1(i)}u_{k_2(i)}\|_\nu \leq \|v_i - r_{k_1(i)}u_{k_2(i)}\|_i < 2^{-i-2} \leq 2^{-\nu-2}$, and similarly for the third term. In order to bound the second term, note that by our choice of $k_1(\nu)$ we have that $d(r_{k_1(i)}, r_{k_1(j)}) < 2^{-2\nu-2-C}$. By Proposition 1, this implies $\|r_{k_1(i)} - r_{k_1(j)}\|_\nu < 2^{-\nu-2-C} < \frac{2^{-\nu-1}}{2K(\nu)(\nu+1)}$. A similar bound holds for $\|u_{k_2(i)} - u_{k_2(j)}\|_\nu$. Putting these in (8) yields $\|r_{k_1(i)}u_{k_2(i)} - r_{k_1(j)}u_{k_2(j)}\|_\nu < 2^{-\nu-1}$. Thus we conclude that $\|v_i - v_j\|_\nu < 2^{-\nu-2} + 2^{-\nu-1} + 2^{-\nu-2} = 2^{-\nu}$, and hence $d(v_i, v_j) < 2^{-\nu}$, i.e. v_i is id-convergent. A similar reasoning proves that v_i converges to ru. Hence the above describes a tracking function for scalar multiplication.

Continuous limit: let $u \in \mathcal{Z}_c$ be represented by the tuple $(x_0, y_0, \ldots y_{N^2})$, where each $x_0, y_j \in \mathcal{X}_c$. We first observe that, for any natural number $n \in \mathbb{N}$, the value of $u(n)$ can be approximated as

$$u(n) \approx \begin{cases} x_{nN} & \text{if } n \leq N; \\ x_{N^2} + (N - n)y_{N^2} & \text{if } n \geq N, \end{cases}$$

where the x_j are again recursively obtained via the trapezoid rule. Consequently, one can devise a computable procedure $\mathbf{eval} : (e, n, \ell) \mapsto e_{\text{eval}}$ such that, given a code e of some element $u = \alpha_{\mathcal{Z}}(e)$ and natural numbers n, ℓ, it produces a code e_{eval} of some element $x = \alpha_{\mathcal{X}}(e_{\text{eval}})$ with $d(x, u(n)) < 2^{-\ell}$; i.e. x approximates $u(n)$ within an error of $2^{-\ell}$.

Now let $c = \langle T, M \rangle$ be a code for an effective Cauchy sequence $u_j = \alpha_{\mathcal{Z}}\{T\}(j)$ in $\mathcal{Z}_c$ converging to a computable element $u \in C^1(\mathbb{T}, \mathcal{X})$. We want to compute a code $c_\infty = \langle T_\infty, M_\infty \rangle$ for an effective Cauchy sequence $x_j = \alpha_{\mathcal{X}}\{T_\infty\}(j)$ in $\mathcal{X}_c$ converging to the limit $x = \mathcal{L}u = \lim_{t \to \infty} u(t) \in \mathcal{X}$.

The idea is to define $\{T_\infty\}(j) = \mathbf{eval}(\{T\}(k_j), t_j, \ell_j)$, for a suitable choice of $\ell_j = j + 3$, $t_j = j + 2$ and $k_j = \{M\}(3j + 5)$. To prove that (x_j) is an id-convergent Cauchy sequence, let $\nu \in \mathbb{N}$ be given, and suppose that $i, j \geq \nu$. By applying the triangular inequality, $d_\mathcal{X}(x_i, x_j)$ is upper bounded as

$$d_\mathcal{X}(x_i, x_j) \leq d_\mathcal{X}(x_i, u_{k_i}(t_i)) + d_\mathcal{X}(u_{k_i}(t_i), u(t_i)) + d_\mathcal{X}(u(t_i), u(t_j))$$
$$+ d_\mathcal{X}(u(t_j), u_{k_j}(t_j)) + d_\mathcal{X}(u_{k_j}(t_j), x_j).$$

By our choice of $\ell_j = j + 3$ we immediately get that $d_\mathcal{X}(x_i, u_{k_i}(t_i)) < 2^{-\nu-3}$ and $d_\mathcal{X}(u_{k_j}(t_j), x_j) < 2^{-\nu-3}$. Since u is an id-convergent Cauchy stream, and by our choice of $t_j = j + 2$, we can also bound $d_\mathcal{X}(u(t_i), u(t_j)) < 2^{-\nu-2}$. Next we need to handle the terms $d_\mathcal{X}(u_{k_i}(t_i), u(t_i))$ and $d_\mathcal{X}(u_{k_j}(t_j), u(t_j))$, which amounts to show that $k_j = \{M\}(3j + 5)$ is suitably large.

Indeed, observe that $d_\mathcal{Z}(u_{k_j}, u) \leq 2^{-3j-5} = 2^{-3t_j+1}$. Using Proposition 1 then yields $\|u_{k_j} - u\|_{t_j} \leq 2^{-2t_j+1}$, and using (3) we have[6]

$$\|u_{k_j}(t_j) - u(t_j)\|_{t_j} \leq \frac{t_j}{2^{t_j-1}} 2^{-t_j} \leq 2^{-t_j}.$$

Once more by Proposition 1 we get $d_\mathcal{X}(u_{k_j}(t_j), u(t_j)) \leq 2^{-t_j} \leq 2^{-\nu-2}$. The same reasoning also gives the bound $d_\mathcal{X}(u_{k_i}(t_i), u(t_i)) \leq 2^{-\nu-2}$. Combining all these bounds yields $d_\mathcal{X}(x_i, x_j) < 2^{-\nu}$, so that (x_j) is an id-convergent Cauchy sequence. In particular, we can take M_∞ to be a code for the identity function.

This construction shows that $c = \langle T, M \rangle \mapsto c_\infty = \langle T_\infty, M_\infty \rangle$ is an effective procedure. We also proved that, for all $j \in \mathbb{N}$, $d_\mathcal{X}(x_j, u(t_j)) < 2^{-j-3} + 2^{-j-2}$, implying that $\lim_j x_j = \lim_t u(t)$; hence c_∞ encodes an effective Cauchy sequence converging to $\mathcal{L}u$ as desired. $\square$

References

1. Bournez, O., Campagnolo, M.L., Graça, D.S., Hainry, E.: Polynomial differential equations compute all real computable functions on computable compact intervals. J. Complex. **23**(3), 317–335 (2007). https://doi.org/10.1016/j.jco.2006.12.005
2. Bush, V.: The differential analyzer. A new machine for solving differential equations. J. Frankl. Inst. **212**(4), 447–488 (1931)
3. Campagnolo, M.L., Moore, C., Costa, J.F.: Iteration, inequalities, and differentiability in analog computers. J. Complex. **16**(4), 642–660 (2000). https://doi.org/10.1006/jcom.2000.0559
4. Coddington, E.A., Levinson, N.: Theory of Ordinary Differential Equations. Tata McGraw-Hill Education, New York (1955)
5. Graça, D.S., Campagnolo, M.L., Buescu, J.: Robust simulations of turing machines with analytic maps and flows. In: Cooper, S.B., Löwe, B., Torenvliet, L. (eds.) CiE 2005. LNCS, vol. 3526, pp. 169–179. Springer, Heidelberg (2005). https://doi.org/10.1007/11494645_21
6. Hartree, D.R.: Calculating Instruments and Machines. Cambridge University Press, Cambridge (1950)

---

[6] Observe that $t_j \leq 2^{t_j-1}$ for any $t_j = j + 2 \geq 2$.

7. Kohlenbach, U., Lambov, B.: Bounds on iterations of asymptotically quasi-nonexpansive mappings. BRICS Rep. Ser. **10**(51) (2003)
8. Kohlenbach, U., Leuştean, L.: Asymptotically nonexpansive mappings in uniformly convex hyperbolic spaces. J. Eur. Math. Soc. **12**(1), 71–92 (2010). https://doi.org/10.4171/JEMS/190
9. Mal'cev, A.I.: Constructive algebras I. Rus. Math. Surv. **16**, 77–129 (1961)
10. Pour-El, M.B., Richards, I.: Computability in Analysis and Physics. Springer, Heidelberg (1989)
11. Poças, D., Zucker, J.: Analog networks on function data streams. Computability **7**(4), 301–322 (2018). https://doi.org/10.3233/COM-170077
12. Poças, D., Zucker, J.: Approximability in the GPAC. Log. Methods Comput. Sci. **15**(3) (2019). https://doi.org/10.23638/LMCS-15(3:24)2019
13. Reed, M., Simon, B.: Methods of Modern Mathematical Physics: Functional Analysis. Academic Press Inc., Cambridge (1980)
14. Rudin, W.: Principles of Mathematical Analysis. International Series in Pure and Applied Mathematics, 3rd edn. McGraw-Hill, New York (1976)
15. Shannon, C.: Mathematical theory of the differential analyser. J. Math. Phys. **20**, 337–354 (1941)
16. Stoltenberg-Hansen, V., Tucker, J.V.: Effective algebras. Handbook of Logic in Computer Science, vol. 4, pp. 357–526. Oxford University Press, Oxford (1995)
17. Stoltenberg-Hansen, V., Tucker, J.V.: Concrete models of computation for topological algebras. Theoret. Comput. Sci. **219**(1–2), 347–378 (1999)
18. Thomson, W., Tait, P.: Treatise on Natural Philosophy, 2nd edn, pp. 479–508. Cambridge University Press, Cambridge (1880)
19. Tucker, J.V., Zucker, J.I.: Abstract versus concrete computation on metric partial algebras. ACM Trans. Comput. Logic (TOCL) **5**(4), 611–668 (2004). https://doi.org/10.1145/1024922.1024924
20. Tucker, J.V., Zucker, J.I.: Computable total functions on metric algebras, universal algebraic specifications and dynamical systems. J. Logic Algebraic Program. **62**(1), 71–108 (2005). https://doi.org/10.1016/j.jlap.2003.10.001
21. Tucker, J.V., Zucker, J.I.: Abstract versus concrete computability: the case of countable algebras. In: Stoltenberg-Hansen, V., Väänänen, J. (eds.) Logic Colloquium '03. Lecture Notes in Logic, pp. 377–408. Cambridge University Press, Cambridge (2006). https://doi.org/10.1017/9781316755785.019
22. Weihrauch, K.: Computable Analysis: An Introduction. Texts in Theoretical Computer Science. Springer, Heidelberg (2000). https://doi.org/10.1007/978-3-642-56999-9

Modal Type Theory Based on the Intuitionistic Modal Logic IEL⁻

Daniel Rogozin[1,2]([⊠]) (iD)

[1] Lomonosov Moscow State University, Moscow, Russia
[2] Serokell OÜ, Tallinn, Estonia
`daniel.rogozin@serokell.io`

Abstract. The modal intuitionistic epistemic logic **IEL⁻** was proposed by Artemov and Protopopescu as the intuitionistic version of belief logic. We construct the modal lambda calculus which is Curry-Howard isomorphic to **IEL⁻** as the type-theoretical representation of applicative computation widely known in functional programming.

Keywords: Intuitionistic epistemic logic · Modal type theory · Functional programming · Applicative functor

1 Introduction

The intuitionistic modal logic **IEL⁻** was proposed by Artemov and Protopopescu [1]. **IEL⁻** provides the logic of beliefs agreed with BHK-semantics of intuitionistic logic. The logic **IEL⁻** is a weaker version of the intuitionistic epistemic logic **IEL**, the logic of intuitionistic knowledge and intuitionistic beliefs, which are provably consistent.

The logic **IEL⁻** is defined by following axioms and derivation rules:

Definition 1. *Intuitionistic epistemic logic* **IEL⁻***:*

1. **IPC** *axioms*
2. $\Box(A \rightarrow B) \rightarrow (\Box A \rightarrow \Box B)$
3. $A \rightarrow \Box A$
4. *Rule: Modus Ponens.*

The last modal axiom is also called *co-reflection*. One may consider this axiom as the principle which connects intuitionistic truth and intuitionistic knowledge.

From a Kripkean point of view, the logic **IEL⁻** is the logic of all frames $\langle W, R, E \rangle$ [1]. Here $\langle W, R \rangle$ is a partial order and E is a binary 'knowledge'relation, a subrelation of $\preceq$. The relation E should satisfy the following conditions:

1. $E(w) \subseteq R(w)$ for each $w \in W$
2. $E(u) \subseteq E(w)$, if wRu

S. Artemov and A. Nerode (Eds.): LFCS 2020, LNCS 11972, pp. 236–248, 2020.
https://doi.org/10.1007/978-3-030-36755-8_15

A model for **IEL**$^-$ is a quadruple $\mathcal{M} = \langle W, R, E, \vartheta \rangle$, an extended intuitionistic Kripke model with the additional forcing relation for modal formulas defined via the relation E.

$$\mathcal{M}, x \Vdash \Box A \Leftrightarrow \forall y \in E(x) \ \mathcal{M}, y \Vdash A.$$

Thus, one has the following theorem proved by Artemov and Protopopescu:

Theorem 1. *Let* $\mathbb{F}$ *be a class of* **IEL**$^-$ *frames defined as above, then* $\mathrm{Log}(\mathbb{F}) =$ **IEL**$^-$.

Krupski and Yatmanov investigated proof-theoretical and algorithmic aspects of the stronger logic **IEL**. In this paper [7], they provided the sequent calculus for **IEL** and proved that the derivability problem of this calculus is PSPACE-complete.

2   Motivation

Let us discuss the motivation of our research briefly.

Functional programming languages such as Haskell, Idris or Purescript have special type classes[1] for computation with computational environment.

By *computational context* (or, *environment*), we mean some, roughly speaking, type-level map f, where f is a "function" from $*$ to $*$: such a type-level map takes a simple type which has kind $*$ and yields another simple type of kind $*$. For more detailed description of the type system with kinds implemented in Haskell see [12].

Here, the underlying type class is `Functor` which is defined as follows:

```
class Functor f where
    fmap :: (a -> b) -> f a -> f b
```

In Haskell, `Functor` is a generalisation of such higher-order function as `map`. This function passes some unary function and the list of elements and yields another one list applying that function to each element of an input list. In other words, `map` returns an image of a list by a given function. Let us take a look at its implementation briefly:

```
map :: (a -> b) -> [a] -> [b]
map f []     = []
map f (x:xs) = f x : (map f xs)
```

The first line declares that `map` is a binary function. The arguments of `map` are a unary function of type $a \to b$ and a list of elements from type a. The output is a list of elements from type b. This line of the piece of code is the so-called type-signature. Type-signature describes the behaviour of the function in terms of input and output types.

---

[1] In Haskell, type class is a general interface for some special group of datatypes.

The next two lines describe a recursive implementation of map. At first, we tell that an image of the empty list is empty. This part is the termination condition of a recursion. After that, we consider the case with a non-empty list. A non-empty list is a list obtained by adding an element to the top of the list. Suppose one has a list xs and x is an element of type a. It is obvious that $x : xs$ is a non-empty list. Then, in the case of non-empty list $x : xs$, one needs to call map recursively on the tail xs. We also apply a given function f to the head x. Finally, we add $f\,x$ to the top of the list map $f\,xs$ which is an image of the tail xs.

A list is one of the simplest examples of a functor in such languages as Haskell. Generally, Functor provides a uniform method to carry unary functions through parametrised type such as a list. In other words, the notion of a functor in functional programming is similar to the functor from category theory.

One may extend a functor to the so-called monad which is a functional programming counterpart of Kleisli triples. In Haskell-like languages, one also has the type class called Monad, a type class of an abstract data type of action in some computational environment. Here we define the Monad type class as follows:

```
class Functor m => Monad m where
   return :: a -> m a
   (>>=)  :: m a -> (a -> m b) -> m b
```

Monad is a type class that extends Functor with two methods called return and (>>=) (a monadic bind).

Monads provide a uniform tool to describe some sequence of actions, where the result of each step depends on the previous ones somehow. In other words, one has so-called monadic binding by which such a sequence of actions with dependencies performs.

In Haskell, one has a parametrised data type called Maybe. The main use of Maybe is redefining some partial function as the total one. Here is the definition:

```
data Maybe a = Nothing | Just a
```

The data type consists of two constructors. Suppose we deal with some computation that might terminate with some failure. Nothing is a flag that claims this failure arose. The second constructor Just stores some value of type a, a successful result of a considered computation.

For example, one needs to extract the first element of some list, but there might be a failure, if the list is empty. This problem could solved with the Maybe data type:

```
safeHead :: [a] -> Maybe a
safeHead []     = Nothing
safeHead (x:xs) = Just x
```

The Maybe instance of Monad is the following one:

```
instance Monad Maybe where
   return = Just
```

(Just x**)** >>= f = f x
Nothing >>= f = **Nothing**

Here, the `return` method merely embeds any value of type a into the type
`Maybe` a via the `Just` constructor. The implementation of a monadic bind for
`Maybe` is also quite simple. Suppose one has a function f of type $a \to$ `Maybe` b
and some value x of type `Maybe` a. Here we match on x. If x is `Nothing`, then
the monadic bind yields `Nothing`. Otherwise, we extract the value of type a and
apply a given function to the extracted value.

The monad interface for `Maybe` allows one to perform sequences of actions,
where some value on some step of execution might be undefined. If all values
are well defined on each step, then the result of an execution is a term of the
form `Just` n. Otherwise, if something went wrong and we have no required value
somewhere, then the computation halts with `Nothing`.

The other examples of `Monad` instances have more or less the same explanation since the monadic interface was proposed for a side-effect processing.

Let us discuss why `Applicative` class was introduced since this class is relatively recent. This class was described by Paterson and McBride to describe
effectful programming in an applicative style [8].

One may consider the `Applicative` type class as the intermediate one
between `Functor` and `Monad`.

The main aim of an applicative functor is a generalisation the action of a
functor for functions of arbitrary arity, for instance:

```
lift A 2
   :: Applicative  f
 ⇒ (a -> b -> c)
 -> f  a -> f  b -> f  c
liftA2  f  x  y = ((pure  f) <*> x) <*> y
```

`liftA2` is a version of `fmap` for arbitrary two-argument function. It is clear that
one may implement `liftA3`, `liftA4`, and `liftAn` for each $n \in \mathbb{N}$. In the case of
lists, `liftA2` passes two-argument function, two lists, and yields the list obtained
by applying to every possible pair the first element of which is an element of the
first list and the second element belongs to the second list.

The case of monadic computation was introduced by Moggi [9] who provided so-called monadic metalanguage, the modal lambda calculus that describe
a computation with an abstract monad. From a proof-theoretical point of view,
this modal extension of the simply-typed lambda calculus is Curry-Howard isomorphic to the lax logic considered by Goldblatt in the context of Grothendieck
topology [3], where a modal operator is understood as a so-called geometric
modality.

In this paper, we consider applicative computation type-theoretically, which
is weaker than the monadic one.

It is not difficult to see that the modal axioms of **IEL⁻** and types of the
`Applicative` methods in Haskell-like languages are syntactically similar to each
other. We investigate the relationship between intuitionistic epistemic logic

IEL⁻ and applicative computation by constructing the type system which is Curry-Howard isomorphic to **IEL⁻**.

This type system consists of the rules for simply-typed lambda-calculus extended via the special modal rules. We assume that our type system will axiomatise the simplest case of an applicative computation. We provide a proof-theoretical view of this kind of computation in functional programming and prove such metatheoretical properties as strong normalisation and confluence.

The initial idea to consider applicative functors type-theoretically belongs to Krishnaswami [6] and we are going to develop his ideas.

3   Typed Lambda-Calculus Based on IEL⁻

The first is to define the natural deduction calculus for **NIEL⁻**. For simplicity, we restrict our language to $\rightarrow$, $\wedge$, and $\square$.

Definition 2. *The natural deduction calculus* **NIEL⁻** *for* **IEL⁻** *is an extension of the intuitionistic natural deduction calculus with the additional inference rules for modality:*

$$\frac{}{\Gamma, A \vdash A} \; \mathbf{ax}$$

$$\frac{\Gamma, A \vdash B}{\Gamma \vdash A \rightarrow B} \rightarrow_I \qquad \frac{\Gamma \vdash A \rightarrow B \quad \Gamma \vdash A}{\Gamma \vdash B} \rightarrow_E$$

$$\frac{\Gamma \vdash A \quad \Gamma \vdash B}{\Gamma \vdash A \wedge B} \wedge_I \qquad \frac{\Gamma \vdash A_1 \wedge A_2}{\Gamma \vdash A_i} \wedge_E, i = 1,2$$

$$\frac{\Gamma \vdash A}{\Gamma \vdash \square A} \square_{I1} \qquad \frac{\Gamma \vdash \square \vec{A} \quad \vec{A} \vdash B}{\Gamma \vdash \square B} \square_{I2}$$

The first modal rule allows one to derive co-reflection and its consequences. The second modal rule is a counterpart of $\square_I$ rule in natural deduction calculus for constructive **K** (see [5]). We will denote $\Gamma \vdash \square A_1, \ldots, \Gamma \vdash \square A_n$ and $A_1, \ldots, A_n \vdash B$ as $\Gamma \vdash \square \vec{A}$ and $\vec{A} \vdash B$ for brevity.

It is straightforward to check that the second modal rule is equivalent to the $\square$-rule à la **K**:

$$\frac{\Gamma \vdash A}{\square \Gamma \vdash \square A}$$

Let us show that one may translate **NIEL⁻** into **IEL⁻** as follows:

Lemma 1. $\Gamma \vdash_{\mathbf{NIEL^-}} A \Rightarrow \mathbf{IEL^-} \vdash \bigwedge \Gamma \rightarrow A$.

Proof. Induction on the derivation. Let us consider the modal cases.

1. If $\Gamma \vdash_{\mathbf{NIEL}^-} A$, then $\mathbf{IEL}^- \vdash \bigwedge \Gamma \to \Box A$.

(1) $\bigwedge \Gamma \to A$	assumption
(2) $A \to \Box A$	co-reflection
(3) $(\bigwedge \Gamma \to A) \to ((A \to \Box A) \to (\bigwedge \Gamma \to \Box A))$	IPC theorem
(4) $(A \to \Box A) \to (\bigwedge \Gamma \to \Box A)$	from (1), (3) and MP
(5) $\bigwedge \Gamma \to \Box A$	from (2), (4) and MP

2. If $\Gamma \vdash_{\mathbf{NIEL}^-} \overrightarrow{\Box A}$ and $\overrightarrow{A} \vdash B$, then $\mathbf{IEL}^- \vdash \bigwedge \Gamma \to \Box B$.

(1) $\bigwedge \Gamma \to \Box A_1, \ldots, \bigwedge \Gamma \to \Box A_n$	assumption
(2) $\bigwedge \Gamma \to \bigwedge_{i=1}^{n} \Box A_i$	$\mathbf{IEL}^-$ theorem
(3) $\bigwedge_{i=1}^{n} \Box A_i \to \Box \bigwedge_{i=1}^{n} A_i$	$\mathbf{IEL}^-$ theorem
(4) $\bigwedge \Gamma \to \Box \bigwedge_{i=1}^{n} A_i$	from (2), (3) and transitivity
(5) $\bigwedge_{i=1}^{n} A_i \to B$	assumption
(6) $(\bigwedge_{i=1}^{n} A_i \to B) \to \Box(\bigwedge_{i=1}^{n} A_i \to B)$	co-reflection
(7) $\Box(\bigwedge_{i=1}^{n} A_i \to B)$	from (5), (6) and MP
(8) $\Box \bigwedge_{i=1}^{n} A_i \to \Box B$	from (7) and normality
(9) $\bigwedge \Gamma \to \Box B$	from (4), (8) and transitivity

$\Box$

Lemma 2. *If $IEL^- \vdash A$, then $\mathbf{NIEL}^- \vdash A$.*

Proof. By straightforward derivation of modal axioms in $\mathbf{NIEL}^-$. We will consider those derivations via terms below. $\Box$

On the next step, we build the typed lambda-calculus based on the $\mathbf{NIEL}^-$ by proof-assignment in inference rules. Let us define terms and types for this modal lambda calculus.

Definition 3. *The set of terms:*
Let $\mathbb{V}$ be the set of variables. The set $\Lambda_{\Box}$ of terms is defined by the grammar:
$$\Lambda_{\Box} :: = \mathbb{V} \mid (\lambda \mathbb{V}.\Lambda_{\Box}) \mid (\Lambda_{\Box}\Lambda_{\Box}) \mid (\langle \Lambda_{\Box}, \Lambda_{\Box} \rangle) \mid (\pi_1 \Lambda_{\Box}) \mid (\pi_2 \Lambda_{\Box}) \mid$$
$$(\mathbf{box}\ \Lambda_{\Box}) \mid (\mathbf{let\ box}\ \mathbb{V}^* = \Lambda_{\Box}^* \ \mathbf{in}\ \Lambda_{\Box})$$

where $\mathbb{V}^*$ and $\Lambda_{\Box}^*$ denote the set of finite sequences of variables $\cup_{i=0}^{\infty} \mathbb{V}^i$ and the set of finite sequences of terms $\cup_{i=0}^{\infty} \Lambda_{\Box}^i$. In the term $(\mathbf{let\ box}\ \overrightarrow{x} = \overrightarrow{M} \ \mathbf{in}\ N)$, the sequence of variables $\overrightarrow{x}$ and the sequence of terms $\overrightarrow{M}$ should have the same length. Otherwise, the term is not well-formed.

As we discuss below, the terms of the form **let box** $\overrightarrow{x} = \overrightarrow{M}$ **in** N correspond to the special local binding.

Definition 4. *The set of types:*

Let $\mathbb{T}$ *be the set of atomic types. The set* $\mathbb{T}_\square$ *of types is generated by the grammar:*

$$\mathbb{T}_\square ::= \mathbb{T} \mid (\mathbb{T}_\square \rightarrow \mathbb{T}_\square) \mid (\mathbb{T}_\square \times \mathbb{T}_\square) \mid (\square\mathbb{T}_\square) \tag{1}$$

A context is defined standardly [10,12] as a sequence of type declarations $\Gamma = \{x_1 : A_1, \ldots, x_n : A_n\}$, where x_i is a variable and A_i is a type for each $i \in \{1, \ldots, n\}$.

Definition 5. *The modal lambda calculus* $\lambda_{\mathbf{IEL}^-}$ *based on* $\mathbf{NIEL}^-_{\wedge,\rightarrow}$:

$$\frac{}{\Gamma, x : A \vdash x : A} \; ax$$

$$\frac{\Gamma, x : A \vdash M : B}{\Gamma \vdash \lambda x.M : A \rightarrow B} \rightarrow_i \qquad \frac{\Gamma \vdash M : A \rightarrow B \qquad \Gamma \vdash N : A}{\Gamma \vdash MN : B} \rightarrow_e$$

$$\frac{\Gamma \vdash M : A \qquad \Gamma \vdash N : B}{\Gamma \vdash \langle M, N \rangle : A \times B} \times_i \qquad \frac{\Gamma \vdash M : A_1 \times A_2}{\Gamma \vdash \pi_i M : A_i} \times_e, \, i = 1, 2$$

$$\frac{\Gamma \vdash M : A}{\Gamma \vdash \mathbf{box} \; M : \square A} \; \square_I \qquad \frac{\Gamma \vdash \overrightarrow{M} : \square\overrightarrow{A} \qquad \overrightarrow{x} : \overrightarrow{A} \vdash N : B}{\Gamma \vdash \mathbf{let \; box} \; \overrightarrow{x} = \overrightarrow{M} \; \mathbf{in} \; N : \square B} \; let_\square$$

$\Gamma \vdash \overrightarrow{M} : \square\overrightarrow{A}$ is a short form for the sequence $\Gamma \vdash M_1 : \square A_1, \ldots, \Gamma \vdash M_n : \square A_n$ and $\overrightarrow{x} : \overrightarrow{A} \vdash N : B$ is a short form for $x_1 : A_1, \ldots, x_n : A_n \vdash N : B$. We use this short form instead of **let box** $x_1, \ldots, x_n = M_1, \ldots, M_n$ **in** N.

The $\square_I$-typing rule is the same as $\bigcirc$-introduction in monadic metalanguage [11]. $\square_I$ injects an object of type A into $\square$. According to this rule, it is clear that the type constructor **box** reflects the Haskell method **pure** in `Applicative` class.

The rule $let_\square$ is similar to the $\square$-rule in typed lambda calculus for intuitionistic normal modal logic **IK**, which is introduced in [4]. Informally, one may read **let box** $\overrightarrow{x} = \overrightarrow{M}$ **in** N as a simultaneous local binding in N, where each free variable of a term N should be binded with term of modalised type from $\overrightarrow{M}$. In other words, we modalise all free variables of term N and 'substitute' them to terms from the sequence $\overrightarrow{M}$.

Our calculus extends the typed lambda calculus for **IK** with $\square_I$-rule with the co-reflection rule which allows one to modalise any type in an arbitrary context.

Here are some examples:

$$\cfrac{\cfrac{\cfrac{x : A \vdash x : A}{x : A \vdash \textbf{box } x : \Box A} \; \Box_I}{\vdash (\lambda x.\textbf{box } x) : A \to \Box A} \; {\to}_I}$$

$$\cfrac{\cfrac{\cfrac{f : \Box(A \to B) \vdash f : \Box(A \to B) \qquad x : \Box A \vdash x : \Box A \qquad \cfrac{g : A \to B \vdash g : A \to B \qquad y : A \vdash y : A}{g : A \to B, y : A \vdash gy : B} \; {\to}_e}{f : \Box(A \to B), x : \Box A \vdash \textbf{let box } g, y = f, x \textbf{ in } gy : \Box B} \; \text{let}_\Box}{f : \Box(A \to B) \vdash \lambda x.\textbf{let box } g, y = f, x \textbf{ in } gy : \Box A \to \Box B} \; {\to}_I}{\vdash \lambda f.\lambda x.\textbf{let box } g, y = f, x \textbf{ in } gy : \Box(A \to B) \to \Box A \to \Box B} \; {\to}_I$$

Here we provided the derivations for modal axioms of **IEL**$^-$. In fact, we proved Lemma 2 using proof-assignment via terms.

Now we define free variables and substitutions:

Definition 6. *The set $FV(M)$ of free variables for a term M:*

1. $FV(x) = \{x\}$;
2. $FV(\lambda x.M) = FV(M) \setminus \{x\}$;
3. $FV(MN) = FV(M) \cup FV(N)$;
4. $FV(\langle M, N \rangle) = FV(M) \cup FV(N)$;
5. $FV(\pi_i M) = FV(M)$, $i = 1, 2$;
6. $FV(\textbf{box } M) = FV(M)$;
7. $FV(\textbf{let box } \overrightarrow{x} = \overrightarrow{M} \textbf{ in } N) = \cup_{i=1}^{n} FV(M)$, *where* $n = |\overrightarrow{M}|$.

Definition 7. *Substitution:*

1. $x[x := N] = N$, $x[y := N] = x$;
2. $(MN)[x := N] = M[x := N]N[x := N]$;
3. $(\lambda x.M)[y := N] = \lambda x.M[y := N]$, $y \in FV(M)$;
4. $(M, N)[x := P] = (M[x := P], N[x := P])$;
5. $(\pi_i M)[x := P] = \pi_i(M[x := P])$, $i = 1, 2$;
6. $(\textbf{box } M)[x := P] = \textbf{box } (M[x := P])$;
7. $(\textbf{let box } \overrightarrow{x} = \overrightarrow{M} \textbf{ in } N)[y := P] = \textbf{let box } \overrightarrow{x} = (\overrightarrow{M}[y := P]) \textbf{ in } N$.

Substitutions and free variable for terms of kind **let box** $\overrightarrow{x} = \overrightarrow{M}$ **in** N are defined similarly to [4]. That is, we do not take into account free variables of N because those variables occur in the list $\overrightarrow{x}$ and are eliminated by the assignment $\overrightarrow{x} = \overrightarrow{M}$.

Now we define the reduction rules:

Definition 8. *β-reduction rules for $\lambda_{\textbf{IEL}^-}$.*

1. $(\lambda x.M)N \to_\beta M[x := N]$
2. $\pi_1 \langle M, N \rangle \to_\beta M$
3. $\pi_2 \langle M, N \rangle \to_\beta N$
4. **let box** $\overrightarrow{x}, y, \overrightarrow{z} = \overrightarrow{M}$, **let box** $\overrightarrow{w} = \overrightarrow{N}$ **in** $Q, \overrightarrow{P}$ **in** $R \to_\beta$
   let box $\overrightarrow{x}, \overrightarrow{w}, \overrightarrow{z} = \overrightarrow{M}, \overrightarrow{N}, \overrightarrow{P}$ **in** $R[y := Q]$
5. **let box** $\overrightarrow{x} = \textbf{box } \overrightarrow{M}$ **in** $N \to_\beta$ **box** $N[\overrightarrow{x} := \overrightarrow{M}]$

6. **let box** __ = __ **in** $M \to_\beta$ **box** M, *where* __ *is an empty sequence of terms*

If M reduces to N by one of these rules, then we will write $M \to_r N$. A multistep reduction $\twoheadrightarrow_r$ is a reflexive transitive closure of $\to_r$.

Now we formulate the standard lemmas for contexts.

Lemma 3. *Generation for $\square_I$.*
  *Let $\Gamma \vdash$ **box** $M : \square A$, then $\Gamma \vdash M : A$.*

Proof. Straightforwardly. □

Lemma 4. *Basic lemmas.*

1. *If $\Gamma \vdash M : A$ and $\Gamma \subseteq \Delta$, then $\Delta \vdash M : A$*
2. *If $\Gamma \vdash M : A$, then $\Delta \vdash M : A$, where $\Delta = \{x : A \,|\, (x : A) \in \Gamma \,\&\, x \in FV(M)\}$*
3. *If $\Gamma, x : A \vdash M : B$ and $\Gamma \vdash N : A$, then $\Gamma \vdash M[x := N] : B$*

Proof. The items 1–2 are proved by induction on the derivation of $\Gamma \vdash M : A$. The item 3 is proved by induction on the derivation of $\Gamma \vdash N : A$. □

Theorem 2. *Subject reduction*
  If $\Gamma \vdash M : A$ and $M \twoheadrightarrow_r N$, then $\Gamma \vdash N : A$.

Proof. Induction on the derivation $\Gamma \vdash M : A$ and on the generation of $\to_\beta$. The general statement follows from transitivity of $\twoheadrightarrow_\beta$. □

Theorem 3. $\twoheadrightarrow_\beta$ *is strongly normalising;*

Proof. Follows from Theorem 5 below, so far as reduction in monadic metalanguage is strongly normalising [2] and $\lambda_{\mathbf{IEL}-}$ is sound with respect to this system. □

Theorem 4. $\twoheadrightarrow_r$ *is confluent.*

Proof. By Newman's lemma [12], if the relation is strongly normalising and locally confluent, then this relation is confluent.

It is sufficient to show that a multistep reduction $\twoheadrightarrow_r$ is locally confluent.

Lemma 5. *Local confluence*
  If $M \to_r N$ and $M \to_r Q$, then there exists some term P, such that $N \twoheadrightarrow_r P$ and $Q \twoheadrightarrow_r P$.

Proof. Let us consider the following critical pairs and show that they are joinable:

1.
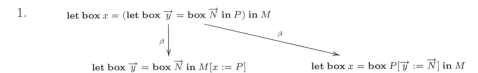

$$\textbf{let box } \overrightarrow{y} = \textbf{box } \overrightarrow{N} \textbf{ in } M[x := P] \rightarrow_\beta$$
$$\textbf{box } M[x := P][\overrightarrow{y} := \overrightarrow{N}]$$
$$\textbf{let box } x = \textbf{box } P[\overrightarrow{y} := \overrightarrow{N}] \textbf{ in } M \rightarrow_\beta$$
$$\textbf{box } M[x := P[\overrightarrow{y} := \overrightarrow{N}]] \equiv$$

Since $x \notin \overrightarrow{y}$

$$\textbf{box } M[x := P][\overrightarrow{y} := \overrightarrow{N}]$$

2. $\textbf{let box } x = (\textbf{let box }\_\_ = \_\_ \textbf{ in } N) \textbf{ in } M$

$$\Big\downarrow \beta \qquad\qquad\qquad\qquad \beta$$

$\textbf{let box }\_\_ = \_\_ \textbf{ in } M[x := N] \qquad\qquad \textbf{let box } x = \textbf{box } N \textbf{ in } M$

$\textbf{let box }\_\_ = \_\_ \textbf{ in } M[x := N] \rightarrow_\beta \textbf{let box } (M[x := N])$
$\textbf{let box } x = \textbf{box } N \textbf{ in } M \rightarrow_\beta \textbf{box } (M[x := N])$

□

Also one may consider four critical pairs which are considered in confluence proof for lambda-calculus based on the intuitionistic normal modal logic **IK** [4]. It is clear that those critical pairs are joinable too.   □

4   Relation with the Monadic Metalanguage

The monadic metalanguage is the modal lambda calculus calculus based on the categorical interpretation of computation proposed by Moggi [9]. As we told above, the monadic metalanguage might be considered as the type-theoretical representation of computation with an abstract data type of action. In fact, the monadic metalanguage is a type-theoretical formulation for monadic computation implemented in Haskell.

Let us show that $\lambda_{\textbf{IEL}^-}$ is sound with respect to the monadic metalanguage.

Definition 9. *The monadic metalanguage*

The monadic metalanguage is an extension of the simply typed lambda calculus with the additional typing rules:

$$\frac{\Gamma \vdash M : A}{\Gamma \vdash \textbf{val } M : \bigcirc A} \bigcirc_I \qquad\qquad \frac{\Gamma \vdash M : \bigcirc A \qquad \Gamma, x : A \vdash N : \bigcirc B}{\Gamma \vdash \textbf{let val } x = M \textbf{ in } N : \bigcirc B} \, let_\bigcirc$$

The reduction rules are the following ones (in addition to the standard rule for abstraction and application):

1. $\textbf{let val } x = \textbf{val } M \textbf{ in } N \rightarrow_\beta N[x := M]$;
2. $\textbf{let val } x = (\textbf{let val } y = N \textbf{ in } P) \textbf{ in } M \rightarrow_\beta \textbf{let val } y = N \textbf{ in } (\textbf{let val } x = P \textbf{ in } M)$;
3. $\textbf{let val } x = M \textbf{ in } x \rightarrow_\eta M$.

Let us define the translation ⌜.⌝ from $\lambda_{\textbf{IEL}^-}$ to the monadic metalanguage:

1. $\ulcorner p_i \urcorner = p_i$, where p_i is atomic;
2. $\ulcorner A \to B \urcorner = \ulcorner A \urcorner \to \ulcorner B \urcorner$;
3. $\ulcorner \Box A \urcorner = \bigcirc \ulcorner A \urcorner$.

1. $\ulcorner x \urcorner = x$, x is a variable;
2. $\ulcorner \lambda x.M \urcorner = \lambda x.\ulcorner M \urcorner$;
3. $\ulcorner M \ N \urcorner = \lceil M \urcorner \ulcorner N \urcorner$;
4. $\ulcorner \mathbf{box} \ M \urcorner = \mathbf{val} \ \ulcorner M \urcorner$;
5. $\ulcorner \mathbf{let \ box} \ \vec{x} = \vec{M} \ \mathbf{in} \ N \urcorner = \mathbf{let \ val} \ \vec{x} = \ulcorner \vec{M} \urcorner \ \mathbf{in} \ \ulcorner N \urcorner$.

where $\mathbf{let \ val} \ \vec{x} = \ulcorner \vec{M} \urcorner \ \mathbf{in} \ N$ denotes $\mathbf{let \ val} \ x_1 = \ulcorner M_1 \urcorner \ \mathbf{in} \ (\dots \ \mathbf{in} \ (\mathbf{let \ val} \ x_n = \ulcorner M_n \urcorner \ \mathbf{in} \ N) \dots)$

It is clear that, if $\Gamma = \{x_1 : A_1, \dots, x_n : A_n\}$ is a context, then $\ulcorner \Gamma \urcorner = \{x_1 : \ulcorner A_1 \urcorner, \dots, x_n : \ulcorner A_n \urcorner\}$

Let us denote $\vdash_{\lambda_{\mathbf{IEL}-}}$ as the derivability relation in $\lambda_{\mathbf{IEL}-}$ in order to distinguish from the derivability in the monadic metalanguage.

Lemma 6. *If $\Gamma \vdash_{\lambda_{\mathbf{IEL}-}} M : A$, then $\ulcorner \Gamma \urcorner \vdash \ulcorner M \urcorner : \ulcorner A \urcorner$ in the monadic metalanguage.*

Proof. By induction on $\Gamma \vdash_{\lambda_{\mathbf{IEL}-}} M : A$. One may prove the cases of $\Box_I$ and $\mathbf{let}_\Box$ as follows:

$$\frac{\ulcorner \Gamma \urcorner \vdash \ulcorner M \urcorner : \ulcorner A \urcorner}{\ulcorner \Gamma \urcorner \vdash \mathbf{val} \ \ulcorner M \urcorner : \bigcirc \ulcorner A \urcorner}$$

$$\frac{\ulcorner \Gamma \urcorner \vdash \ulcorner \vec{M} \urcorner : \bigcirc \ulcorner \vec{A} \urcorner \qquad \dfrac{\vec{x} : \ulcorner \vec{A} \urcorner \vdash \ulcorner N \urcorner : \ulcorner B \urcorner}{\vec{x} : \ulcorner \vec{A} \urcorner \vdash \mathbf{val} \ \ulcorner N \urcorner : \bigcirc \ulcorner B \urcorner}}{\ulcorner \Gamma \urcorner \vdash \mathbf{let \ val} \ \vec{x} = \ulcorner \vec{M} \urcorner \ \mathbf{in} \ \mathbf{val} \ \ulcorner N \urcorner : \bigcirc \ulcorner B \urcorner}$$

$\qquad\qquad\qquad\qquad\qquad\qquad\qquad\qquad\qquad\qquad\qquad\qquad\qquad\qquad\qquad\qquad\qquad$ $\square$

Now one may formulate the following lemma:

Lemma 7. *1. $\ulcorner M[x := N] \urcorner = \ulcorner M \urcorner [x := \ulcorner N \urcorner]$;*
2. $M \twoheadrightarrow_r N \Rightarrow \ulcorner M \urcorner \twoheadrightarrow_\beta \ulcorner N \urcorner$;

Proof. 1. Induction on the structure of M.
2. By the induction on $\twoheadrightarrow_r$:
    (a) For simplicity, we will consider the case with only one variable in $\mathbf{let \ box}$ local binding, that can be easily extended to an arbitrary number of valiables in local binding:

$\ulcorner \mathbf{let \ box} \ x = (\mathbf{let \ box} \ \vec{y} = \vec{N} \ \mathbf{in} \ P) \ \mathbf{in} \ M \urcorner =$
$\quad \mathbf{let \ val} \ x = (\mathbf{let \ val} \ \vec{y} = \ulcorner \vec{N} \urcorner \ \mathbf{in} \ \mathbf{val} \ \ulcorner P \urcorner) \ \mathbf{in} \ \mathbf{val} \ \ulcorner M \urcorner \twoheadrightarrow_\beta$
$\quad \mathbf{let \ val} \ \vec{y} = \ulcorner \vec{N} \urcorner \ \mathbf{in} \ (\mathbf{let \ val} \ x = \ulcorner P \urcorner \ \mathbf{in} \ \mathbf{val} \ \ulcorner M \urcorner) \twoheadrightarrow_\beta$
$\quad \mathbf{let \ val} \ \vec{y} = \ulcorner \vec{N} \urcorner \ \mathbf{in} \ \mathbf{val} \ \ulcorner M \urcorner [x := \ulcorner P \urcorner] =$
$\ulcorner \mathbf{let \ box} \ \vec{y} = \vec{N} \ \mathbf{in} \ M[x := P] \urcorner$

(b)
$$\begin{aligned}
&\ulcorner \textbf{let box } \overrightarrow{x} = \textbf{box } \overrightarrow{N} \textbf{ in } M \urcorner = \\
&\quad \textbf{let val } \overrightarrow{x} = \textbf{val } \ulcorner \overrightarrow{N} \urcorner \textbf{ in val} \ulcorner M \urcorner \to_\beta \\
&\quad \textbf{val } \ulcorner M \urcorner [\overrightarrow{x} := \ulcorner \overrightarrow{N} \urcorner] = \\
&\ulcorner \textbf{box } M[\overrightarrow{x} := \overrightarrow{N}] \urcorner
\end{aligned}$$

(c)
$$\begin{aligned}
&\ulcorner \textbf{let box } x = M \textbf{ in } x \urcorner = \\
&\quad \textbf{let val } x = \ulcorner M \urcorner \textbf{ in val } x \to_\eta \\
&\ulcorner M \urcorner
\end{aligned}$$

$\square$

Theorem 5. IEL$^-$ *is sound with respect to the monadic metalanguage.*

Proof. Follows from the lemmas above. $\square$

5   Summary

In this paper, we built the modal lambda-calculus based on the intuitionistic epistemic logic **IEL**$^-$ and proved such metatheoretical properties as confluence and strong normalisation. We investigated the connection between applicative computation and **IEL**$^-$ in order to consider this kind of computation widely used in functional programming type-theoretically.

Also, we proved that the obtained system has such properties as subject reduction, confluence, and strong normalisation.

Acknowledgment. The author is grateful to Neel Krishnaswami, Vladimir Krupski, Valerii Plisko, and Vladimir Vasyukov for consulting and advice.

The author thanks anonymous peer-reviewers for valuable and considerable comments and remarks that improved the paper significantly.

The research described in this paper was supported by Russian Foundation for Basic Research (grant 16-03-00364).

References

1. Artemov, S., Protopopescu, T.: Intuitionistic epistemic logic. Rev. Symb. Log. **9**(2), 266–298 (2016). https://doi.org/10.1017/S1755020315000374
2. Benton, P.N., Bierman, G.M., de Paiva, V.: Computational types from a logical perspective. J. Funct. Program. **8**(2), 177–193 (1998). https://doi.org/10.1017/S0956796898002998
3. Goldblatt, R.I.: Grothendieck topology as geometric modality. Math. Log. Q. **27**(3135), 495–529 (1981). https://doi.org/10.1002/malq.19810273104
4. Kakutani, Y.: Call-by-name and call-by-value in normal modal logic. In: Shao, Z. (ed.) APLAS 2007. LNCS, vol. 4807, pp. 399–414. Springer, Heidelberg (2007). https://doi.org/10.1007/978-3-540-76637-7_27
5. Kavvos, G.A.: The many worlds of modal λ-calculi: I. curry-howard for necessity, possibility and time. CoRR abs/1605.08106 (2016). http://arxiv.org/abs/1605.08106

6. Krishnaswami, N.: A computational lambda calculus for applicative functors. http://semantic-domain.blogspot.com/2012/08/a-computational-lambda-calculus-for.html

7. Krupski, V.N., Yatmanov, A.: Sequent calculus for intuitionistic epistemic logic IEL. In: Artemov, S., Nerode, A. (eds.) LFCS 2016. LNCS, vol. 9537, pp. 187–201. Springer, Cham (2016). https://doi.org/10.1007/978-3-319-27683-0_14

8. McBride, C., Paterson, R.: Applicative programming with effects. J. Funct. Program. **18**(1), 1–13 (2008). https://doi.org/10.1017/S0956796807006326

9. Moggi, E.: Notions of computation and monads. Inform. Comput. **93**(1), 55–92 (1991). https://doi.org/10.1016/0890-5401(91)90052-4. http://www.sciencedirect.com/science/article/pii/0890540191900524, selections from 1989 IEEE Symposium on Logic in Computer Science

10. Nederpelt, R., Geuvers, H.: Type Theory and Formal Proof: An Introduction, 1st edn. Cambridge University Press, New York (2014)

11. Pfenning, F., Davies, R.: A judgmental reconstruction of modal logic. Math. Struct. Comput. Sci. **11**(4), 511–540 (2001). https://doi.org/10.1017/S0960129501003322

12. Sorensen, M.H., Urzyczyn, P.: Lectures on the Curry-Howard Isomorphism. Elsevier Science, Amsterdam (2006)

Lifting Recursive Counterexamples to Higher-Order Arithmetic

Sam Sanders[✉][iD]

Department of Mathematics, TU Darmstadt, Darmstadt, Germany
sasander@me.com
https://sasander.wixsite.com/academic

Abstract. In classical computability theory, a *recursive counterexample* to a theorem shows that the latter does not hold when restricted to computable objects. These counterexamples are highly useful in the *Reverse Mathematics* program, where the aim of the latter is to determine the minimal axioms needed to prove a given theorem of ordinary mathematics. Indeed, recursive counterexamples often (help) establish the 'reverse' implication in the typical equivalence between said minimal axioms and the theorem at hand. The aforementioned is generally formulated in the language of second-order arithmetic and we show in this paper that recursive counterexamples are readily modified to provide similar implications in higher-order arithmetic. For instance, the higher-order analogue of 'sequence' is the topological notion of 'net', also known as 'Moore-Smith sequence'. Our results on metric spaces suggest that the latter can only be reasonably studied in weak systems via representations (aka codes) in the language of second-order arithmetic.

Keywords: Recursive counterexamples · Higher-order arithmetic · Specker sequences · Nets

1 Introduction

Computability theory has its roots in the seminal work of Turing, providing an intuitive notion of computation based on what we nowadays call *Turing machines* ([34]). Now, *classical* (resp. higher-order) computability theory deals with the computability of sets of natural numbers (resp. higher-order objects). In classical computability theory, a *recursive counterexample* to a theorem (formulated in an appropriate language) shows that the latter does not hold when restricted to computable sets. An historical overview may be found in the introduction of [7].

Recursive counterexamples are also highly useful in the *Reverse Mathematics* program (RM hereafter; see Sect. 2.1). Indeed, the aim of RM is to determine the minimal axioms needed to prove a given theorem of ordinary mathematics, often resulting in an *equivalence* between these axioms and the theorem; recursive counterexamples often (help) establish the 'reverse' implication from the theorem at hand to the minimal axioms (see e.g. [29, p. 1368] for this opinion).

S. Artemov and A. Nerode (Eds.): LFCS 2020, LNCS 11972, pp. 249–267, 2020.
https://doi.org/10.1007/978-3-030-36755-8_16

As is well-known, both (classical) RM and classical recursion theory are (essentially) restricted to the language of second-order arithmetic, i.e. natural numbers and sets thereof. It is then a natural, if somewhat outlandish, question whether recursive counterexamples (and the associated implications in classical RM) yield any interesting results in higher-order RM and computability theory. In this paper, we show that recursive counterexamples are readily modified to provide interesting implications in higher-order arithmetic. We shall treat the following theorems: montone convergence theorem/Specker sequences (Sect. 3.1), compactness of metric spaces (Sect. 3.2), the Rado selection lemma (Sect. 3.3), and the ordering of fields (Sect. 3.4).

We shall work in Kohlenbach's higher-order RM ([16]; see Sect. 2.1). We do not claim that the above results are always optimal or new; we even provide a counterexample in Sect. 3.1. Our aim is to show that *with little modification* recursive counterexamples, second-order as they may be, also establish results in higher-order arithmetic. As a bonus, our results pertaining to metric spaces suggest that the latter can only be reasonably studied in weak systems via representations (aka codes) in the language of second-order arithmetic.   ·

Finally, the reader is welcome to their own interpretation of the aforementioned results, *as long it accords with all the facts*. In our opinion, one *reasonable* interpretation is that second- and higher-order arithmetic are not *as* different as sometimes claimed, and that recursive counterexamples and reversals provide a bridge of sorts between the two. However, the results in this paper *do not* support the argument that higher-order arithmetic contains 'nothing new' compared to second-order arithmetic, as discussed in detail in Sect. 4.2.

2   Preliminaries

We introduce *Reverse Mathematics* in Sect. 2.1, as well as its generalisation to *higher-order arithmetic*, and the associated base theory RCA_0^ω. We introduce some essential axioms in Sect. 2.2.

2.1   Reverse Mathematics

Reverse Mathematics is a program in the foundations of mathematics initiated around 1975 by Friedman ([8,9]) and developed extensively by Simpson ([28]). The aim of RM is to identify the minimal axioms needed to prove theorems of ordinary, i.e. non-set theoretical, mathematics. In almost all cases, these minimal axioms are also *equivalent* to the theorem at hand (over a weak logical system). The reversal, i.e. the derivation of the minimal axioms from the theorem, is often proved based on recursive counterexample to the latter (see [29, p. 1368]).

We refer to [32] for an introduction to RM and to [28] for an overview of RM. We expect basic familiarity with RM, but do sketch some aspects of Kohlenbach's *higher-order* RM ([16]) essential to this paper, including the base theory RCA_0^ω (Definition 1). As will become clear, the latter is officially a type theory but can accommodate (enough) set theory via Definition 4.

First of all, in contrast to 'classical' RM based on *second-order arithmetic* Z_2, higher-order RM uses L_ω, the richer language of *higher-order arithmetic*. Indeed, while L_2, the language of Z_2, is restricted to natural numbers and sets of natural numbers, higher-order arithmetic can accommodate sets of sets of natural numbers, sets of sets of sets of natural numbers, et cetera. To formalise this idea, define the collection of *all finite types* $\mathbf{T}$ by the two clauses:

(i) $0 \in \mathbf{T}$ and (ii) if $\sigma, \tau \in \mathbf{T}$ then $(\sigma \to \tau) \in \mathbf{T}$,

where 0 is the type of natural numbers, and $\sigma \to \tau$ is the type of mappings from objects of type σ to objects of type τ. In this way, $1 \equiv 0 \to 0$ is the type of functions from numbers to numbers, and $n+1 \equiv n \to 0$ maps type n objects to numbers. Viewing sets as characteristic functions, we note that Z_2 only includes objects of type 0 and 1.

Secondly, the language L_ω includes variables $x^\rho, y^\rho, z^\rho, \ldots$ of any finite type $\rho \in \mathbf{T}$. Types may be omitted when they can be inferred from context. The constants of L_ω include the type 0 objects $0, 1$ and $<_0, +_0, \times_0, =_0$ which are intended to have their usual meaning as operations on $\mathbb{N}$. Equality at higher types is defined in terms of '$=_0$' as follows: for any objects x^τ, y^τ, we have

$$[x =_\tau y] \equiv (\forall z_1^{\tau_1} \ldots z_k^{\tau_k})[x z_1 \ldots z_k =_0 y z_1 \ldots z_k], \tag{1}$$

if the type τ is composed as $\tau \equiv (\tau_1 \to \ldots \to \tau_k \to 0)$. Furthermore, L_ω includes the *recursor constant* $\mathbf{R}_\sigma$ for any $\sigma \in \mathbf{T}$, which allows for iteration on type σ-objects as in the special case (2). Formulas and terms are defined as usual.

Definition 1. The base theory RCA_0^ω consists of the following axioms.

a. Basic axioms expressing that $0, 1, <_0, +_0, \times_0$ form an ordered semi-ring with equality $=_0$.
b. Basic axioms defining the well-known Π and Σ combinators (aka K and S in [1]), which allow for the definition of λ-*abstraction*.
c. The defining axiom of the recursor constant $\mathbf{R}_0$: For m^0 and f^1:

$$\mathbf{R}_0(f, m, 0) := m \text{ and } \mathbf{R}_0(f, m, n+1) := f(n, \mathbf{R}_0(f, m, n)). \tag{2}$$

d. The *axiom of extensionality*: for all $\rho, \tau \in \mathbf{T}$, we have:

$$(\forall x^\rho, y^\rho, \varphi^{\rho \to \tau})[x =_\rho y \to \varphi(x) =_\tau \varphi(y)]. \tag{$\mathsf{E}_{\rho,\tau}$}$$

e. The induction axiom for quantifier-free[1] formulas of L_ω.
f. QF-AC^{1,0}: The quantifier-free Axiom of Choice as in Definition 2.

Definition 2. The axiom QF-AC consists of the following for all $\sigma, \tau \in \mathbf{T}$:

$$(\forall x^\sigma)(\exists y^\tau)A(x, y) \to (\exists Y^{\sigma \to \tau})(\forall x^\sigma)A(x, Y(x)), \tag{QF-AC$^{\sigma,\tau}$}$$

for any quantifier-free formula A in the language of L_ω.

---

[1] To be absolutely clear, variables (of any finite type) are allowed in quantifier-free formulas of the language L_ω: only quantifiers are banned.

We let IND be the induction axiom for all formulas in L_ω.

As discussed in [16, §2], RCA_0^ω and RCA_0 prove the same sentences 'up to language' as the latter is set-based and the former function-based. Recursion as in (2) is called *primitive recursion*; the class obtained from $\mathbf{R}_\rho$ for all $\rho \in \mathbf{T}$ is called *Gödel's system T* of all (higher-order) primitive recursive functionals.

We use the usual notations for natural, rational, and real numbers, and the associated functions, as introduced in [16, pp. 288–289].

Definition 3 (Real numbers and related notions in RCA_0^ω)

a. Natural numbers correspond to type zero objects, and we use 'n^0' and '$n \in \mathbb{N}$' interchangeably. Rational numbers are defined as signed quotients of natural numbers, and '$q \in \mathbb{Q}$' and '$<_\mathbb{Q}$' have their usual meaning.

b. Real numbers are coded by fast-converging Cauchy sequences $q_{(\cdot)} : \mathbb{N} \to \mathbb{Q}$, i.e. such that $(\forall n^0, i^0)(|q_n - q_{n+i}| <_\mathbb{Q} \frac{1}{2^n})$. We use Kohlenbach's 'hat function' from [16, p. 289] to guarantee that every q^1 defines a real number.

c. We write '$x \in \mathbb{R}$' to express that $x^1 := (q_{(\cdot)}^1)$ represents a real as in the previous item and write $[x](k) := q_k$ for the k-th approximation of x.

d. Two reals x, y represented by $q_{(\cdot)}$ and $r_{(\cdot)}$ are *equal*, denoted $x =_\mathbb{R} y$, if $(\forall n^0)(|q_n - r_n| \leq 2^{-n+1})$. Inequality '$<_\mathbb{R}$' is defined similarly. We sometimes omit the subscript '$\mathbb{R}$' if it is clear from context.

e. Functions $F : \mathbb{R} \to \mathbb{R}$ are represented by $\Phi^{1\to1}$ mapping equal reals to equal reals, i.e. satisfying $(\forall x, y \in \mathbb{R})(x =_\mathbb{R} y \to \Phi(x) =_\mathbb{R} \Phi(y))$.

f. The relation '$x \leq_\tau y$' is defined as in (1) but with '$\leq_0$' instead of '$=_0$'. Binary sequences are denoted '$f^1, g^1 \leq_1 1$', but also '$f, g \in C$' or '$f, g \in 2^\mathbb{N}$'. Elements of Baire space are given by f^1, g^1, but also denoted '$f, g \in \mathbb{N}^\mathbb{N}$'.

g. For a binary sequence f^1, the associated real in $[0,1]$ is $\mathsf{r}(f) := \sum_{n=0}^\infty \frac{f(n)}{2^{n+1}}$.

h. Sets of type ρ objects $X^{\rho\to0}, Y^{\rho\to0}, \dots$ are given by their characteristic functions $F_X^{\rho\to0} \leq_{\rho\to0} 1$, i.e. we write '$x \in X$' for $F_X(x) =_0 1$.

The following special case of item (h) is singled out, as it will be used frequently.

Definition 4 [RCA_0^ω]. A 'subset D of $\mathbb{N}^\mathbb{N}$' is given by its characteristic function $F_D^2 \leq_2 1$, i.e. we write '$f \in D$' for $F_D(f) = 1$ for any $f \in \mathbb{N}^\mathbb{N}$. A 'binary relation $\preceq$ on a subset D of $\mathbb{N}^\mathbb{N}$' is given by some functional $G_\preceq^{(1\times1)\to0}$, namely we write '$f \preceq g$' for $G_\preceq(f, g) = 1$ and any $f, g \in D$. Assuming extensionality on the reals as in item (e), we obtain characteristic functions that represent subsets of $\mathbb{R}$ and relations thereon. Using pairing functions, it is clear we can also represent sets of finite sequences (of reals), and relations thereon.

Next, we mention the highly useful ECF-interpretation.

Remark 5 (The ECF-interpretation). The (rather) technical definition of ECF may be found in [33, p. 138, §2.6]. Intuitively, the ECF-interpretation $[A]_{\mathsf{ECF}}$ of a formula $A \in L_\omega$ is just A with all variables of type two and higher replaced by countable representations of continuous functionals. Such representations are also (equivalently) called 'associates' or 'RM-codes' (see [15, §4]). The ECF-interpretation connects RCA_0^ω and RCA_0 (see [16, Prop. 3.1]) in that if RCA_0^ω

proves A, then RCA_0 proves $[A]_{\mathsf{ECF}}$, again 'up to language', as RCA_0 is formulated using sets, and $[A]_{\mathsf{ECF}}$ is formulated using types, namely only using type zero and one objects.

In light of the widespread use of codes in RM and the common practise of identifying codes with the objects being coded, it is no exaggeration to refer to ECF as the *canonical* embedding of higher-order into second-order RM. For completeness, we also list the following notational convention for finite sequences.

Notation 6 (Finite sequences). We assume a dedicated type for 'finite sequences of objects of type ρ', namely ρ^*. Since the usual coding of pairs of numbers goes through in RCA_0^ω, we shall not always distinguish between 0 and 0^*. Similarly, we do not always distinguish between 's^ρ' and '$\langle s^\rho \rangle$', where the former is 'the object s of type ρ', and the latter is 'the sequence of type ρ^* with only element s^ρ'. The empty sequence for the type ρ^* is denoted by '$\langle \rangle_\rho$', usually with the typing omitted.

Furthermore, we denote by '$|s| = n$' the length of the finite sequence $s^{\rho^*} = \langle s_0^\rho, s_1^\rho, \ldots, s_{n-1}^\rho \rangle$, where $|\langle \rangle| = 0$, i.e. the empty sequence has length zero. For sequences s^{ρ^*}, t^{ρ^*}, we denote by '$s*t$' the concatenation of s and t, i.e. $(s*t)(i) = s(i)$ for $i < |s|$ and $(s*t)(j) = t(|s| - j)$ for $|s| \leq j < |s| + |t|$. For a sequence s^{ρ^*}, we define $\overline{s}N := \langle s(0), s(1), \ldots, s(N - 1) \rangle$ for $N^0 < |s|$. For a sequence $\alpha^{0 \to \rho}$, we also write $\overline{\alpha}N = \langle \alpha(0), \alpha(1), \ldots, \alpha(N - 1) \rangle$ for *any* N^0. By way of shorthand, $(\forall q^\rho \in Q^{\rho^*})A(q)$ abbreviates $(\forall i^0 < |Q|)A(Q(i))$, which is (equivalent to) quantifier-free if A is.

2.2 Some Axioms of Higher-Order RM

We introduce some functionals which constitute the counterparts of some of the Big Five systems, in higher-order RM. We use the formulation from [16] and [19]. First of all, ACA_0 is readily derived from:

$$(\exists \mu^2)(\forall f^1)\big[(\exists n)(f(n) = 0) \to [(f(\mu(f)) = 0) \wedge (\forall i < \mu(f))f(i) \neq 0] \tag{μ^2}$$
$$\wedge [(\forall n)(f(n) \neq 0) \to \mu(f) = 0]\big],$$

and $\mathsf{ACA}_0^\omega \equiv \mathsf{RCA}_0^\omega + (\mu^2)$ proves the same sentences as ACA_0 by [13, Theorem 2.5]. The (unique) functional μ^2 in (μ^2) is called *Feferman's μ* ([1]), and is clearly *discontinuous* at $f =_1 11\ldots$; in fact, (μ^2) is equivalent to the existence of $F : \mathbb{R} \to \mathbb{R}$ such that $F(x) = 1$ if $x >_\mathbb{R} 0$, and 0 otherwise ([16, §3]), and to

$$(\exists \varphi^2 \leq_2 1)(\forall f^1)\big[(\exists n)(f(n) = 0) \leftrightarrow \varphi(f) = 0\big]. \tag{$\exists^2$}$$

Finally, we list the following comprehension axiom, first introduced in [24].

Definition 7 [BOOT]. $(\forall Y^2)(\exists X^1)(\forall n^0)\big[n \in X \leftrightarrow (\exists f^1)(Y(f, n) = 0)\big].$

Clearly, BOOT is inspired by the following axiom:

$$(\exists E^3 \leq_3 1)(\forall Y^2)\big[(\exists f^1)(Y(f) = 0) \leftrightarrow E(Y) = 0\big], \tag{$\exists^3$}$$

yielding full second-order arithmetic Z_2, while $Z_2^\Omega \equiv \mathsf{RCA}_0^\omega + (\exists^3)$ is a conservative extension of the latter (see [13]). No comprehension axiom weaker than $(\exists^3)$ can prove BOOT by the results in [19], [20], [25]. Nonetheless, one readily shows that $[\mathsf{BOOT}]_{\mathsf{ECF}}$ is equivalent to ACA_0 and we finish this section with a conceptual remark on how ECF connects second- and higher-order arithmetic.

Remark 8 (The nature of ECF). We discuss the meaning of the words 'A is converted into B by the ECF-translation'. Such statement is obviously not to be taken literally, as e.g. $[\mathsf{BOOT}]_{\mathsf{ECF}}$ is not verbatim ACA_0. Nonetheless, $[\mathsf{BOOT}]_{\mathsf{ECF}}$ follows from ACA_0 by noting that $(\exists f^1)(Y(f, n) = 0) \leftrightarrow (\exists \sigma^{0^*})(Y(\sigma * 00, n) = 0)$ for continuous Y^2 (see [24, §3]). Similarly, let HBU be the Heine-Borel theorem for *uncountable* covers of Cantor space as studied in [19]. Then $[\mathsf{HBU}]_{\mathsf{ECF}}$ is not verbatim the Heine-Borel theorem for countable covers, but the latter does imply the former by noting that for continuous functions, the associated canonical cover has a trivial countable sub-cover enumerated by the rationals in $[0, 1]$.

In general, that continuous objects have countable representations is the very foundation of the formalisation of mathematics in L_2, and identifying continuous objects and their countable representations is routinely done. Thus, when we say that A is converted into B by the ECF-translation, we mean that $[A]_{\mathsf{ECF}}$ is about a class of continuous objects to which B is immediately seen to apply, with a possible intermediate step involving representations. Since this kind of step forms the bedrock of classical RM, it is therefore harmless in this context.

3   Main Results

We establish the results sketched in Sect. 1. In each section, we study a known recursive counterexample and show that it lifts to higher-order arithmetic *with minimal effort*.

3.1   Specker Nets

In Sect. 3.1, we lift the implication involving the monotone convergence theorem *for sequences* and arithmetical comprehension to higher-order arithmetic. This results in an implication involving the *monotone convergence theorem* for *nets* indexed by Baire space and the comprehension axiom BOOT from Sect. 2.2. Nets and associated concepts are introduced in Sect. 3.1.

In more detail, the proof that the monotone convergence theorem implies ACA_0 from [28, III.2] is based on a recursive counterexample by Specker ([31]), who proved the existence of a computable increasing sequence of rationals in the unit interval that does not converge to any computable real number. We show that these results lift to the higher-order setting in that *essentially the same proof* yields that the monotone convergence theorem *for nets* indexed by Baire space implies BOOT. In particular, the notion of *Specker sequence* readily generalises to *Specker net*. We provide full details for this case, going as far as comparing the original and 'lifted' proof side-by-side. A much less detailed proof was first published in [24].

Nets: Basics and Definitions. We introduce the notion of net and associated concepts. Intuitively speaking, nets are the generalisation of sequences to (possibly) uncountable index sets; nets are essential for convergence in topology and domain theory. On a historical note, Moore-Smith and Vietoris independently introduced these notions about a century ago in [18] and [35], which is why nets are also called *Moore-Smith sequences*. Nets and filters yield the same convergence theory, but e.g. third-order nets are represented by fourth-order filters, i.e. nets are more economical in terms of type complexity (see [2]).

We use the following definition from [14, Ch. 2].

Definition 9 [Nets]. A set $D \neq \emptyset$ with a binary relation '$\preceq$' is *directed* if

a. The relation $\preceq$ is transitive, i.e. $(\forall x, y, z \in D)([x \preceq y \wedge y \preceq z] \rightarrow x \preceq z)$.
b. For $x, y \in D$, there is $z \in D$ such that $x \preceq z \wedge y \preceq z$.
c. The relation $\preceq$ is reflexive, i.e. $(\forall x \in D)(x \preceq x)$.

For such $(D, \preceq)$ and topological space X, any mapping $x : D \rightarrow X$ is a *net* in X. We denote $\lambda d.x(d)$ as 'x_d' or '$x_d : D \rightarrow X$' to suggest the connection to sequences. The directed set $(D, \preceq)$ is not always explicitly mentioned together with a net x_d.

We only use directed sets that are subsets of $\mathbb{N}^{\mathbb{N}}$, i.e. as given by Definition 4. Similarly, we only study nets $x_d : D \rightarrow \mathbb{R}$ where D is a subset of $\mathbb{N}^{\mathbb{N}}$. Thus, a net x_d in $\mathbb{R}$ is just a type $1 \rightarrow 1$ functional with extra structure on its domain D provided by '$\preceq$' as in Definition 4, i.e. part of third-order arithmetic.

The definitions of convergence and increasing net are of course familiar.

Definition 10 [Convergence of nets]. If x_d is a net in X, we say that x_d *converges* to the limit $\lim_d x_d = y \in X$ if for every neighbourhood U of y, there is $d_0 \in D$ such that for all $e \succeq d_0$, $x_e \in U$.

Definition 11 [Increasing nets]. A net $x_d : D \rightarrow \mathbb{R}$ is *increasing* if $a \preceq b$ implies $x_a \leq_{\mathbb{R}} x_b$ for all $a, b \in D$.

Many (convergence) notions concerning sequences carry over to nets *mutatis mutandis*. A rather general RM study of nets may be found in [26], [27], [24], [25]. We shall study the monotone convergence theorem for nets as follows.

Definition 12 [$\mathsf{MCT}_{\mathrm{net}}^{[0,1]}$]. Any increasing net in $[0, 1]$ indexed by $\mathbb{N}^{\mathbb{N}}$ converges.

The 'original' monotone convergence theorem for *sequences* as in [28, III.2] is denoted $\mathsf{MCT}_{\mathrm{seq}}^{[0,1]}$. Following Remark 8, we say that $\mathsf{MCT}_{\mathrm{seq}}^{[0,1]}$ is the ECF-interpretation of $\mathsf{MCT}_{\mathrm{net}}^{[0,1]}$. The implications $\mathsf{MCT}_{\mathrm{seq}}^{[0,1]} \leftarrow \mathsf{ACA}_0$ and $\mathsf{MCT}_{\mathrm{net}}^{[0,1]} \leftarrow \mathsf{BOOT}$ are in fact proved in exactly the same way.

Finally, sequences are nets with index set $(\mathbb{N}, \leq_{\mathbb{N}})$ and theorems pertaining to nets therefore apply to sequences. However, some care is advised as e.g. a sub-net of a sequence is not necessarily a *sub-sequence* (see [25, §3]).

Specker Nets and Comprehension. In this section, we show that $\mathsf{MCT}_{\mathsf{net}}^{[0,1]} \rightarrow$ BOOT using a minor variation of the well-known proof $\mathsf{MCT}_{\mathsf{seq}}^{[0,1]} \rightarrow \mathsf{ACA}_0$ from [28, III.2.2] involving Specker sequences.

First of all, we distill the essence of the latter proof, as follows.

i. We prove $\mathsf{MCT}_{\mathsf{seq}}^{[0,1]} \rightarrow$ range, where the latter states that the range exists for any function, i.e. $(\forall f^1)(\exists X \subset \mathbb{N})(\forall k \in \mathbb{N})\big(k \in X \leftrightarrow (\exists m^0)(f(m) = k)\big)$.

ii. Fix f^1 and define the Specker sequence $c_n := \sum_{i=0}^{n} 2^{-f(i)}$.

iii. Note that $\mathsf{MCT}_{\mathsf{seq}}^{[0,1]}$ applies and let c be $\lim_{n \to \infty} c_n$.

iv. Establish the following equivalence:

$$(\exists m^0)(f(m) = k) \leftrightarrow (\forall n^0)\big(|c_n - c| < 2^{-k} \rightarrow (\exists i \leq n)(f(i) = k)\big), \qquad (3)$$

v. Apply Δ_1^0-comprehension to (3), yielding the set X needed for range.

We now show how to lift the previous steps to higher-order arithmetic, resulting in a proof of $\mathsf{MCT}_{\mathsf{net}}^{[0,1]} \rightarrow$ BOOT in Theorem 15.

Regarding item (v), to lift proofs involving Δ_1^0-comprehension to the higher-order framework, we introduce the following comprehension axiom:

$$(\forall Y^2, Z^2)\big[(\forall n^0)\big((\exists f^1)(Y(f,n) = 0) \leftrightarrow (\forall g^1)(Z(g,n) = 0)\big) \qquad (\Delta\text{-CA})$$
$$\rightarrow (\exists X^1)(\forall n^0)(n \in X \leftrightarrow (\exists f^1)(Y(f,n) = 0))\big].$$

A snippet of countable choice suffices to prove Δ-comprehension and we observe that the ECF-translation converts Δ-comprehension into Δ_1^0-comprehension while $\mathsf{QF\text{-}AC}^{0,1}$ becomes $\mathsf{QF\text{-}AC}^{0,0}$, all following Remark 8.

Theorem 13. *The system* $\mathsf{RCA}_0^\omega + \mathsf{QF\text{-}AC}^{0,1}$ *proves* $\Delta\text{-CA}$.

Proof. The antecedent of Δ-comprehension implies the following

$$(\forall n^0)(\exists g^1, f^1)(Z(g,n) = 0 \rightarrow Y(f,n) = 0). \qquad (4)$$

Applying $\mathsf{QF\text{-}AC}^{0,1}$ to (4) yields $\Phi^{0 \rightarrow 1}$ such that

$$(\forall n^0)\big((\forall g^1)(Z(g,n) = 0) \rightarrow Y(\Phi(n), n) = 0\big), \qquad (5)$$

and by assumption an equivalence holds in (5), and we are done. $\qquad \square$

The previous is not *spielerei*: the crux of numerous reversals $T \rightarrow \mathsf{ACA}_0$ is that the theorem T somehow allows for the reduction of Σ_1^0-formulas to Δ_1^0-formulas, while Δ_1^0-comprehension -included in RCA_0- then yields the required Σ_1^0-comprehension, and ACA_0 follows. Additional motivation for Δ-CA is provided by Theorem 30.

Regarding item (i), lifting range to the higher-order framework is fairly basic: we just consider the existence of the range of *type two* functionals (rather than type one functions), as in RANGE below.

Theorem 14. *The system* RCA_0^ω *proves that* BOOT *is equivalent to*

$$(\forall G^2)(\exists X^1)(\forall n^0)\big[n \in X \leftrightarrow (\exists f^1)(G(f) = n)\big]. \tag{RANGE}$$

Proof. The forward direction is immediate. For the reverse direction, define G^2 as follows for n^0 and g^1: put $G(\langle n \rangle * g) = n+1$ if $Y(g,n) = 0$, and 0 otherwise. Let $X \subseteq \mathbb{N}$ be as in RANGE and note that

$$(\forall m^0 \geq 1)(m \in X \leftrightarrow (\exists f^1)(G(f) = m) \leftrightarrow (\exists g^1)(Y(g, m-1) = 0)). \tag{6}$$

which is as required for BOOT after trivial modification. $\qquad\square$

This theorem was first proved as [24, Theorem 3.19]. Again, the previous is not a gimmick: reversals involving ACA_0 are often established using range, and those yield implications involving RANGE, for instance as follows.

Theorem 15. *The system* $\mathsf{RCA}_0^\omega + \Delta\text{-}\mathsf{CA}$ *proves* $\mathsf{MCT}_{net}^{[0,1]} \to \mathsf{BOOT}$.

Proof. In case $\neg(\exists^2)$, note that $\mathsf{MCT}_{net}^{[0,1]}$ implies $\mathsf{MCT}_{seq}^{[0,1]}$ as sequences are nets with directed set $(\mathbb{N}, \leq_\mathbb{N})$. By [28, III.2], ACA_0 is available, which readily implies BOOT for continuous Y^2, but all functions on $\mathbb{N}^\mathbb{N}$ are continuous by [16, §3].

In case $(\exists^2)$, we shall establish RANGE and obtain BOOT by Theorem 14, which mimics the above item (i). We let $(D, \preceq_D)$ be a directed set with D consisting of the finite sequences in $\mathbb{N}^\mathbb{N}$ and $v \preceq_D w$ if $(\forall i < |v|)(\exists j < |w|)(v(i) =_1 w(j))$ for any v^{1^*}, w^{1^*}. Note that $(\exists^2)$ is necessary for this definition.

Following item (ii), fix some Y^2 and define the 'Specker net' $c_w : D \to [0,1]$ as $c_w := \sum_{i=0}^{|w|-1} 2^{-Y(w(i))}$. Clearly, c_w is increasing as in Definition 11 and let c be the limit provided by $\mathsf{MCT}_{net}^{[0,1]}$, following item (iii). Following item (iv), consider the following generalisation of (3), for any $k \in \mathbb{N}$:

$$(\exists f^1)(Y(f) = k) \leftrightarrow (\forall w^{1^*})\big(|c_w - c| < 2^{-k} \to (\exists g \in w)(Y(g) = k)\big), \tag{7}$$

for which the reverse direction is trivial thanks to $\lim_w c_w = c$. For the forward direction in (7), assume the left-hand side holds for $f = f_1^1$ and fix some $w_0^{1^*}$ such that $|c - c_{w_0}| < \frac{1}{2^k}$. Since c_w is increasing, we also have $|c - c_w| < \frac{1}{2^k}$ for $w \succeq_D w_0$. Now there must be f_0 in w_0 such that $Y(f_0) = k$, as otherwise $w_1 = w_0 * \langle f_1 \rangle$ satisfies $w_1 \succeq_D w_0$ but also $c_{w_1} > c$, which is impossible.

Note that (7) has the right form to apply $\Delta\text{-}\mathsf{CA}$ (modulo obvious coding), and the latter provides the set required by RANGE, following item (v). $\qquad\square$

We refer to the net c_w from the proof as a *Specker net* following the concept of *Specker sequence* pioneered in [31]. We hope that the reader agrees that the previous proof is *exactly* the final part of the proof of [28, III.2.2] as in items (i)–(v), save for the replacement of sequences by nets and functions by functionals. The aforementioned 'reuse' comes at a cost however: the proof of $\mathsf{MCT}_{net}^{[0,1]} \leftrightarrow \mathsf{BOOT}$ in [24, §3.2] does not make use of countable choice or $\Delta\text{-}\mathsf{CA}$. Moreover, from the proof of this equivalence, once can essentially 'read off' that a realiser for $\mathsf{MCT}_{net}^{[0,1]}$ computes $\exists^3$ in the sense of Kleene's S1-S9, and vice versa (see also [25, §3.1]). It seems one cannot obtain this S1-S9 result from the above proof because of Δ-CA.

Finally, Theorem 15 readily generalises by increasing the size of the index sets to any set of objects of finite type. The case of nets indexed by $\mathcal{N} \equiv \mathbb{N}^{\mathbb{N}} \to \mathbb{N}$ may be found in [24, Theorem 3.38]. In particular, the monotone convergence theorem for nets indexed by $\mathcal{N}$ in $[0,1]$ implies the following axiom:

$$(\forall G^3)(\exists X^1)(\forall n^0)[n \in X \leftrightarrow (\exists Y^2)(G(Y) = n)], \qquad (\mathsf{RANGE}^1)$$

which states the existence of the range of type three functionals.

3.2   Compactness of Metric Spaces

Complete separable metric spaces are represented (or: coded) in second-order arithmetic by countable dense subsets with a pseudo-metric (see e.g. [4, 28]). Various notions of compactness can then be formulated and their relations have been analysed in detail (see e.g. [4]). In this section, we lift some of these results to higher-order arithmetic; in doing so, we shall observe that the development of metric spaces in weak systems must proceed via codes, lest strong comprehensions or countable choice be needed in basic cases.

Our starting point is [4, Theorem 3.13], which establishes the equivalence between ACA_0 and *every (countable) Heine-Borel compact complete metric space is totally bounded*. The reverse implication is established via range and we lift this result to higher-order arithmetic. We make use of the standard definition of metric spaces, which does not use coding and can be found verbatim in [22, 23].

Definition 16. A *complete metric space* $\tilde{D}$ over $\mathbb{N}^{\mathbb{N}}$ consists of $D \subseteq \mathbb{N}^{\mathbb{N}}$, an equivalence[2] relation $=_D$, and $d : (D \times D) \to \mathbb{R}$ such that for all $e, f, g \in D$:

a. $d(e, f) =_{\mathbb{R}} 0 \leftrightarrow e =_D f$,
b. $0 \leq_{\mathbb{R}} d(e, f) =_{\mathbb{R}} d(f, e)$,
c. $d(f, e) \leq_{\mathbb{R}} d(f, g) + d(g, e)$,

and such that every Cauchy sequence in D converges to some element in D.

To be absolutely clear, the final condition regarding $\tilde{D}$ in the definition means that if $\lambda n.f_n$ is a sequence in D such that $(\forall k^0)(\exists N)(\forall m^0, n^0 \geq N)(d(f_n, f_m) <_{\mathbb{R}} \frac{1}{2^k})$, then there is $g \in D$ such that $(\forall k^0)(\exists n^0)(\forall m \geq n)(d(f_m, g) < \frac{1}{2^k})$. A *point* in $\tilde{D}$ is just any element in D. Two points $e, f \in \tilde{D}$ are said to be *equal* if $e =_D f$. Note that the 'hat function' from [16] readily yields $\mathbb{R}$ as a metric space over $\mathbb{N}^{\mathbb{N}}$.

We use standard notation like $B(e, r)$ for the open ball $\{f \in D : d(f, e) <_{\mathbb{R}} r\}$. The first item in Definition (16) expresses a kind of extensionality property and we tacitly assume that every mapping with domain D respects '$=_D$'.

---

[2] An *equivalence relation* is a binary, reflexive, transitive, and symmetric relation.

Definition 17 [Heine-Borel]. A complete metric space $\tilde{D}$ over $\mathbb{N}^{\mathbb{N}}$ is *Heine-Borel compact* if for any $Y : D \to \mathbb{R}^+$, the cover $\cup_{e \in D} B(e, Y(e))$ has a finite sub-cover.

We define *countable* Heine-Bore compactness as the previous definition restricted to *countable* covers of D.

Definition 18 [Totally bounded]. A complete metric space $\tilde{D}$ over $\mathbb{N}^{\mathbb{N}}$ is *totally bounded* if there is a sequence of finite sequences $\lambda n.x_n^{0 \to 1^*}$ of points in $\tilde{D}$ such that for any $x \in D$ there is $n \in \mathbb{N}$ such that $d(x, x_n(i)) < 2^{-n}$ for some $i < |x_n|$.

We now obtain the following theorem by lifting the proof of [4, Theorem 3.13].

Theorem 19. $\mathsf{RCA}_0^\omega + \mathsf{IND}$ *proves that either of the following items:*

a. *a Heine-Borel compact complete metric space over* $\mathbb{N}^{\mathbb{N}}$ *is totally bounded,*
b. *item (a) restricted to countable Heine-Borel compactness,*
c. *item (a) with sequential compactness instead of Heine-Borel compactness,*

implies the comprehension axiom BOOT.

Proof. We derive RANGE from item (a), and BOOT is therefore immediate; the implication involving item (b) is then immediate. Fix some Y^2 and define $D = \mathbb{N}^{\mathbb{N}} \cup \{0_D\}$ where 0_D is some special element. Define $f =_D e$ as $Y(f) =_0 Y(e)$ for any $e, f \in D \setminus \{0_D\}$, while $0_D = 0_D$ is defined as true and $f =_D 0_D$ is defined as false for $d \in D \setminus \{0_D\}$.

Define $d : D^2 \to \mathbb{R}$ as follows: $d(f, g) = |2^{-Y(f)} - 2^{-Y(g)}|$ if $f, g \neq_D 0$, $d(0_D, 0_D) = 0$, and $d(0_D, f) = d(f, 0_D) = 2^{-Y(f)}$ for $f \neq_D 0$. Clearly, this is a metric space in the sense of Definition 16 and the 'zero element' 0_D satisfies $\lim_{n \to \infty} d(0_D, f_n) =_\mathbb{R} 0$, assuming Y is unbounded on $\mathbb{N}^{\mathbb{N}}$ and $\lambda n.f_n$ is a sequence in D witnessing this, i.e. $Y(f_n) \geq n$ for any $n \in \mathbb{N}$.

Now, given a Cauchy sequence $\lambda n.f_n$ in D, either it converges to 0_D or $d(0_D, f_n)$ is eventually constant, i.e. the completeness property of $\tilde{D}$ is satisfied. Moreover, the Heine-Borel property as in Definition 17 is also straightforward, as any neighbourhood of 0_D covers all but finitely many $2^{-Y(f)}$ for $f \in \mathbb{N}^{\mathbb{N}}$ by definition. One seems to need IND to form the finite sub-cover. Let $\lambda n.x_n$ be the sequence provided by item (a) that witnesses that $\tilde{D}$ is totally bounded. Now define $X \subseteq \mathbb{N}$ as:

$$n \in X \leftrightarrow (\exists i < |x_{n+1}|)[2^{-n} =_\mathbb{R} d(0_D, x_{n+1}(i))], \tag{8}$$

and one readily shows that $n \in X \leftrightarrow (\exists f^1)(Y(f) = n)$, i.e. RANGE follows. Note that one can remove '$=_\mathbb{R}$' from (8) in favour of a decidable equality.

Regarding item (c), if a sequence $\lambda n.f_n$ is 'unbounded' as in $(\forall m^0)(\exists n^0)(Y(f_n) > m)$, then there is an obvious sub-sequence that converges to 0_D. In case we have $(\exists m^0)(\forall n^0)(Y(f_n) \leq m)$, there is a constant sub-sequence, and the space $\tilde{D}$ is clearly sequentially compact. □

In light of the considerable logical hardness of BOOT, it is clear that *for developing mathematics in weak systems, one must avoid items (a)–(c) and therefore*

Definition 16, i.e. the use of codes for metric spaces would seem to be essential for this development. This is particularly true since item (b) only deals with *countable* covers, i.e. the only higher-order object is the metric space, and the same for item (c). For those still not entirely convinced, Corollary 20 below shows that countable choice can be derived from item (a), i.e. the non-constructive nature of the latter is rampant compared to the version involving codes, namely [4, Theorem 3.13.ii].

Now, Definition 18 is used in RM (see e.g. [4,28]) and is sometimes referred to as *effective* total boundedness as there is a sequence that enumerates the finite sequences of approximating points. As it turns out, this extra information yields countable choice in the higher-order setting. Note that the monotone convergence theorem for nets *with a modulus of convergence* similarly yields $\mathsf{BOOT}+\mathsf{QF}\text{-}\mathsf{AC}^{0,1}$ by [24, §3.3]; obtaining countable choice in this context therefore seems normal.

Corollary 20. *The system* RCA_0^ω *proves the implication* [item (a) → $\mathsf{QF}\text{-}\mathsf{AC}^{0,1}$].

Proof. In light of $n \in X \leftrightarrow (\exists f^1)(Y(f) = n)$ and (8), one of the $x_{n+1}(i)$ for $i < |x_{n+1}|$ provides a witness to $(\exists f^1)(Y(f) = n)$ if such there is. $\qquad\square$

One can show that item (b) implies the associated second-order statement in case $\neg(\exists^2)$; the usual proof of [4, Theorem 3.13] can then be used. Thus, the ECF-translation (more or less) converts item (a) to the original second-order theorem. Intuitively speaking, assuming $D \subseteq \mathbb{N}^\mathbb{N}$ has a continuous characteristic function, it can be replaced with an enumeration of all finite sequences σ^{0^*} such that $\sigma * 00 \in D$.

Finally, one can generalise the previous to higher types. For instance, Definition 16 obviously generalises *mutatis mutandis* to yield the definition of complete metric spaces $\tilde{D}$ over $\mathcal{N} \equiv \mathbb{N}^\mathbb{N} \to \mathbb{N}$, and the same for any finite type. As opposed to nets indexed by function spaces like $\mathcal{N}$, a metric space based on the latter is quite standard. The proof of Theorem 19 can then be relativised with ease.

Corollary 21. $\mathsf{RCA}_0^\omega + \mathsf{IND}$ *proves that the following:*

(d) *a countable Heine-Borel compact complete metric space over* $\mathcal{N}$ *is totally bounded,*

implies the comprehension axiom RANGE^1.

Note that RANGE^1 was first introduced in [24, §3.7] and follows from the monotone convergence theorem for nets *indexed by* $\mathcal{N}$. In fact, the usual proof of the monotone convergence theorem involving Specker sequences immediately generalises to Specker nets indexed by $\mathcal{N}$, as discussed in Sect. 3.1.

In conclusion, it seems the development of metric spaces in weak systems must proceed via codes, lest strong comprehensions or countable choice be needed in basic cases. Indeed, BOOT is not provable from any comprehension axiom weaker than $(\exists^3)$, and 'larger' metric spaces require even stronger comprehension axioms (see Corollary 21). Moreover, countable choice is also lurking around the corner by Corollary 20, implying codes are the only way we can reasonably study general metric spaces in weak logical systems.

3.3   Rado Selection Lemma

We study the Rado selection lemma, introduced in [21]. The countable version of this lemma is equivalent to ACA_0 by [28, III.7.8], while a proof based on range can be found in [12, §3]. We shall lift the reversal to higher-order arithmetic, making use of RANGE. We first need some definitions.

Definition 22. A *choice function* f for a collection of non-empty A_i indexed by I, is such that $f(i) \in A_i$ for all $i \in I$.

A collection of finite subsets of $\mathbb{N}$ indexed by $\mathbb{N}^{\mathbb{N}}$ is of course given by a mapping $Y^{1\to 0^*}$. In case the latter is continuous, the index set is actually countable.

Definition 23 [(Rado($\mathbb{N}^{\mathbb{N}}$)]. Let A_i be a collection of finite sets indexed by $\mathbb{N}^{\mathbb{N}}$ and let F_J^2 be a choice function for the collection A_j for $j \in J$, for any finite set $J \subset \mathbb{N}^{\mathbb{N}}$. Then there is a choice function F^2 for the entire collection A_j such that for all finite $J \subset \mathbb{N}^{\mathbb{N}}$, there is a finite $K \supseteq J$ such that for $j \in J$, $F(j) =_0 F_K(j)$.

The following theorem is obtained by lifting the proof of [12, Theorem 3.30] to higher-order arithmetic.

Theorem 24. *The system* RCA_0^ω *proves* Rado($\mathbb{N}^{\mathbb{N}}$) $\to$ BOOT.

Proof. We assume that finite sequences in $\mathbb{N}^{\mathbb{N}}$ are coded by elements of $\mathbb{N}^{\mathbb{N}}$ in the usual way. We will prove RANGE, i.e. fix some G^2. For any w^{1^*}, define $A_w := \{0, 1\}$ and the associated choice function $F_w^2(h^1) := 1$ if $(\exists g \in w)(G(g) = h(0))$, and zero otherwise. For F^2 as in Rado($\mathbb{N}^{\mathbb{N}}$), we have the following implications for any $n \in \mathbb{N}$ and where $\widetilde{n} := \langle n \rangle * \langle n \rangle * \ldots$ is a sequence:

$$(\exists g^1)(G(g) = n) \to (\exists w_0^{1^*})(F_{w_0}(\widetilde{n}) = 1) \to F(\widetilde{n}) = 1 \to (\exists g^1)(G(g) = n). \quad (9)$$

The first implication in (9) follows by definition, while the others follow by the properties of F^2. Hence, RANGE follows, yielding BOOT. $\square$

The previous proof, does not make use of countable choice or Δ-CA. Thus, for larger collections indexed by subsets of type n objects, one readily obtains e.g. RANGE^1 as in Corollary 21, but without extra choice or comprehension. Finally, a reversal in Theorem 24 seems to need BOOT plus choice.

Hirst introduces a version of the Rado selection lemma in [12, §3] involving a bounding function, resulting in a reversal to WKL_0. A similar bounding function could be introduced, restricting $\mathbb{N}^{\mathbb{N}}$ to some compact sub-space while obtaining (only) the Heine-Borel theorem for uncountable covers as in HBU from [19].

3.4   Fields and Order

We lift the following implication to higher-order arithmetic: *that any countable formally real field is orderable implies weak König's lemma* (see [28, IV.4.5]). This result is based on a recursive counterexample by Ershov from [6], as (cheerfully) acknowledged in [10, p. 145].

First of all, the aforementioned implication is obtained via an intermediate principle involving the separation of disjoint ranges of functions (see [28, IV.4.4]). The generalisation of the latter to higher-order arithmetic and type two functionals is SEP^1 as follows:

$$(\forall Y^2, Z^2)\big[(\forall f^1, g^1)(Y(f) \neq Z(g)) \to (\exists X^1)(\forall g^1)(Y(g) \in X \wedge Z(g) \notin X)\big].$$

Modulo $\mathsf{QF\text{-}AC}^{0,1}$, SEP^1 is equivalent to the Heine-Borel theorem for uncountable covers of $[0, 1]$ as in HBU from [19]. We need the following standard definitions.

Definition 25 [Field over Baire space]. A field K over Baire space consists of a set $|K| \subseteq \mathbb{N}^{\mathbb{N}}$ with distinguished elements 0_K and 1_K, an equivalence relation $=_K$, and operations $+_K$, $-_K$ and $\times_K$ on $|K|$ satisfying the usual field axioms.

Definition 26. A field K is *formally real* if there is no sequence $c_0, \ldots, c_n \in |K|$ such that $0 =_K \sum_{i=0}^{n} c_i^2$.

Definition 27. A field K over $\mathbb{N}^{\mathbb{N}}$ is orderable if there exists an binary relation '$<_K$' on $|K|$ satisfying the usual axioms of ordered field.

As in [28], we sometimes identify K and $|K|$. With these definitions, the following theorem is a generalisation of [28, IV.4.5.2].

Definition 28. [ORD] A formally real field over $\mathbb{N}^{\mathbb{N}}$ is orderable.

We have the following theorem where the proof is obtained by lifting the proof of [28, IV.4.5] to higher-order arithmetic.

Theorem 29. *The system* $\mathsf{RCA}_0^\omega + \Delta\text{-}\mathsf{CA}$ *proves that* $\mathsf{ORD} \to \mathsf{SEP}^1$.

Proof. Let p_k be an enumeration of the primes and fix some Y^2, Z^2 as in the antecedent of SEP^1. By [28, II.9.7], the algebraic closure of $\mathbb{Q}$, denoted $\overline{\mathbb{Q}}$, is available in RCA_0. For w^{1^*}, define K_w as the sub-field of $\overline{\mathbb{Q}}(\sqrt{-1})$ generated by:

$$\big\{\sqrt[4]{p_{Y(w(i))}} : i < |w|\big\} \cup \big\{\sqrt{-\sqrt{p_{Z(w(j))}}} : j < |w|\big\} \cup \big\{\sqrt{p_k} : k < |w|\big\}.$$

Note that one can define such a sub-field from a *finite* set of generators in RCA_0 (see [28, IV.4]). Unfortunately, this is not possible for *infinite* sets and we need a different approach, as follows. By the proof of Theorem 14 (and (6) in particular), there is a functional G^2 such that:

$$\big(\forall b \in \overline{\mathbb{Q}}(\sqrt{-1})\big)\big[(\exists w^{1^*})(b \in K_w) \leftrightarrow (\exists v^{1^*})(G(v) = b)\big]. \tag{10}$$

Intuitively, we now want to define the field $\cup_{f\in\mathbb{N}^{\mathbb{N}}}K_{\langle f\rangle}$, but the latter cannot be (directly) defined as a set in weak systems. We therefore take the following approach: we define a field K over Baire space using G from (10), as follows: for w^{1^*}, v^{1^*}, define $w +_K v$ as that u^{1^*} such that $G(u) = G(v) +_{\overline{\mathbb{Q}}(\sqrt{-1})} G(w)$. This u^{1^*} is found by removing from $v * w$ all elements from $G(v)$ and $G(w)$ that sum to 0 in $G(v) +_{\overline{\mathbb{Q}}(\sqrt{-1})} G(w)$. Multiplication $\times_K$ is defined similarly, while $-_K w^{1^*}$ provides an extra label such that $G(-_K w) = -b$ if $G(w) = b$ and the 'inverse function' of $\times_K$ is defined similarly. Using (10) for $w = \langle\rangle$, 0_K and 1_K are given by those finite sequences v_0 and v_1 such that $G(v_0) = 0_{\overline{\mathbb{Q}}(\sqrt{-1})}$ and $G(v_1) = 1_{\overline{\mathbb{Q}}(\sqrt{-1})}$. Finally, $v^{1^*} =_K w^{1^*}$ is defined as the decidable equality $G(w) =_{\mathbb{Q}(\sqrt{-1})} G(v)$.

We call the resulting field K and proceed to show that it is formally real. To this end, note that K_w can be embedded into $\overline{\mathbb{Q}}$ by mapping $\sqrt{p_{Y(w(i))}}$ to $\sqrt{p_{Y(w(i))}}$ and $\sqrt{p_k}$ to $-\sqrt{p_k}$ for $k \neq Y(w(i))$ for $i < |w|$. Hence, K_w is formally real for every w^{1^*}. As a result, K is also formally real because a counterexample to this property would live in K_v for some v^{1^*}. Applying ORD, K now has an order $<_K$. Since $\sqrt{p_{Y(f)}}$ has a square root in $K_{\langle f\rangle}$, namely $\sqrt[4]{p_{Y(f)}}$, we have $u^{1^*} >_K 0_K$ if $G(u) = \sqrt{p_{Y(f)}}$, using the basic properties of the ordered field K. One similarly obtains $v^{1^*} <_K 0_K$ if $G(v) = \sqrt{p_{Z(g)}}$. Intuitively speaking, the order $<_K$ thus allows us to separate the ranges of Y and Z. To this end, consider the following equivalence, for every k^0:

$$(\exists u^{1^*})(u >_K 0_K \wedge G(u) = \sqrt{p_k}) \leftrightarrow (\forall v^{1^*})\big(G(v) = \sqrt{p_k} \to v >_K 0_K\big). \quad (11)$$

The forward direction in (11) is immediate in light of the properties of $=_K$ and $<_K$. For the reverse direction in (11), fix k_0 and find $w_0^{1^*}$ such that $|w_0| > k_0$. Since $\sqrt{p_{k_0}} \in K_{w_0}$, (10) yields $v_0^{1^*}$ such that $G(v_0) = \sqrt{p_{k_0}}$. The right-hand side of (11) implies $v_0 >_K 0_K$, and the left-hand side of (11) follows.

Finally, apply Δ-comprehension to (11) and note that the resulting set X satisfies $(\forall f^1)(Y(f) \in X \wedge Z(f) \notin X)$. Indeed, fix f_1^1 and put $k_1 := Y(f_1)$. Clearly, $\sqrt{p_{k_1}} \in K_{\langle f_1\rangle}$, yielding v_1 such that $G(v_1) = \sqrt{p_{k_1}}$ by (10). As noted above, the latter number has a square root, implying $v_1 >_K 0$, and $Y(f_1) = k_1 \in X$ by definition. Similarly, $k_2 := Z(f_1)$ satisfies $\sqrt{p_{k_2}} \in K_{\langle f_1\rangle}$ and (10) yields v_2 such that $G(v_2) = \sqrt{p_{k_2}}$. Since $-\sqrt{p_{k_2}}$ has a square root in K, $v_2 <_K 0_K$ follows, and $k_2 \notin X$, again by the definition of X. $\qquad\square$

Next, the following theorem yield further motivation for Δ-CA.

Theorem 30. *The system* RCA_0^ω *proves* $\mathsf{SEP}^1 \to \Delta\text{-CA}$.

Proof. To establish this implication, let G^2, H^2 be such that

$$(\forall k^0)\big[(\exists f^1)(G(f) = k) \leftrightarrow (\forall g^1)(H(g) \neq k)\big]. \quad (12)$$

By definition, G, H satisfy the antecedent of SEP^1. Let X be the set obtained by applying the latter and consider:

$$(\exists f^1)(G(f) = k) \to k \in X \to (\forall g^1)(H(g) \neq k) \to (\exists f^1)(G(f) = k),$$

where the final implication follows from (12); for the special case (12), X is now the set required by Δ-CA. For $Y^{(1\times 0)\to 0}$, define G^2 as follows for n^0 and g^1: put $G(\langle n\rangle * g) = n+1$ if $Y(g,n) = 0$, and 0 otherwise. Note that for $k \geq 1$, we have

$$(\exists f^1)(Y(f,k) = 0) \leftrightarrow (\exists g^1)(G(g) = k).$$

Hence, (12) is 'general enough' to obtain full Δ-CA, and we are done. $\square$

4   Concluding Remarks

We finish this paper with some conceptual remarks.

4.1   Future Work and Alternative Approaches

In this section, we discuss future work and alternative approaches.

First of all, we list some topics that can be lifted to higher-order arithmetic in the same way as in the previous sections.

a. Algebraic closures of countable fields (see [28, III.3 and IV.4]).
b. Maximal ideals of countable fields (see [28, III.3] or [11]).
c. Ordering countable groups (see [30]).
d. Persistence of real numbers (see [5]).
e. Closed and separably closed sets in $\mathbb{R}$ (see [3, §2]).

In each case, the theorem at hand allows one to define the range of functions, or separate the disjoint ranges of functions. The proofs in the indicated references then generalise as in the previous sections.

Secondly, we discuss possible alternative approaches, and why they are not fruitful. Now, recursive counterexamples often give rise to *Brouwerian counterexamples*, and vice versa (see [7, p. xii] for this opinion). A Brouwerian counterexample to a theorem shows that the latter is rejected in (a certain strand of) constructive mathematics (see [17] for details). We choose to use recursive counterexamples (and the associated RM results) because those are formulated in a formal system, which enables us to lift the associated proofs without too much trouble. The same would not be possible for Brouwerian counterexamples, due to the lack of an explicit/unified choice of formal system for e.g. Bishop's constructive mathematics. To be absolutely clear, there is nothing *wrong* with constructive mathematics in general; *however*, the lack of an explicit/unified formal system for constructive mathematics means that we cannot 'lift' Brouwerian counterexamples with the same ease (or at all).

4.2   Implications and Interpretations

While we initially had no intention of discussing the implications of the above results in this paper, at least two colleagues have provided interpretations that do not do the justice to the bigger picture provided by [19] and [24]. This section is an attempt at avoiding further misconceptions.

In our opinion, one reasonable interpretation of the results in this paper is that second- and higher-order arithmetic are not *as* different as sometimes claimed, and that recursive counterexamples and reversals provide a bridge of sorts between the two. However, the results in this paper *do not* support the argument that higher-order arithmetic contains 'nothing new' compared to second-order arithmetic. The exact opposite is the case, as follows; associated results may be found in [19] and [24].

In a nutshell, Kohlenbach's higher-order RM (see Sect. 2.1) is based on *comprehension* and *discontinuity*, while the aforementioned principles BOOT, HBU, and SEP^1 cannot be captured well in this hierarchy, necessitating a new 'continuity' hierarchy based on the *neighbourhood function principle* NFP, as first developed in [24]. Let us discuss the previous sentence in a lot more detail.

First of all, as noted in Sect. 2.2, Kohlenbach's counterpart of arithmetical comprehension ACA$_0$ is given by $(\exists^2)$, and the functional in the latter is clearly *discontinuous*. As shown in [16, §3], this axiom is also equivalent to e.g. the existence of a discontinuous function on $\mathbb{R}$. As expected, ECF converts $(\exists^2)$ to '$0 = 1$', as a discontinuous function does not have a continuous representation by an RM-code. The same holds for the higher-order versions of Π_k^1-CA$_0$ *mutatis mutandis*; the higher-order version of Π_1^1-CA$_0$ is given by the *Suslin functional*. In a nutshell, Kohlenbach's higher-order comprehension is based on discontinuity.

Secondly, one of the main (conceptual) results of [19] is that higher-order comprehension does not capture e.g. Heine-Borel compactness well at all. Indeed, while Z_2^Ω proves HBU, the system $Z_2^\omega \equiv \cup_k \Pi_k^1$-CA$_0^\omega$ does not; note that Π_k^1-CA$_0^\omega$ is RCA$_0^\omega$ plus the existence of a type two functional deciding Π_k^1-formulas (allowing for type zero and one parameters only). The same holds for BOOT and related theorems like the Lindelöf lemma for Baire space (see [19] for the latter). Since all the aforementioned axioms have relatively weak first-order strength (at most ACA$_0$) in isolation, higher-order comprehension seems unsuitable for capturing them. *By contrast*, we show in [24] that these axioms are equivalent to natural fragments of the *neighbourhood function principle* NFP, which is as follows:

$$(\forall f^1)(\exists n^0)A(\overline{f}n) \rightarrow (\exists \gamma \in K_0)(\forall f^1)A(\overline{f}\gamma(f)), \tag{13}$$

for any formula $A \in \mathsf{L}_\omega$ and where '$\gamma \in K_0$' means that γ^1 is a total associate/RM code on $\mathbb{N}^\mathbb{N}$. Thus, (13) expresses the existence of a continuous choice function. In a nutshell, NFP is based on *continuity* and ECF converts NFP, BOOT, and HBU to resp. Z_2, ACA$_0$, and WKL$_0$.

In conclusion, while comprehension is generally a great axiom schema for classifying theorems in RM (of any order), principles like BOOT and HBU do not have a nice classification based on Kohlenbach's higher-order comprehension. By contrast, the latter do have a nice classification based on NFP. Now, higher-order comprehension amounts to *discontinuity*, while NFP is a *continuity* schema, stating as it does the existence of a continuous choice function. In this light, higher-order arithmetic includes (at least) two 'orthogonal' scales (one based on continuity, one based on discontinuity) for classifying theorems. The 'liftings' in this paper generalise second-order theorems to higher-order theorems from the NFP-scale *with little modification.* One disadvantage is that the 'lifted' proofs from this paper often use more comprehension or countable choice than the known proofs, as discussed in Sect. 3.1.

Acknowledgement. Our research was supported by the John Templeton Foundation via the grant *a new dawn of intuitionism* with ID 60842. We express our gratitude towards this institution. We thank Anil Nerode and Paul Shafer for their valuable advice. We also thank the anonymous referees for the helpful suggestions. Opinions expressed in this paper do not necessarily reflect those of the John Templeton Foundation.

References

1. Avigad, J., Feferman, S.: Gödel's functional ("dialectica") interpretation. In: Handbook of Proof Theory. Studies in Logic and the Foundations of Mathematics, vol. 137, pp. 337–405 (1998)
2. Bartle, R.G.: Nets and filters in topology. Am. Math. Mon. **62**, 551–557 (1955)
3. Brown, D.K.: Notions of closed subsets of a complete separable metric space in weak subsystems of second-order arithmetic. In: Logic and Computation, Pittsburgh, PA (1987). Contemporary Mathematics, vol. 106, pp. 39–50. American Mathematical Society, Providence, RI (1990)
4. Brown, D.K.: Notions of compactness in weak subsystems of second order arithmetic. In: Simpson, S. (ed.) Reverse Mathematics 2001. Perspectives in Logic, pp. 47–66. Association for Symbolic Logic, Urbana, IL (2005)
5. Dorais, F.G., Hirst, J.L., Shafer, P.: Reverse mathematics, trichotomy and dichotomy. J. Log. Anal. **4**, 14 (2012). Paper 13
6. Ershov, Y.L.: Theorie der numerierungen III. Z. Math. Logik Grundlagen Math. **23**(4), 289–371 (1977)
7. Ershov, Y.L., Goncharov, S.S., Nerode, A., Remmel, J.B., Marek, V.W. (eds.): Handbook of Recursive Mathematics. Volume 138 of Studies in Logic and the Foundations of Mathematics, Vol. 1. North-Holland, Amsterdam (1998). Recursive Model Theory
8. Friedman, H.: Some systems of second order arithmetic and their use. In: Proceedings of the International Congress of Mathematicians (Vancouver, B.C., 1974), vol. 1, pp. 235–242 (1975)
9. Friedman, H.: Systems of second order arithmetic with restricted induction, I & II (abstracts). J. Symb. Log. **41**, 557–559 (1976)
10. Friedman, H., Simpson, S.G., Smith, R.: Countable algebra and set existence axioms. Ann. Pure Appl. Logic **25**(2), 141–181 (1983)
11. Hatzikiriakou, K.: Minimal prime ideals and arithmetic comprehension. J. Symb. Log. **56**(1), 67–70 (1991)
12. Hirst, J.L.: Combinatorics in subsystems of second order arithmetic, Ph.D. thesis, The Pennsylvania State University, ProQuest LLC, Ann Arbor, MI (1987)
13. Hunter, J.: Higher-order reverse topology, Ph.D. thesis, The University of Wisconsin-Madison, ProQuest LLC, Ann Arbor, MI (2008)
14. Kelley, J.L.: General Topology. Springer, New York (1975). Reprint of the 1955 edition; Graduate Texts in Mathematics, no. 27

15. Kohlenbach, U.: Foundational and mathematical uses of higher types. In: Reflections on the Foundations of Mathematics (Stanford, CA, 1998). Lecture Notes in Logic 15, pp. 92–116. Association for Symbolic Logic, Urbana, IL (2002)
16. Kohlenbach, U.: Higher order reverse mathematics. In: Simpson, S. (ed.) Reverse Mathematics 2001. Perspectives in Logic, pp. 281–295. Association for Symbolic Logic, Urbana, IL (2005)
17. Mandelkern, M.: Brouwerian counterexamples. Math. Mag. **62**(1), 3–27 (1989)
18. Moore, E.H., Smith, H.: A general theory of limits. Am. J. Math. **44**, 102–121 (1922)
19. Normann, D., Sanders, S.: On the mathematical and foundational significance of the uncountable. J. Math. Log. (2018). https://doi.org/10.1142/S0219061319500016
20. Normann, D., Sanders, S.: Pincherle's theorem in Reverse Mathematics and computability theory. arXiv:1808.09783 (2018, submitted)
21. Rado, R.: Axiomatic treatment of rank in infinite sets. Can. J. Math. **1**, 337–343 (1949)
22. Royden, H.L.: Real Analysis, 3rd edn. Macmillan Publishing Company, New York (1988)
23. Rudin, W.: Real and Complex Analysis, 3rd edn. McGraw-Hill, New York (1987)
24. Sanders, S.: Plato and the foundations of mathematics. arxiv:1908.05676, pp. 44 (2019, submitted)
25. Sanders, S.: Nets and reverse mathematics: a pilot study, to appear in Computability, pp. 34 (2019)
26. Sanders, S.: Nets and reverse mathematics. In: Manea, F., Martin, B., Paulusma, D., Primiero, G. (eds.) CiE 2019. LNCS, vol. 11558, pp. 253–264. Springer, Cham (2019). https://doi.org/10.1007/978-3-030-22996-2_22
27. Sanders, S.: Reverse mathematics and computability theory of domain theory. In: Iemhoff, R., Moortgat, M., de Queiroz, R. (eds.) WoLLIC 2019. LNCS, vol. 11541, pp. 550–568. Springer, Heidelberg (2019). https://doi.org/10.1007/978-3-662-59533-6_33
28. Simpson, S.G.: Subsystems of Second Order Arithmetic. Perspectives in Logic, 2nd edn. Cambridge University Press, Cambridge (2009)
29. Simpson, S.G., Rao, J.: Reverse algebra. In: Handbook of Recursive Mathematics, Volume 2, Studies in Logic and the Foundations of Mathematics, vol. 139, pp. 1355–1372. North-Holland (1998)
30. Solomon, D.R.: Reverse mathematics and ordered groups, Ph.D. thesis, Cornell University, ProQuest LLC (1998)
31. Specker, E.: Nicht konstruktiv beweisbare Sätze der Analysis. J. Symb. Log. **14**, 145–158 (1949)
32. Stillwell, J.: Reverse Mathematics, Proofs from the Inside Out. Princeton University Press, Princeton (2018)
33. Troelstra, A.S.: Metamathematical Investigation of Intuitionistic Arithmetic and Analysis. Lecture Notes in Mathematics, vol. 344. Springer, Berlin (1973). https://doi.org/10.1007/BFb0066739
34. Turing, A.: On computable numbers, with an application to the Entscheidungsproblem. Proc. Lond. Math. Soc. **42**, 230–265 (1936)
35. Vietoris, L.: Stetige Mengen (German). Monatsh. Math. Phys. **31**, 173–204 (1921)

On the Tender Line Separating Generalizations and Boundary-Case Exceptions for the Second Incompleteness Theorem Under Semantic Tableaux Deduction

Dan E. Willard$^{(\boxtimes)}$

Computer Science and Mathematics Departments, University at Albany,
Albany, USA
dwillard@albany.edu

Abstract. Our previous research has studied the semantic tableaux deductive methodology, of Fitting and Smullyan, and observed that it permits boundary-case exceptions to the Second Incompleteness Theorem, when multiplication is viewed as a 3-way relation (rather than as a total function). It is known that tableaux methodologies do prove a schema of theorems, verifying all instances of the Law of the Excluded Middle. But yet we show that if one promotes this schema of theorems into formalized logical axioms, then the meaning of the pronoun "I" in our self-referencing engine changes, and our partial evasions of the Second Incompleteness Theorem come to a complete halt.

Keywords: Semantic tableaux · Hilbert's Consistency Program · Second Incompleteness Theorem · Law of Excluded Middle

1 Introduction

The existence of a significant chasm separating the goals of Hilbert's consistency program from the implications of the Second Incompleteness Theorem was evident immediately after Gödel published [20]'s seminal result. We exhibited in [45–47,49–53] a large number of articles about generalizations and boundary case exceptions to the Second Incompleteness Theorem, starting with our 1993 article [45]. These papers, which included six papers published in the JSL and APAL, showed that every extension α of Peano Arithmetic can be mapped onto an axiom system α^* that can recognize its own consistency and prove analogs of all α's Π_1 theorems (in a slightly different language, called L^*).

These formalisms were called "Self-Justifying" systems. They were able to verify their own consistency by containing a built-in self-referencing axiom which declared *"I am consistent"* (as will be explained later). In particular,

© Springer Nature Switzerland AG 2020
S. Artemov and A. Nerode (Eds.): LFCS 2020, LNCS 11972, pp. 268–286, 2020.
https://doi.org/10.1007/978-3-030-36755-8_17

our axiom systems α^* used the Fixed-Point Theorem to assure α^*'s self-referencing analogs of the pronoun "I" would enable it to refer to itself in the context of its *"I am consistent"* axiomatic declaration.

It turns out that such a self-referencing mechanism will produce unacceptable Gödel-style diagonalizing contradictions, when either α^* or its particular employed definition of consistency are too strong. This is because our methodologies *only* become contradiction-free *when* α^* uses sufficiently weak underlying structures.

These weak structures obviously have significant disadvantages. Their virtue is that their formalisms α^* can be arranged to prove more Π_1 like theorems than Peano Arithmetic, while offering *some type* of *partial* knowledge about their own consistency. We will call such formalisms **"Declarative Exceptions"** to the Second Incompleteness Theorem.

An alternative type of exception to the Second Incompleteness Theorem, which we will call an **"Infinite-Ranged Exception"**, was recently developed by Sergei Artemov [4] (It is related to the works of Beklemishev [6] and Artemov-Beklemishev [5].) Artemov has observed Peano Arithmetic can verify its own consistency, from a special infinite-ranging perspective. This means PA will generate an infinite set of theorems T_1, T_2, T_3 ... where each T_i shows some subset S_i of PA is unable to prove $0 = 1$ and where PA equals the formal union of these special selected S_i satisfying the inclusion property of $S_1 \subset S_2 \subset S_3 \subset ...$.

This perspective is also not a panacea. Thus, the abstract in [4] cautiously used the adjective of "somewhat" to describe how it sought to partially achieve the goals sought by Hilbert's Consistency Program (with an infinite collection of theorems T_1, T_2, T_3 ... replacing Hilbert's intended goal of finding one unifying formal consistency theorem).

Our "Declarative"' exceptions to the Second Incompleteness Theorem and Artemov's "Infinite Ranging" exceptions are rigorous results that are nicely compatible with each other. This is because each acknowledged that the Second Incompleteness Theorem is a strong result, that *will admit no full-scale exceptions*. Also, these results are of interest because Gödel conjectured that Hilbert's Consistency Program would ultimately, reach, *some levels of partial success* (see next section). We will explain, herein, how Gödel's conjecture can be *partially justified,* due to an unusual consequence of the Law of the Excluded Middle.

More specifically, we shall focus on the semantic tableaux deductive mechanisms of Fitting and Smullyan [15,39] and their special properties from the perspective of our JSL-2005 article [49]. Each instance of the Law of the Excluded Middle has been treated by most tableaux mechanisms as a provable theorem, rather than as a built-in logical axiom. This may, at first, appear to be an insignificant distraction because most deductive methodologies do not have their consistency reversed when a theorem is promoted into becoming a logical axiom.

Our self-justifying axiom systems are *different,* however, because their built-in self-referencing *"I am consistent"* axioms have their meanings change, fundamentally, when their self-referencing concept of "I" involves promoting a schema of theorems verifying the Law of Excluded Middle *into formal logical axioms*.

This effect is counterintuitive because similar distinctions exist almost nowhere else in Logic. However some confusion, that has surrounded our prior work, can be clarified when one realizes that *an interaction* between the self-referencing concept of "I" with the Law of Excluded Middle causes the Second Incompleteness Theorem to become activated *precisely when* the Law of Excluded Middle *is promoted* into becoming a schema of logical axioms.

The intuitive reason for this unusual effect is that the transforming of derived theorems *into* logical axioms can shorten proofs under the Fitting-Smullyan semantic tableaux formalism. In a context where our special axiom systems in Sect. 3 use a self-referencing *"I am consistent"* axiom *and view* multiplication as a 3-way relation (rather than as a total function), this compression will be capable of enacting the power of the Second Incompleteness Theorem. Moreover, the next chapter will explain how this issue is germane to central questions raised by Gödel and Hilbert about feasible boundary-case exceptions to the Second Incompleteness Effect.

2   Revisiting Some Intuitions of Gödel and Hilbert

Interestingly, neither Gödel (unequivocally) nor Hilbert (after learning about Gödel's work) would dismiss the possibility of a compromise solution, whereby a *fragment* of the goals of Hilbert's Consistency Program would remain intact. Thus, Hilbert never withdrew [26]'s statement ∗ for justifying this program:

∗ *"Let us admit that the situation in which we presently find ourselves with respect to paradoxes is in the long run intolerable. Just think: in mathematics, this paragon of reliability and truth, the very notions and inferences, as everyone learns, teaches, and uses them, lead to absurdities. And where else would reliability and truth be found if even mathematical thinking fails?"*

Gödel was, also, cautious (especially during the early 1930's) not to speculate whether all facets of Hilbert's Consistency program would come to a termination. He thus inserted the following hesitant caveat into his famous 1931 paper [20]:

∗∗ *"It must be expressly noted that Theorem XI"* (e.g. the Second Incompleteness Theorem) *"represents no contradiction of the formalistic standpoint of Hilbert. For this standpoint presupposes only the existence of a consistency proof by finite means, and there might conceivably be finite proofs which cannot be stated in P or in …"*

Several biographies of Gödel [11, 22, 55] have noted that Gödel's intention (prior to 1930) was to establish Hilbert's proposed objectives, before he formalized his famous result that led in the opposite direction. Moreover, Yourgrau's biography [55] of Gödel records how von Neumann found it necessary during the early 1930's to *"argue against Gödel himself"* about the definitive termination of Hilbert's consistency program, which *"for several years"* after [20]'s publication, Gödel *"was cautious not to prejudge"*. It is known that Gödel had hinted that the Second Incompleteness Theorem was more significant during a 1933 Vienna lecture [21].

Yet despite this endorsement, a YouTube talk by Gerald Sacks [38] explicitly recalled Gödel telling Sacks, during 1961–1962, that some type of revival of Hilbert's Consistency Program would eventually become feasible (as explained in detail by footnote[1]). This recent Year-2014 YouTube lecture by Gerald Sacks had caught many scholars by surprise because Gödel published fewer than 85 pages in his life. Thus, Gödel never explicitly recorded, during the second half of his life, his partial reluctance about the relevance of the Second Incompleteness Theorem, as Sacks did recall in [38].

The research that has followed Gödel's seminal 1931 discovery has mainly focused on studying generalizations of the Second Incompleteness Theorem (instead of also examining its boundary-case exceptions). Many of these generalizations of the Second Incompleteness Theorem [2,3,7–10,13,16,23–25,29,32–36,40–44,46–48,50] are quite beautiful. The author of this paper is especially impressed by a generalization of the Second Incompleteness Effect, arrived at by the combined work of Pudlák and Solovay together with added research by Nelson and Wilkie-Paris [31,35,41,44]. These results, which also have been more recently discussed in [10,23,42,46], have noted the Second Incompleteness Theorem does not require the presence of the Principle of Induction to apply to most formalisms that use a Hilbert-Frege type of deduction. (The Remark 1 of the next chapter will offer a detailed summary of this helpful generalization of the Second Incompleteness Theorem.)

3   Main Notation and Background Literature

Let us call an ordered pair (α, D) a **Generalized Arithmetic Configuration** (abbreviated as a **"GenAC"**) when its first and second components are defined as follows:

1. The **Axiom Basis** "α" for a GenAC is defined as its set of proper axioms.
2. The second component "D" of a GenAC, called its **Deductive Apparatus**, is defined as the union of its logical axioms "L_D" with its rules for obtaining inferences.

Example 1. This notation allows us to separate the logical axioms L_D, associated with (α, D), from its "basis axioms" α. It also allows us to compare different deductive apparatuses from the literature. Thus, the D_E apparatus, from Enderton's textbook [12], uses only modus ponens as a rule of inference, but it deploys a complicated 4-part schema of logical axioms. This differs from the D_M and D_H apparatuses of the Mendelson [30] and Hájek-Pudlák [25] textbooks, which use a more reduced set of logical axioms but require two rules of inference (modus ponens and generalization). The D_F apparatus, from Fitting's and Smullyan's textbooks [15,39], actually uses *no logical axioms,* but

---

[1] Some quotes from Sacks's YouTube talk [38] are that Gödel *"did not think"* the objectives of Hilbert's Consistency Program *"were erased"* by the Incompleteness Theorem, and Gödel believed (according to Sacks) it left Hilbert's program *"very much alive and even more interesting than it initially was"*.

it instead employs a broader "tableaux style" rule of inference. **AN IMPOR-TANT POINT** is that while proofs have different lengths under different apparatuses, all the common apparatuses will produce the same set of final theorems from an initial common "axiom basis" of α (as explained in footnote[2]).

Definition 1. Let α again denote an axiom basis, D designate a deduction apparatus, and (α, D) denote their GenAC. Henceforth, (α, D)'s will be called **Self Justifying** when

   i. one of (α, D)'s theorems (or possibly one of α's axioms) states that the deduction method D, applied to the basis system α, produces a consistent set of theorems, and

  ii. the GenAC formalism (α, D) is actually, in fact, consistent.

Example 2. Using Definition 1's notation, our prior research [45, 46, 49, 50, 53] constructed GenAC pairs (α, D) that were "Self Justifying". We also proved that the Incompleteness Theorem implies specific limits beyond which self-justifying formalisms simply cannot transgress. For any (α, D), all our articles observed it was easy to construct a system $\alpha^D \supseteq \alpha$ that satisfies the Part-i condition (in an isolated context *where the Part-ii condition is not also satisfied*). In essence, α^D could consist of all of α's axioms plus the added "**SelfRef**(α, D)" sentence, defined below:

   $\oplus$ There is no proof (using D's deduction method) of $0 = 1$ from the *union* of the axiom system α with *this* sentence "SelfRef(α, D)" (looking at itself).

Kleene [28] was the first to show how to encode analogs of SelfRef(α, D)'s above statement, which we often call an "**I AM CONSISTENT**" **axiom**. Each of Kleene, Rogers and Jeroslow [27, 28, 37] emphasized α^D may be inconsistent (e.g. violate Part-ii of self-justification's definition *despite* the assertion in SelfRef(α, D)'s particular statement). This is because if the pair (α, D) is too strong then a quite conventional Gödel-style diagonalization argument can be applied to the axiom basis of $\alpha^D = \alpha +$ SelfRef(α, D), where the added presence of the statement SelfRef(α, D) will cause this extended version of α, ironically, to become automatically inconsistent. Thus, an encoding for "SelfRef(α, D)" is relatively easy, via an application of the Fixed Point Theorem, but this sentence is *potentially devastating*.

Definition 2. Let $Add(x, y, z)$ and $Mult(x, y, z)$ denote two 3-way predicates, specifying $x + y = z$ and $x * y = z$, for which the associative, commutative, identity and distributive properties have Π_1 style encodings provable under an axiom system of α. Then we will say that α **recognizes** successor, addition and multiplication as **Total Functions** iff it can prove all of (1)–(3) as theorems:

---

[2] This is because all the common apparatuses satisfy the requirement of Gödel's Completeness Theorem.

$$\forall x \exists z \; Add(x, 1, z) \tag{1}$$

$$\forall x \; \forall y \; \exists z \; Add(x, y, z) \tag{2}$$

$$\forall x \; \forall y \; \exists z \; Mult(x, y, z) \tag{3}$$

We will call the GenAC system (α, D) a **Type-M** formalism iff it includes (1)–(3) as theorems, **Type-A** if it includes only (1) and (2) as theorems, and it will be called **Type-S** if it contains only (1) as a theorem. Also, (α, D) will be called **Type-NS** iff it can prove none of (1)–(3).

Remark 1. The separation of GenAC systems into the categories of Type–NS, Type-S, Type-A and Type-M systems helps summarize the prior literature about generalizations and boundary-case exceptions for the Second Incompleteness Theorem. This is because:

i. The combined research of Pudlák, Solovay, Nelson and Wilkie-Paris [31, 35, 41, 44], as formalized by Theorem ++, implies that no natural Type–S system (α, D) can recognize its own consistency (and thereby be self-justifying) when D is one of Example 1's three examples of Hilbert-Frege style deduction operators of D_E, D_H or D_M. It thus establishes the following result:

   ++ *(Solovay's modification [41] of Pudlák [35]'s formalism using some of Nelson and Wilkie-Paris [31,44]'s methods):* Let (α, D) denote a Type-S GenAC system which assures the successor operation will provably satisfy both $x' \neq 0$ and $x' = y' \Leftrightarrow x = y$. Then (α, D) cannot verify its own consistency whenever simultaneously D is some type of a Frege-Hilbert deductive apparatus and α treats addition and multiplication as 3-way relations, satisfying their usual associative, commutative distributive and identity axioms.

   Essentially, Solovay [41] privately communicated to us in 1994 an analog of theorem ++. Many authors have noted Solovay has been reluctant to publish his nice privately communicated results on many occasions [10, 25, 31, 33, 35, 44]. Thus, approximate analogs of ++ were explored subsequently by Buss-Ignjatović, Hájek and Švejdar in [10, 23, 42], as well as in Appendix A of our paper [46] and in [48] Also, Pudlák's initial 1985 article [35] captured the majority of ++'s essence, chronologically before Solovay's observations, Also, Friedman did some closely related work in [16].

ii. Part of what makes ++ interesting is that [46, 49, 50] presented two cases of self-justifying GenAC systems, whose natural hybrid is precluded by ++. Specifically, these results involve using Example 2's self-referencing *"I am consistent"* axiom (from statement $\oplus$). Thus, they established that some (not all) Type-NS systems [46, 50] can verify their own consistency under a Hilbert-Frege style deductive apparatus[3], and some (not all) Type-A systems [45, 46, 49, 51] can, likewise, corroborate their consistency under a more

---

[3] The Example 1 had provided three examples of Hilbert-Frege style deduction operators, called D_E, D_H and D_M. It explained how these deductive operators differ from a tableaux-style deductive apparatus by containing a modus ponens rule.

restrictive semantic tableaux apparatus. Also, we observed in [47,52] how one could refine ++ with Adamowicz-Zbierski's methods [2] to show most Type-M systems cannot recognize their semantic tableaux consistency.

Remark 2. Several of our papers, starting with our 1993 article [45], have used Example 2's *"I am consistent"* axiomatic declaration ⊕ for evading the Second Incompleteness Effect. Other possible types of evasions rest on the cut-free methods of Gentzen and Kreisel-Takeuti [19,29], an interpretational approach (such as what Adamowicz, Bigorajska, Friedman, Nelson, Pudlák and Visser had applied in [1,17,31,35,43]), or Artemov's Infinite-Range perspective [4] (where an infinite schema of theorems replaces one single unified consistency theorem). We encourage the reader to examine all these articles, each of which has their own separate virtues. Our focus, in this paper, will be primarily on the next section's Theorems 1 and 2. They will show that some types of partial *(and not full)* evasions of the Second Incompleteness Effect are possible under a semantic tableaux deductive apparatus.

4   Main Theorems and Related Notation

A function F is called **Non-Growth** when $F(a_1, \ldots, a_j) \leq Maximum(a_1, \ldots a_j)$ does hold. Six examples of non-growth functions are *Integer Subtraction* (where $x - y$ is defined to equal zero when $x \leq y$), *Integer Division* (where $x \div y$ equals x when $y = 0$, and it equals $\lfloor x/y \rfloor$ otherwise), $Maximum(x, y)$, $Logarithm(x)$, $Root(x, y) = \lceil x^{1/y} \rceil$ and $Count(x, j)$ (which designates the number of physical "1" bits that are stored among x's rightmost j bits). The term **U-Grounding Function** will refer to either one of one of these six functions or the *growth-oriented* Addition and $Double(x) = x + x$ operations. Our language L^*, introduced in [49], was built out of these eight functions plus the primitives of "0", "1", "=" and "≤".

This language L^* differs from a conventional arithmetic by **EXCLUDING** a formal multiplication function symbol. It CAN VIEW multiplication as a 3-way relation (via the use of its Division primitive.) This revised notation will lead us to a surprisingly strong evasion of the Second Incompleteness Effect.

Let t be any term. The v-based quantifiers used by the wffs $\forall v \leq t \; \Psi(v)$ and $\exists \, v \leq t \, \Psi(v)$ will be called *bounded quantifiers*. Any formula in our language L^*, all of whose quantifiers are similarly bounded, will be called a Δ_0^* formula. The Π_n^* and Σ_n^* formulae are defined by usual rules except they **DO NOT** contain multiplication function symbols. These rules are that:

1. Every Δ_0^* formula will also be a "Π_0^*" and "Σ_0^*" formula.
2. A wff will be called Π_n^* when it is encoded as $\forall v_1 \ldots \forall v_k \; \Phi$ with Φ being Σ_{n-1}^*.
3. A wff will be called Σ_n^* when it is encoded as $\exists v_1 .. \exists v_k \; \Phi$, with Φ being Π_{n-1}^*.

Also, the sentence Ψ will be called a **Rank-1*** statement iff it can be encoded in either a Π_1^* or Σ_1^* format. (The reader is reminded that our definitions for Π_1^* or

Σ_1^* formulae differ from Arithmetic's counterparts by excluding multiplication function symbols.)

There will be three variants of formal deductive apparatus methods, which we will now compare. The first is *semantic tableaux*. It will receive the abbreviated name of **"Tab"** and correspond to Fitting's textbook formalism from [15]. (It is also summarized by us in the attached Appendix A.) Thus, a Tab-proof for a theorem Ψ, from an axiom basis α is a tree-like structure that begins with the sentence $\neg\Psi$ stored inside the tree's root and whose every root$-$to$-$leaf path establishes a contradiction by containing some pair of contradictory nodes that will "close" its path. The rules for generating internal nodes, along each root$-$to$-$leaf path, are that each node must be *either* a proper axiom of α *or* a deduction from an ancestor node via one of the Appendix A's six stated "elimination" rules for the $\wedge$, $\vee$, $\rightarrow$, $\neg$, $\forall$, and $\exists$ symbols.

Our second explored deductive apparatus is called *Extended Tableaux*, and shall be abbreviated as **"Xtab"**. Its definition is identical to **Tab**-deduction, except that for any sentence ϕ in our language L^*, the sentence $\phi \vee \neg\phi$ is allowed as an internal node in an Xtab proof tree. (In other words, *Xtab*-deduction differs from *Tab*-deduction by allowing all instances of the Law of Excluded Middle to appear as permitted logical axioms. In contrast, *Tab*-deduction will view these instances only as derived theorems.)

Our third deductive apparatus was called **Tab-1** in [49]. It is, essentially, a compromise between Tab and Xtab, where a "Tab-1" proof for Ψ from an axiom basis α corresponds to a set of ordered pairs $(p_1, \phi_1), (p_2, \phi_2), ..(p_k, \phi_k)$ where

1. $\phi_k = \Psi$
2. Each p_j is a Tab-proof of what we have called a Rank-1* sentence ϕ_j from the union of α with the preceding Rank-1* sentences of ϕ_1, ϕ_2, .. ϕ_{j-1} .

We emphasize Tab-1 deduction can be *substantially* less efficient than Xtab because *the former requires ϕ_j be a Rank-1* sentence, while Xtab *does not impose* a similar Rank-1* constraint upon ϕ, when it invokes its permitted axiom of $\phi \vee \neg\phi$.

Let us say that an axiom system α owns a **Level-1** appreciation of its own self-consistency (under a deductive apparatus D) iff it can verify that D produces no two simultaneous proofs for a Π_1^* sentence and its negation. Within this context, where β denotes any basis axiom system using L^*'s U-Grounding language, $\mathrm{IS}_D(\beta)$ was defined in [49] to be an axiomatic formalism capable of recognizing all of β's Π_1^* theorems and corroborating its own Level-1 consistency under D's deductive apparatus. It consisted of the following four groups of axioms:

Group-Zero: Two of the Group-zero axioms will define the constant-symbols, $\bar{c}_0$ and $\bar{c}_1$, designating the integers of 0 and 1. The Group-zero axioms will also define the growth functions of Addition and $Double(x) = x + x$. (They will enable our formalism to define any integer $n \geq 2$ using fewer than $3 \cdot \lceil \mathrm{Log}\, n \rceil$ logic symbols.)

Group-1: This axiom group will consist of a finite set of Π_1^* sentences, denoted as F, which can prove any Δ_0^* sentence that holds true under the standard model of the natural numbers. (Any finite set of Π_1^* sentences F, with this property, may be used to define Group-1, as [49] had noted.)

Group-2: Let $\ulcorner \Phi \urcorner$ denote Φ's Gödel Number, and $\text{HilbPrf}_\beta(\ulcorner \Phi \urcorner, p)$ denote a Δ_0^* formula indicating that p is a Hilbert-Frege styled proof of theorem Φ from axiom system β. For each Π_1^* sentence Φ, the Group-2 schema will contain the below axiom (4). (Thus $\text{IS}_D(\beta)$ can trivially prove all β's Π_1^* theorems.)

$$\forall p \quad \{\text{HilbPrf}_\beta(\ulcorner \Phi \urcorner, p) \;\Rightarrow\; \Phi\} \tag{4}$$

Group-3: The final part of $\text{IS}_D(\beta)$ will be a self-referencing Π_1^* axiom, that indicates $\text{IS}_D(\beta)$ is "Level-1 consistent" under D's deductive apparatus. It thus amounts to the following declaration:

> \# *No two proofs exist for a Π_1^* sentence and its negation, when D's deductive apparatus is applied to an axiom system, consisting of the union of Groups 0, 1 and 2 with* **this sentence** *(looking at itself).*

One encoding for \# as a self-referencing Π_1^* axiom, had appeared in [49]. Thus, (5) is a Π_1^* representation for \# where: (1) $\text{Prf}_{\text{IS}_D(\beta)}(a, b)$ is a Δ_0^* formula indicating that b is a proof of a theorem a from the axiom basis $\text{IS}_D(\beta)$ under D's deductive apparatus, and (2) $\text{Pair}(x, y)$ is a Δ_0^* formula indicating that x is a Π_1^* sentence and y represents x's negation.

$$\forall x \, \forall y \, \forall p \, \forall q \quad \neg \, [\; \text{Pair}(x, y) \,\wedge\, \text{Prf}_{\text{IS}_D(\beta)}(x, p) \,\wedge\, \text{Prf}_{\text{IS}_D(\beta)}(y, q)] \tag{5}$$

For the sake of brevity, we will not provide the exact details, here, about how (5) can be encoded via the Fixed Point Theorem. Adequate details were provided by us in [46, 49].

Definition 3. Let "D" denote any one of the *Tab*, *Xtab* or *Tab-1* deductive apparatus. Then we will say that the resulting mapping of $\text{IS}_D(\bullet)$ is **Consistency Preserving** iff $\text{IS}_D(\beta)$ is automatically consistent whenever all the axioms of β hold true under the standard model of the natural numbers.

The preceding definition raises questions about whether the mappings of $\text{IS}_{Tab}(\bullet)$, $\text{IS}_{Tab-1}(\bullet)$, and $\text{IS}_{Xtab}(\bullet)$ are consistency preserving. It turns out that Theorem 1 will show the first two of these mappings are consistency preserving, while Theorem 2 explores how the Law of the Excluded Middle conflicts with $\text{IS}_{Xtab}(\bullet)$'s Group-3 axiom.

Theorem 1. *The $\text{IS}_{Tab-1}(\bullet)$ and $\text{IS}_{Tab}(\bullet)$ mapping are consistency preserving. (I.e. the axiom systems $\text{IS}_{Tab-1}(\beta)$ and $\text{IS}_{Tab}(\beta)$ are automatically consistent whenever all β's axioms hold true under the standard model of the Natural Numbers.)*

Theorem 2. *In contrast, $IS_{Xtab}(\bullet)$ fails to be consistency-preserving mappings. (More specifically, $IS_{Xtab}(\beta)$ is automatically inconsistent whenever β proves some conventional Π_1^* theorems stating that addition and multiplication satisfy their usual associative, commutative, distributive and identity properties.)*

The proofs of Theorems 1 and 2 would be quite lengthy, if they were derived from first principles. Fortunately, it is unnecessary for us to do so here because we gave a detailed justification of Theorem 1's result for $IS_{Tab-1}(\bullet)$ in [49], and one can incrementally modify the Remark 1's special Invariant of $++$ to justify Theorem 2. Thus, it will be possible for the next two sections of this paper to adequately summarize the intuition behind Theorems 1 and 2, without delving into all the formal details.

Part of the reason Theorems 1 and 2 are of interest is because of their surprising contrast. Thus, some historians have wondered whether Hilbert and Gödel were entirely incorrect when their statements $*$ and $**$ suggested some form of the Consistency Program should likely be viable. Moreover Gerald Sacks's Year-2014 YouTube lecture [38] has reinforced this question by recording how Gödel had repeated analogs of $**$'s speculation during 1961–1962. The contrast between Theorems 1 and 2 will thus provide a plausible suggestion that some portion of what Hilbert and Gödel advocated may be part-way feasible.

This extended abstract will not have the page space to go into all the details. However, the next three sections will, be sufficient to communicate the main gist behind the proofs of Theorems 1 and 2,

5   Intuition Behind Theorem 1

Let us recall the acronym **"Tab"** stands for semantic tableaux deduction. This was defined by Fitting [14, 15] to be a tree-like proof of a theorem Ψ from an axiom basis α, whose root consists of the temporary negated assumption of $\neg\,\Psi$ and whose every root–to–leaf path establishes a contradiction by containing some pair of contradictory nodes that "closes" its path. Each internal node along these paths must *either* be a proper axiom of α *or be* a deduction from an ancestor node via one of the "elimination" rules associated with the logic symbols of $\wedge$, $\vee$, $\rightarrow$, $\neg$, $\forall$, or $\exists$ (that are itemized in the attached Appendix A).

Example 3. Let $IS_{Tab}^M(\bullet)$ denote a mapping transformation identical to Theorem 1's formalism of $IS_{Tab}(\bullet)$, *except that* IS_{Tab}^M shall contain a further multiplication function operation and, accordingly, have its Group-3 "I am consistent" axiom statements *updated* to recognize multiplication as a total function. It turns out this change will cause $IS_{Tab}^M(\bullet)$ to stop satisfying the consistency-preservation property, which Theorem 1 attributed to $IS_{Tab}(\bullet)$.

The intuition behind this change can be roughly summarized if we let $x_0, x_1, x_2, \dots$ and $y_0, y_1, y_2, \dots$ denote the sequences defined by:

$$x_0 = 2 = y_0 \tag{6}$$

$$x_i = x_{i-1} + x_{i-1} \tag{7}$$

$$y_i = y_{i-1} * y_{i-1} \tag{8}$$

For $i > 0$, let ϕ_i and ψ_i denote the sentences in (7) and (8) respectively. Also, let ϕ_0 and ψ_0 denote (6)'s sentence. Then $\phi_0, \phi_1, \ldots \phi_n$ imply $x_n = 2^{n+1}$, and $\psi_0, \psi_1, \ldots \psi_n$ imply $y_n = 2^{2^n}$. Thus, the latter sequence shall grow at an exponentially faster rate than the former. It turns out that this change in growth speed will cause the $\mathrm{IS}_{Tab}^M(\bullet)$, and $\mathrm{IS}_{Tab}(\bullet)$ to have opposite self-justification properties.

In particular, let the quantities $\mathrm{Log}(y_n) = 2^n$ and $\mathrm{Log}(x_n) = n+1$ represent the lengths for the binary codings for y_n and x_n. Thus, y_n's coding will have a length 2^n, which is *much larger* than the $n+1$ steps that $\psi_0, \psi_1, \ldots \psi_n$ uses to define y_n's existence. In contrast, x_n's binary encoding will have a smaller length of $n+1$. These observations are helpful because every proof of the Incompleteness Theorem involves a Gödel number z encoding a capacity to self-reference its own definition.

The faster growing series $y_0, y_1, \ldots y_n$ should, intuitively, have this self-referencing capacity because y_n's binary encoding has a 2^{n+1} length that greatly exceeds the size of the $O(n)$ steps used to define its value. Leaving aside many of [47,52]'s further details, this fast growth explains roughly why a Type-M logic, such as IS_{Tab}^M, satisfies the semantic tableaux version of the Second Incompleteness Theorem, unlike IS_{Tab}.

Our paradigm also explains why IS_{Tab}'s Type-A formalisms produce boundary-case exceptions to the semantic tableaux version of the Second Incompleteness Theorem. This is because [49] showed that it was unable to construct numbers z that can self-reference their own definitions (when only the *more slowly growing* addition primitive is available). In particular assuming only two bits are needed to encode each sentence in the sequence $\phi_0, \phi_1, \ldots \phi_n$, the length $n+1$ for x_n's binary encoding is insufficient for encoding this sequence.

Leaving aside many of [49]'s details, this short length for x_n explains the core intuition behind [49]'s evasion of the Second Incompleteness Theorem under IS_{Tab}. It arises essentially because of the difference between the growth rates of the sequences $x_1, x_2, x_3 \ldots$ and $y_1, y_2, y_3 \ldots$.

There is obviously insufficient space for this extended abstract to provide more details, here. A full detailed proof of Theorem 1 can be found in [49]. It establishes (see[4]) that Peano Arithmetic can prove β's consistency implies *both* the consistency and also the self-justifying properties for $\mathrm{IS}_{Tab-1}(\beta)$. Our more modest goal, within the present abbreviated paper, has been *merely to* summarize the intuition behind Theorem 1's surprising evasion of the Second Incompleteness effect.

---

[4] The *exact* meaning of this implication is subtle. This is because Peano Arithmetic (PA) cannot know whether β is consistent when $\beta = PA$. Thus, *unlike* the quite different formalism of $\mathrm{IS}_{Tab-1}(PA)$, the system of PA shall linger in a state of self-doubt, about whether both PA and $\mathrm{IS}_{Tab-1}(PA)$ are consistent.

It arises, intuitively, because of the difference in growth rates between the $x_1, x_2, x_3...$ and $y_1, y_2, y_3...$ series.

6   Summary of the Justification For Theorem 2

The proof of Theorem 2 is complex, but it can be nicely summarized because it is related to the justification for Invariant $++$, which Remark 1 had credited to the combined work of Pudlák, Solovay, Nelson and Wilkie-Paris. The crucial aspect of the Frege-Hilbert methodologies, explored by $++$, is that modus ponens assures that a proof of a theorem ψ from an axiom system α has a length no greater than the sum of the proof-lengths needed to derive ϕ and $\phi \rightarrow \psi$ from α. This **"Linear-Sum Effect"** does not apply, actually, also to *Tab*-deduction because it owns no analog of a modus ponens rule (for assuring that ψ's proof length is bounded by the sum of the proof lengths for ϕ and $\phi \rightarrow \psi$).

The *Xtab* methodology, however, differs from *Tab*-deduction by allowing any node of its proof-tree to store a sentence of the form $\phi \vee \neg \phi$, as an application of its allowed use of the Law of Excluded Middle. This added feature will allow an *Xtab* proof for ψ to have a length proportional to the sum of the proof lengths for ϕ and $\phi \rightarrow \psi$ (i.e. it can roughly simulate the actions of modus ponens). In particular, the relevant *Xtab* proof for ψ will consist of the following four steps:

1. The root of an Xtab proof for ψ will be the usual temporary negated hypothesis of $\neg \psi$ (which the remainder of the proof tree will show is impossible to hold).
2. The child of this root node will be an allowed invocation of the Law of the Excluded Middle of the *particular* form $\phi \vee \neg \phi$.
3. The relevant Xtab proof tree will next employ the Appendix A's branching rule for allowing the two sibling nodes of ϕ and $\neg \phi$ to descend from Item 2's node.
4. Finally, our Xtab proof will insert below (3)'s left sibling node of ϕ a subtree that is no longer than a proof for $\phi \rightarrow \psi$, and likewise insert a proof for ϕ below (3)'s right sibling of $\neg \phi$.

The point is that the last step of the above 4-part proof has a length no greater than the sum of the two proof lengths for ϕ and $\phi \rightarrow \psi$ (similar to the proof compressions resulting from a modus ponens operation). Its first three steps shall produce inconsequential effects that increase the overall proof length by no more than a *very tiny* amount proportional to the length of the particular sentence "$\phi \rightarrow \psi$".

We can apply the preceeding "Linear-Sum Effect" to construct an analog of Remark 1's earlier Theorem $++$ that now applies to Xtab deduction. Saving several details for a longer article, the intuition behind this analog is that modus ponens *is the only rule of inference* used by [12]'s classic textbook-style first-order logic system, and Xtab can apply its Linear-Sum Effect to essentially simulate modus ponens. Our natural analog of $++$ will, thus, assure that any axiom system $\mathcal{A}$, using Xtab deduction, is *automatically inconsistent* when:

I. $\mathcal{A}$ can verify Successor is a total function (as is formalized by Line (1)).

II. $\mathcal{A}$ can prove addition and multiplication (viewed as 3-way relations) satisfy their usual associative, commutative, distributive and identity-operator properties.

III. $\mathcal{A}$ proves an added theorem (which turns out to be false) affirming its own consistency when the Xtab deductive apparatus is used.

The preceding paragraph has nicely captured the essence of Theorem 2's proof. It has noted $\mathrm{IS}_{Xtab}(\beta)$ *is automatically inconsistent* when it satisfies analogs of the above three requirements (see footnote[5]). This is intuitively because $\mathrm{IS}_{Xtab}(\beta)$ can roughly simulate the power of modus ponens, whose crucial Linear-Sum Effect from modus ponens will be imitated by $\mathrm{IS}_{Xtab}(\beta)$ (when the latter views the Law of Excluded Middle as a schema of logical axioms). Thus, an analog of Remark 1's invariant $++$ shall apply, consequently, to $\mathrm{IS}_{Xtab}(\beta)$.

From a pedagogical perspective, one obvious drawback to this type of justification for Theorem 2 is that it assumes the reader is familiar with either the combined work of Pudlák, Solovay, Nelson and Wilkie-Paris in [31,35,41,44], or related work (also mentioned in Remark 1) by Buss-Ignjatović, Friedman, Hájek, Švejdar [10,16,23,42] and in Appendix A of [46] (or in Willard's closely related paper of [48]).

Many readers will, of course be unfamiliar with any of this material because it relies upon a much more sophisticated version of the Second Incompleteness Theorem than does appear in most introductory logic textbooks. We have therefore inserted a brief Appendix B into this article. It explains the main idea behind the Theorem $++$, which the Remark 1 credited to a recurring theme appearing in [10,16,23,25,31,35,41,42,44,46,48].

This appendix will not be sufficiently detailed to formulate the precise nature of our reductionistic argument for justifying Theorem 2. (The latter will be saved for a longer version of this article.) Our Appendix B will, however be adequate to provide the reader with a rough intuitive grasp of the type of extension of Theorem $++$, which is needed to establish Theorem 2's generalization of the Second Incompleteness Effect.

7   On the Significance of Theorems 1 and 2

The main topic of this paper is surprising because it is quite unusual for an initially consistent formalism α *to become inconsistent* when its initial schema of theorems (establishing the universal validity of the Law of the Excluded Middle) is essentially transformed into becoming a formal schema of logical axioms.

The reason for this unusual effect is that the meaning of a Group-3 *"I am consistent"* axiom changes, *quite substantially,* when theorems are transformed into logical axioms. This is because unacceptable diagonalizing contradictions

---

[5] $\mathrm{IS}_{Xtab}(\beta)$ actually satisfies a requirement stronger than Item I because it recognizes addition as total.

can occur (as summarized by footnote[6]) when such a transition *significantly alters* the meaning of an *"I am consistent"* axiom.

The resulting contrast between Theorems 1 and 2 is helpful in explaining how Hilbert and Gödel could simultaneously fully appreciate the significance of the Second Incompleteness Theorem, but yet also allow their statements ∗ and ∗∗ to question whether its paradigm could be *partially* evaded. Moreover, Gödel's remark ∗∗ should not be ignored when Gerald Sacks's year-2014 YouTube lecture [38] has recalled how a middle-aged 55-year-old Kurt Gödel had repeated analogs of his 1931 remark ∗∗ during the 1961–1962 period. (It is also noteworthy that Harvey Friedman recorded a You-Tube lecture [18] in 2014 where he indicated he was tentatively open to the possibility that the Second Incompleteness Theorems might permit some type of limited forms of partial exceptions to it.)

Thus, while there is no doubt that the Second Incompleteness Theorem will, certainly, always be remembered for its seminal impact on 20th century Logic, its part-way exceptions should also be seen as significant. This is because futuristic high-tech computers will better understand their self-capacities if they own some *partial* awareness about their own consistency.

There is no page space to go into all the details here. However, the distinction between the initial "IS(A)" system from our 1993 and 2001 papers [45, 46] with the more sophisticated IS $_{Tab-1}(\beta)$ formalism in our year-2005 article [49] should, also, be briefly mentioned. Our older "IS(A)" formalism was actually simpler, but it was substantially weaker because it only recognized the non-existence of a proof of $0 = 1$ from itself. In contrast, IS $_{Tab-1}(\beta)$'s Group-3 axiom can corroborate that *no two simultaneous proofs* exist for a Rank-1* sentence and its negation. This is an important distinction, because the First Incompleteness Theorem indicates no decision procedure can exist for separating all true from false Rank-1* sentences. (See also [50, 51, 53, 54] for other particular refinements for our "IS(A)" formalism.)

In summary, the main purpose of this article has been to explore the contrast between the opposing Theorems 1 and 2. The latter theorem, thus, provides *another helpful reminder* about the millenial importance of Gödel's seminal Second Incompleteness Theorem. Yet at the same time, Theorem 1 illustrates how some *partial exceptions* to Gödel's result do arise, as Hilbert and Gödel had predicted both in their statements ∗ and ∗∗ and in Gödel's private communications with Gerald Sacks [38].

The 2-way contrast between Theorems 1 and 2 may be as significant as their individual actual results. This is because the Second Incompleteness Theorem is fundamental to Logic. Many historians have, thus, been quite perplexed by the *partial* reluctance that Hilbert and Gödel had expressed about it in ∗ and ∗∗. A partial reason for this reluctance is, perhaps, significantly related to the contrast between these two opposing theorems.

---

[6] The point is that proofs are compressed when theorems are transformed into logical axioms, and such compressions can produce diagonalizing contradictions under some Type-A logics using *"I am consistent"* axioms.

Acknowledgment. I thank Seth Chaiken for several helpful comments about how to improve the presentation.

Appendix A: Formal Definition of a Tableaux Proof

Our definition of a semantic tableaux proof is similar to analogs in the textbooks by Fitting and Smullyan [15, 39]. A tableaux proof of a theorem Ψ from a set of proper axioms, denoted as α, will be a tree structure whose root contains the temporary contradictory assumption of $\neg\Psi$ and whose every descending root-to-leaf branch affirms a contradiction by containing both some sentence ϕ and its negation $\neg\phi$. Each internal node in this tree will be either a proper axiom of α or a deduction from a higher ancestor in this tree using one of the following six elimination rules for the logical connective symbols of $\wedge$, $\vee$, $\rightarrow$, $\neg$, $\forall$ and $\exists$. These rules use a notation where "$\mathbf{A} \Longrightarrow \mathbf{B}$" is an abbreviation for a sentence $\mathbf{B}$ being an allowed deduction from its ancestor of $\mathbf{A}$.

1. $\Upsilon \wedge \Gamma \Longrightarrow \Upsilon$ and $\Upsilon \wedge \Gamma \Longrightarrow \Gamma$.
2. $\neg\neg\Upsilon \Longrightarrow \Upsilon$. Other rules for the "$\neg$" symbol are: $\neg(\Upsilon \vee \Gamma) \Longrightarrow \neg\Upsilon \wedge \neg\Gamma$, $\neg(\Upsilon \rightarrow \Gamma) \Longrightarrow \Upsilon \wedge \neg\Gamma$, $\neg(\Upsilon \wedge \Gamma) \Longrightarrow \neg\Upsilon \vee \neg\Gamma$, $\neg\exists v\,\Upsilon(v) \Longrightarrow \forall v\neg\,\Upsilon(v)$ and $\neg\forall v\,\Upsilon(v) \Longrightarrow \exists v\neg\Upsilon(v)$
3. A pair of sibling nodes Υ and Γ is allowed when their ancestor is $\Upsilon \vee \Gamma$.
4. A pair of sibling nodes $\neg\Upsilon$ and Γ is allowed when their ancestor is $\Upsilon \rightarrow \Gamma$.
5. $\forall v\,\Upsilon(v) \Longrightarrow \Upsilon(t)$ where t may denote any term.
6. $\exists v\,\Upsilon(v) \Longrightarrow \Upsilon(p)$ where p is a newly introduced parameter symbol.

A minor additional comment about our notation is that we treat "$\forall\,v \le s\ \Phi(v)$" as an abbreviation for $\forall v\ \{v \le s \rightarrow \Phi(v)\}$ and likewise "$\exists\,v \le s\ \Phi(v)$" as an abbreviation for $\exists v\ \{v \le s\ \wedge\ \Phi(v)\}$. In our year-2005 article [49], we thus applied Rules 5 and 6 to derive the following further hybrid rules for processing the bounded universal and also the bounded existential quantifiers:

a. $\forall v \le s\,\Upsilon(v) \Longrightarrow t \le s \rightarrow \Upsilon(t)$ where t may be any arithmetic term.
b. $\exists v \le s\,\Upsilon(v) \Longrightarrow p \le s \wedge \Upsilon(p)$ where p is a new parameter symbol.

Appendix B: More Details About Theorem 2's Proof

The most surprising aspect of Theorem 2 is the sharp contrast between its result with the opposing property of Theorem 1. Our goal in this appendix will be to intuitively explain why the Invariant $++$ (from Remark 1) ushers in a machinery that applies only to Theorem 2.

During our discussion, we will employ our U-Grounding language L^* that treats multiplication as a 3-way relation (rather than as a functional operation). Its 3-way predicate $Mult(\mathrm{x,y,z})$, for formalizing multiplication, is defined as follows:

$$[(x = 0 \vee y = 0) \Rightarrow z = 0]\ \wedge\ [(x \ne 0 \wedge y \ne 0) \Rightarrow\ (\frac{z}{x} = y\ \wedge\ \frac{z-1}{x} < y)] \quad (9)$$

We will say that an axiom basis α is **Regular** iff

1. It presumes all the U-Grounding operations are total functions (including the Addition and Doubling primitives).
2. It can prove all true Δ_0^* sentences, and α is also consistent.
3. It can prove a Π_1^* theorem showing addition and multiplication, viewed as 3-way relations, satisfy their usual associative, commutative, distributive and identity-operator properties.

Also in this appendix, we will employ a notation where for any $j \geq 0$, the symbol $\omega_j(x)$ will be recursively defined by the following rules:

1. $\omega_0(x) = x^2$.
2. $\omega_{j+1}(x) = 2^{\omega_j(2 \cdot Log_2(x+1))}$

These two rules imply that $\omega_{j+1}(x) > \omega_j(x)$ and $\omega_1(x) \geq x^x$.

Clarification About Notation: Since our U-Grounding language L^* does not permit using any function symbols to grow as fast as multiplication, it does not technically allow us to use any of the ω_j primitive symbols. One can, however, use techniques from [25]'s textbook to construct a Δ_0^* formula $\psi_j(x,y)$ that satisfies (10)'s invariant for all standard numbers. It will, thus, capture most of ω_j's salient features.

$$\forall x \, \forall y \quad \psi_j(x,y) \; \Leftrightarrow \; \omega_j(x) = y \tag{10}$$

Definition 4. A formula $\Phi(x)$ will be called **Locally-J-Closed** relative to the axiom basis α iff α can prove the following three assertions about $\Phi(x)$:

A. All of $\Phi(0)$, $\Phi(1)$ and $\Phi(2)$ are true.
B. The predicate $\Phi(x)$ is operationally closed under the growth operation ω_j. (Line (11) formally encodes this closure condition, using the preceding paragraph's notation.)

$$\forall x \, \forall y \quad \{[\psi_j(x,y) \wedge \Phi(x)] \; \Rightarrow \; \Phi(y)\} \tag{11}$$

C. The predicate $\Phi(x)$ is also closed under (12)'s decrement operation.

$$\forall x \, \forall y < x \; \{\Phi(x) \; \Rightarrow \; \Phi(y)\} \tag{12}$$

Theorem 3. *For each regular axiom basis α (that is consistent) and for each fixed integer $J \geq 1$, there exists a corresponding formula $\Phi(x)$ where α can prove that $\Phi(x)$ is Locally-J-Closed.*

Due to a lack of page space, a formal proof of Theorem 3 will be postponed until a longer version of this article. Theorem 3 is related to various intermediate results that were used to establish Remark 1's Invariant $++$ and [10, 16, 23, 25, 31, 35, 41, 42, 44, 46, 48]'s closely related results.

The fascinating feature of Theorem 3 is that it can explain why Theorems 1 and 2 display nearly opposite effects with regards to Hilbert's Second Open Question. This is because the needed diagonalization for producing Theorem 2's variations of the Second Incompleteness Effect become feasible only[7] when $IS_{Xtab}(\beta)$'s Linear-Sum Effect is applied to the intermediate results produced by its possible derived theorems (which include the formalisms that are illustrated by lines (11) and (12)). On the other hand, no such similar types of nicely compressed constructed proofs are available under Theorem 1's $IS_{Tab-1}(\beta)$ formalism (because all instances of the Law of Excluded Middle are excluded by it from becoming logical axioms). This is the intuitive reason that Theorems 1 and 2 display such sharply contrasting results.

References

1. Adamowicz, Z., Bigorajska, T.: Existentially closed structures and Gödel's second incompleteness theorem. JSL **66**(1), 349–356 (2001)
2. Adamowicz, Z., Zbierski, P.: On Herbrand consistency in weak theories. Arch. Math. Log. **40**(6), 399–413 (2001)
3. Artemov, S.: Explicit provablity and constructive semantics. Bull. Symb. Log. **7**(1), 1–36 (2001)
4. Artemov, S.: The provability of consistency. Cornell Archives arXiv Report 1902.07404v4 (2019)
5. Artemov, S., Beklemishev, L.D.: Provability logic. In: Handbook on Philosophical Logic, 2nd edn, pp. 189-360 (2005)
6. Beklemishev, L.D.: Reflection principles and provability algebras in formal arithmetic. Russ. Math. Surv. **60**(2), 197–268 (2005)
7. Beklemishev, L.D.: Positive provability for uniform reflection principles. APAL **165**(1), 82–105 (2014)
8. Bezboruah, A., Shepherdson, J.C.: Gödel's second incompleteness theorem for Q. JSL **41**(2), 503–512 (1976)
9. Buss, S.R.: Bounded arithmetic. Studies in Proof Theory, Lecture Notes 3, disseminated by Bibliopolis as revised version of Ph.D. thesis (1986)
10. Buss, S.R., Ignjatović, A.: Unprovability of consistency statements in fragments of bounded arithmetic. APAL **74**(3), 221–244 (1995)
11. Dawson, J.W.: Logical Dilemmas: The Life and Work of Kurt Gödel. AKPeters Press (1997)
12. Enderton, H.B.: A Mathematical Introduction to Logic. Academic Press (2001)
13. Feferman, S.: Arithmetization of mathematics in a general setting. FundMath **49**, 35–92 (1960)
14. Fitting, M.: Tableau methods of proofs for modal logics. Notre Dame J. Form. Log. **13**(2), 237–247 (1972)
15. Fitting, M.: First Order Logic and Automated Theorem Proving. Springer, New York (1996). https://doi.org/10.1007/978-1-4612-2360-3
16. Friedman, H.M.: On the consistency, completeness and correctness problems. Ohio State Tech Report (1979). See also Pudlák [36]'s summary of this result

---

[7] Actually, we will only need the "Locally 1-Closure" property to prove that $IS_{Xtab}(\beta)$. cannot possibly be self-justifying.

17. Friedman, H.M.: Translatability and relative consistency. Ohio State Tech Report (1979). See also Pudlák [36]'s summary of this result
18. Friedman, H.M.: Gödel's blessing and Gödel's curse. (This is "Lecture 4" within a 5-part Ohio State YouTube lecture series, dated 14 March 2014)
19. Gentzen, G.: Die Wiederpruchsfreiheit der reinen Zahlentheorie. Math. Ann. **112**, 439–565 (1936)
20. Gödel, K.: Über formal unentscheidbare sätze der Principia Mathematica und verwandte systeme I. Monatshefte für Mathematik und Physik **38**, 349–360 (1931)
21. Gödel, K.: The present situation in the foundations of mathematics. In: Feferman, S., Dawson, J. W., Goldfarb, W., Parsons, C., Solovay, R.M. (eds.) Collected Works Volume III: Unpublished Essays and Lectures, pp. 45–53, Oxford University Press (2004)
22. Goldstein, R.: Incompleteness: The Proof and Paradox of Kurt Gödel. Norton Press (2005)
23. Hájek, P.: Mathematical fuzzy logic and natural numbers. FundMath **81**, 155–163 (2007)
24. Hájek, P.: Towards metamathematics of weak arithmetics over fuzzy logics. Log. J. IPL **19**, 467–475 (2011)
25. Hájek, P., Pudlák, P.: Metamathematics of First Order Arithmetic. Springer, Heidelberg (1991)
26. Hilbert, D.: Über das unendliche. Math. Ann. **95**, 161–191 (1926)
27. Jeroslow, R.: Consistency statements in formal theories. FundMath **72**, 17–40 (1971)
28. Kleene, S.C.: On notation for ordinal numbers. JSL **3**(1), 150–155 (1938)
29. Kreisel, G., Takeuti, G.: Formally self-referential propositions for cut-free classical analysis. Dissertationes Mathematicae **118**, 1–50 (1974)
30. Mendelson, E.: Introduction to Mathematical Logic. CRC Press (2010)
31. Nelson, E.: Predicative Arithmetic. Mathematics Notes. Princeton University Press, Princeton (1986)
32. Parikh, R.: Existence and feasibility in arithmetic. JSL **36**(3), 494–508 (1971)
33. Paris, J.B., Dimitracopoulos, C.: A note on the undefinability of cuts. JSL **48**(3), 564–569 (1983)
34. Parsons, C.: On n–quantifier elimination. JSL **37**(3), 466–482 (1972)
35. Pudlák, P.: Cuts, consistency statements and interpretations. JSL **50**(2), 423–441 (1985)
36. Pudlák, P.: On the lengths of proofs of consistency. In: Pudlák, P. (ed.) Collegium Logicum: 1996 Annals of the Kurt Gödel Society, vol. 2, pp. 65–86. Springer, NewYork (1996). https://doi.org/10.1007/978-3-7091-9461-4_5
37. Rogers, H.A.: Theory of Recursive Functions and Effective Compatibility. McGrawHill (1967)
38. Sacks, G.: Some detailed recollections about Kurt Gödel during a June 2, 2014 YouTube lecture given by Gerald Sacks at the University of Pennsylvania. This lecture records the experiences of Sacks when he was a 1-year assistant for Gödel at the Institute of Advanced Studies. Some quotes from this talk are that Gödel "did not think" the goals of Hilbert's Consistency Program "were erased" by the Incompleteness Theorem, and that Gödel believed (according to Sacks) that it left Hilbert's program "very much alive and even more interesting than it initially was"
39. Smullyan, R.: First Order Logic. Dover Books (1995)
40. Solovay, R.M.: Injecting Inconsistencies into models of PA. APAL **44**(1–2), 102–132 (1989)

41. Solovay, R.M.: Private Telephone conversations during April of 1994 between Dan Willard and Robert M. Solovay. During those conversations Solovay described an unpublished generalization of one of Pudlák's theorems [35], using some methods of Nelson and Wilkie-Paris [31,44]. (The Appendix A of [46] offers a 4-page summary of our interpretation of Solovay's remarks. Several other articles [10,25,31,33,35,44] have also noted that Solovay often has chosen to privately communicate noteworthy insights that he has elected not to formally publish)

42. Švejdar, V.: An interpretation of Robinson arithmetic in its Grzegorczjk's weaker variant. Fundamenta Informaticae **81**, 347–354 (2007)

43. Visser, A.: Faith and falsity. APAL **131**(1–3), 103–131 (2005)

44. Wilkie, A.J., Paris, J.B.: On the scheme of induction for bounded arithmetic. APAL **35**, 261–302 (1987)

45. Willard, E.: Self-verifying axiom systems. In: Gottlob, G., Leitsch, A., Mundici, D. (eds.) KGC 1993. LNCS, vol. 713, pp. 325–336. Springer, Heidelberg (1993). https://doi.org/10.1007/BFb0022580

46. Willard, D.E.: Self-verifying systems, the incompleteness theorem and the tangibiltiy reflection principle. JSL **66**(2), 536–596 (2001)

47. Willard, D.E.: How to extend the semantic tableaux and cut-free versions of the second incompleteness theorem almost to Robinson's arithmetic Q. JSL **67**(1), 465–496 (2002)

48. Willard, D.E.: A version of the second incompleteness theorem for axiom systems that recognize addition but not multiplication as a total function. In: Hendricks, V., Neuhaus, F., Pedersen, S.A., Scheffler, U., Wansing, H. (eds.) First Order Logic Revisited, pp. 337–368, Logos Verlag, Berlin (2004)

49. Willard, D.E.: An exploration of the partial respects in which an axiom system recognizing solely addition as a total function can verify its own consistency. JSL **70**(4), 1171–1209 (2005)

50. Willard, D.E.: A generalization of the second incompleteness theorem and some exceptions to it. APAL **141**(3), 472–496 (2006)

51. Willard, D.E.: On the available partial respects in which an axiomatization for real valued arithmetic can recognize its consistency. JSL **71**(4), 1189–1199 (2006)

52. Willard, D.E.: Passive induction and a solution to a Paris-Wilkie open question. APAL **146**(2), 124–149 (2007)

53. Willard, D.E.: Some specially formulated axiomizations for $I\Sigma_0$ manage to evade the Herbrandized version of the second incompleteness theorem. Inf. Comput. **207**(10), 1078–1093 (2009)

54. Willard, D.E.: On how the introducing of a new θ function symbol into arithmetic's formalism is germane to devising axiom systems that can appreciate fragments of their own Hilbert consistency. Cornell Archives arXiv Report 1612.08071 (2016)

55. Yourgrau, P.: A World Without Time: The Forgotten Legacy of Gödel and Einstein. (See page 58 for the passages we have quoted.) Basic Books (2005)

Author Index

Printed in the United States
By Bookmasters